Harmonic Analysis Method *for* Nonlinear Evolution EQUATIONS, I

Harmonic Analysis Method *for* Nonlinear Evolution EQUATIONS, I

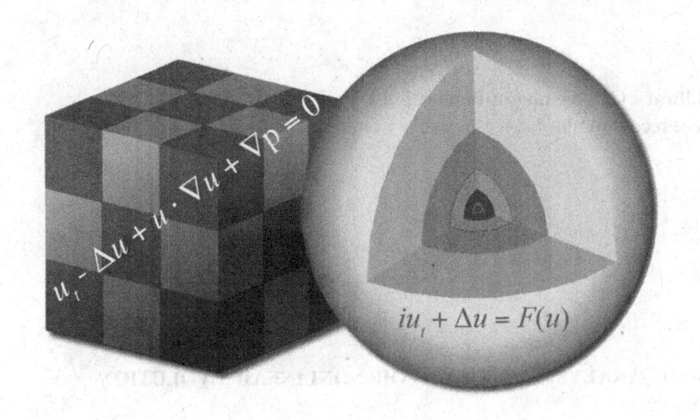

Baoxiang Wang
Peking University, China

Zhaohui Huo
Chinese Academy of Sciences, China

Chengchun Hao
Chinese Academy of Sciences, China

Zihua Guo
Peking University, China

NEW JERSEY · LONDON · SINGAPORE · BEIJING · SHANGHAI · HONG KONG · TAIPEI · CHENNAI

Published by

World Scientific Publishing Co. Pte. Ltd.

5 Toh Tuck Link, Singapore 596224

USA office: 27 Warren Street, Suite 401-402, Hackensack, NJ 07601

UK office: 57 Shelton Street, Covent Garden, London WC2H 9HE

British Library Cataloguing-in-Publication Data
A catalogue record for this book is available from the British Library.

HARMONIC ANALYSIS METHOD FOR NONLINEAR EVOLUTION EQUATIONS, I

ISBN-13 978-981-4360-73-9
ISBN-10 981-4360-73-2

Printed in Singapore.

To our parents

Preface

The sole purpose of science is the glory of human spirit. —— C. G. J. Jacobi

During the past thirty years, the nonlinear evolution equations (NLE) have made a great progress by using the harmonic analysis techniques. This book is devoted to introduce the harmonic analysis method for NLE, which is based on lectures given by B. X. Wang at the Peking University during the past seven years.

In order to solve an NLE, one needs to choose suitable function spaces as working spaces. It is well known that most function spaces are established by using the Lebesgue integrals as basic tools. However, the Lebesgue integration was not recognized to be of importance in its early stage. As long as Lebesgue attempted to attend a conference, some mathematicians working in analysis said to him: "This is not interested for you, we are discussing the differentiable functions.", and some experts engaged in geometry told him: "We are talking the surface which has a tangent plane." One can imagine how Lebesgue was hurt inside. New ideas are usually not so easy to grow up and mathematics brings mathematicians too much sadness and blessedness, which is hard to express by languages.

However, the authors believe that, at least from the local history point of view, many important progresses for nonlinear evolution equations have brought us a series of perpetual surprises over the recent thirty years. The harmonic analysis techniques of NLE can be gone back to the pioneering work of R. S. Strichartz in 1977 who discovered the time-space decay of the solutions of the linear wave equation, so called the Strichartz inequality, which is now a fundamental tool in the study of nonlinear dispersive equations. Since 1980s, the NLE, especially the nonlinear dispersive equations have gained a great development by using the harmonic analysis techniques.

On the other hand, the theory of harmonic analysis is also promoted by the study of NLE. J. Bourgain in 1993 systematically developed the $X^{s,b}$ method which is powerful in handling the derivative nonlinearity in nonlinear dispersive equations, where the $X^{s,b}$ space was previously discovered by J. Rauch, M. Reed and M. Beals for the wave equation. J. Bourgain in 1999 invented the method, so called "separation of the localized energy", to study the energy scattering of nonlinear Schrödinger equations. The "I-method" was introduced by T. Tao's group to study the global wellposedness of nonlinear dispersive equations for a class of initial data with lower regularities. C. E. Kenig and F. Merle developed the concentration compactness method to study the sharp well posedness and scattering for the focusing nonlinear dispersive equations. Recently, the frequency-uniform decomposition techniques are also applied in the study of NLE. A poem describes the current status of the harmonic analysis techniques of NLE:

Resonance to a Book[1]

—*Xi Zhu*

A square of pool likes an opening mirror
Blue sky and white cloud map in it, freely wander
Why is the pond crystal and bottomed out
It is from the source of alive water

More precisely, we will study a class of nonlinear dispersive equations, such as nonlinear Schrödinger equations, nonlinear Klein-Gordon eqautions, KdV equations, as well as the Navier-Stokes equations and the Boltzmann equation. As the first book of this series we will mainly study the local and global wellposedness to the Cauchy problem for those equations. In Chapter 1 we briefly introduce the theories of various function spaces, say Besov and Triebel-Lizorkin spaces. In Chapter 2 we study the Navier-Stokes equation by using the Littlewood-Paley decomposition to establish some time-space estimates for the linear heat equation. Strichartz type estimates for a class of linear dispersive equations will be systematically set up in Chapter 3. Applying those Strichartz' inequalities, in Chapter 4 we will consider the wellposedness of the solutions of nonlinear Schrödinger and Klein-Gordon equations. In Chapter 5 we introduce the $X^{s,b}$-method and study the KdV and derivative Schrödinger equations. In Chapter 6 we introduce the frequency-uniform decomposition method. It is known that

[1] A pool indicates a book.

Morawetz' estimates are basic tools for the scattering theory of nonlinear dispersive equations, which will be summarized in Chapter 7 for nonlinear Schrödinger equations. Finally, in Chapter 8 we introduce some fundamental results for the Boltzmann equation. Chapters 1, 2, 3, 4 and 6 are written by B. X. Wang, Chapter 5 is written by Z. H. Guo and Z. H. Huo, Chapter 7 is written by C. C. Hao and Chapter 8 is written by Z. H. Huo. Many results in Chapters 2, 3, 4, 5 and 6 have been reproved or simplified by the authors.

The authors would like to genuinely thank Professors Yulin Zhou, Hesheng Sun, Boling Guo, Ling Hsiao, Lizhong Peng and Carlos E. Kenig for their constant supports and they are grateful to Professors Kong Ching Chang, Weiyue Ding and Gang Tian for their helps. B. X. Wang thanks Professors Zhouping Xin and Jiecheng Chen for their invitation to give a series of lectures based on the book at Hong Kong and Hangzhou, respectively. B. X. Wang deeply cherishes the memory of Professor Tingfu Wang for his rudimental guidance in mathematics. Thanks are also due to Ms Ji Zhang for her excellent editorial work.

BX Wang

Beijing, May 1, 2011

Contents

Chapter 1

Fourier multiplier, function space $X^s_{p,q}$

The intensive study to the nature is the most plentiful source for the mathematical discovery. —— J. Fourier[1]

In this chapter we begin by considering the Fourier transform on the Schwartz space $\mathscr{S}$ and its dual space $\mathscr{S}'$, the material on Fourier transforms is standard and our treatment here will be sketched, see Stein [202], Yosida [256] for instance. On the basis of the Fourier transform, we introduce the Fourier multiplier space M_p, Besov spaces $B^s_{p,q}$ and Triebel spaces $F^s_{p,q}$ and then discuss their elementary properties. The embeddings between these function spaces are very useful in the study of partial differential equations. For the convenience to the readers, we give a self-contained treatment to the theory of function spaces. For some further results on function spaces, one can refer to [13; 224]. Some notations are well-known for readers and not stated in the text, which will be listed in the Appendix A. For convenience to readers, we would like to remind some important relations between Triebel-Lizorkin, Bessel potential spaces H^s_p ($H^s = H^s_2$) and Sobolev spaces W^m_p with $m \in \mathbb{Z}_+$:

$$F^s_{p,2} = H^s_p, \quad H^m_p = W^m_p, \quad F^s_{2,2} = H^s, \quad \forall\, s \in \mathbb{R},\ m \in \mathbb{Z}_+,\ 1 < p < \infty.$$

1.1 Schwartz space, tempered distribution, Fourier transform

Let $\alpha = (\alpha_1, ..., \alpha_n)$ be a multi-index, denote

$$D^\alpha = \partial^{\alpha_1}_{x_1} ... \partial^{\alpha_n}_{x_n}, \quad \partial^{\alpha_i}_{x_i} = \partial^{\alpha_i}/\partial x_i^{\alpha_i}, \quad |\alpha| = \alpha_1 + ... + \alpha_n, \quad x^\alpha = x_1^{\alpha_1} ... x_n^{\alpha_n}.$$

[1] Joseph Fourier (1768–1830), a French mathematician and physicist, is best known for initiating the investigation of Fourier series and their application to problems of the heat transfer. He is also generally credited with the discovery of the greenhouse effect. He is one of the authority in French analysis school.

We denote by $C^\infty(\mathbb{R}^n)$ the set of all infinitely differentiable functions defined on $\mathbb{R}^n$. Denote

$$\mathscr{S} = \{\phi \in C^\infty(\mathbb{R}^n) : p_k(\phi) < \infty\},$$
$$p_k(\phi) = \sup_{x\in\mathbb{R}^n} (1+|x|^2)^{k/2} \sum_{|\alpha|\leqslant k} |D^\alpha \phi(x)|. \tag{1.1}$$

It is easy to see that $p_k(\cdot)$ is a norm on $\mathscr{S}$ and so, it is a semi-norm on $\mathscr{S}$. From the theory of the linear topology, $\mathscr{S}$ generates a locally convex linear topological space according to the semi-norm system $\{p_k\}_{k=0}^\infty$, which is said to be the Schwartz space, any function in $\mathscr{S}$ is said to be a Schwartz function. The base of zero neighborhoods of $\mathscr{S}$ is

$$B_{k,\varepsilon} = \{\phi \in \mathscr{S} : p_k(\phi) < \varepsilon\}, \ \ k \in \mathbb{Z}_+ = \{0,1,2,...\}, \ \varepsilon > 0.$$

We denote by $\mathscr{S}' := \mathscr{S}'(\mathbb{R}^n)$ the dual space of $\mathscr{S}$ and it is a locally convex linear topological space, which is said to be a tempered distribution space. The base of zero neighborhoods of $\mathscr{S}'$ is

$$U_{B,\varepsilon} = \left\{ f \in \mathscr{S} : \sup_{\phi\in B} |f(\phi)| < \varepsilon \right\},$$

where B is a bounded set in $\mathscr{S}$. Let $\phi, \psi \in \mathscr{S}$, the Fourier (inverse) transform is defined as follows.

$$\widehat{\phi}(\xi) = (\mathscr{F}\phi)(\xi) = (2\pi)^{-n/2} \int_{\mathbb{R}^n} e^{-\mathrm{i}x\cdot\xi} \phi(x) dx,$$
$$\check{\psi}(x) = (\mathscr{F}^{-1}\psi)(x) = (2\pi)^{-n/2} \int_{\mathbb{R}^n} e^{\mathrm{i}x\cdot\xi} \psi(\xi) d\xi, \tag{1.2}$$

where $x \cdot \xi = x_1\xi_1 + ... + x_n\xi_n$ (if there is no confusion, we will write $x\cdot\xi = x\xi$). $\mathscr{S}$ plays a crucial role in the theory of the Fourier analysis, which is compatible with the Fourier transform. In fact, we have the following

Proposition 1.1. *Let $\phi \in \mathscr{S}$. Then we have*

$$\widehat{D^\alpha\phi}(\xi) = \mathrm{i}^{|\alpha|}\xi^\alpha \widehat{\phi}(\xi), \quad D^\alpha \widehat{\phi}(\xi) = (-\mathrm{i})^{|\alpha|} \widehat{x^\alpha \phi}(\xi),$$
$$\mathscr{F}^{-1}(\mathscr{F}\phi) = \phi.$$

Moreover, $\mathscr{F}$ ($\mathscr{F}^{-1}$) : $\mathscr{S} \to \mathscr{S}$ is a continuously linear bijection, i.e., an isomorphism.

Proposition 1.2. *Let $\phi, \psi \in \mathscr{S}$. Then*

$$\int_{\mathbb{R}^n} \phi(x)\widehat{\psi}(x) dx = \int_{\mathbb{R}^n} \widehat{\phi}(x)\psi(x) dx.$$

We define the convolution $\phi * \psi$ on $\mathscr{S}$ in the following way

$$\phi * \psi(x) = \int_{\mathbb{R}^n} \phi(x-y)\psi(y)dy.$$

Proposition 1.3. *Let $\phi, \psi \in \mathscr{S}$. Then*

$$\widehat{\phi * \psi} = (2\pi)^{n/2}\widehat{\phi} \cdot \widehat{\psi}, \quad \widehat{\phi \cdot \psi} = (2\pi)^{-n/2}\widehat{\phi} * \widehat{\psi}.$$

We denote by σ_λ and τ_h the dilation and the translation operators:

$$\sigma_\lambda \phi = \phi(\lambda \cdot), \quad \lambda \in \mathbb{R} \setminus \{0\},$$
$$\tau_h \phi = \phi(\cdot - h), \quad h \in \mathbb{R}^n.$$

Proposition 1.4. *Let $\phi \in \mathscr{S}$, $\lambda \in \mathbb{R} \setminus \{0\}$, $h \in \mathbb{R}^n$. Then*

$$\widehat{\tau_h \phi} = e^{-ih\xi}\widehat{\phi},$$
$$\widehat{e^{ihx}\phi} = \tau_h \widehat{\phi},$$
$$\widehat{\sigma_\lambda \phi} = |\lambda|^{-n}(\sigma_{1/\lambda}\widehat{\phi}).$$

For the proofs of the above propositions, except that the proof of $\mathscr{F}^{-1}\widehat{\phi} = \phi$ needs a special technique, say

$$\widehat{e^{-|x|^2/2}} = e^{-|\xi|^2/2},$$

the other proofs are easily obtained.

In the following we consider the Fourier transform of $f \in \mathscr{S}'$. Inspired by Proposition 1.2, we define $\widehat{f}$ as follows.

$$(\mathscr{F} f)(\phi) = \widehat{f}(\phi) = f(\widehat{\phi}), \quad \phi \in \mathscr{S},$$
$$(\mathscr{F}^{-1} f)(\phi) = \check{f}(\phi) = f(\check{\phi}), \quad \phi \in \mathscr{S}.$$

Proposition 1.5. *$\mathscr{F} : \mathscr{S}' \to \mathscr{S}'$ is a continuously linear bijection map, i.e., an isomorphism.*

Now we can define the convolution of $f \in \mathscr{S}'$ and $\phi \in \mathscr{S}$. Let us recall that for any $f, \phi, \psi \in \mathscr{S}$, we have

$$\int_{\mathbb{R}^n} (f * \phi)(x)\psi(x)dx = \int_{\mathbb{R}^n} f(x)(\tilde{\phi} * \psi)(x)dx,$$

where $\tilde{\phi} = \phi(-\cdot)$. Following this idea, we define

$$(f * \phi)(\psi) = f(\tilde{\phi} * \psi), \quad f \in \mathscr{S}', \ \phi, \psi \in \mathscr{S}.$$

Taking notice of the definition of the Fourier transform and the convolution operator on $\mathscr{S}'$, we can generalize the other operators to $\mathscr{S}'$. We first investigate the corresponding operation for Schwartz functions, which can be converted to an integral operation. Then taking into account the abstract form of this integration operation, we can generalize the operation to $\mathscr{S}'$. Let $f \in \mathscr{S}'$, we define

$$D^\alpha f(\phi) := f((-D)^\alpha \phi),$$
$$\tau_h f(\phi) := f(\tau_{-h}\phi), \quad h \in \mathbb{R}^n,$$
$$\sigma_\lambda f(\phi) := f(\lambda^{-n}\sigma_{1/\lambda}\phi), \quad \lambda > 0.$$

It is known that $f \in \mathscr{S}'$ does not mean that f is a function, the following result indicates that the convolution of f and a Schwartz function must be an infinitely differentiable function.

Proposition 1.6. *Let $f \in \mathscr{S}'$ and $\phi \in \mathscr{S}$. Then $f * \phi$ is an infinitely differentiable function and*

$$(f * \phi)(x) = f(\tau_x \tilde{\phi}).$$

Moreover, $f * \phi$ grows at most in a polynomial way, i.e., for any $\alpha \in \mathbb{Z}^n_+$ there exists a polynomial P_α, such that $|D^\alpha(f * \phi)(x)| \leqslant P_\alpha(x)$, $x \in \mathbb{R}^n$.

Proposition 1.7 (Paley-Wiener-Schwartz theorem).

(1) *$\phi \in \mathscr{S}$ and $\operatorname{supp} \widehat{\phi} \subset \{\xi : |\xi| \leqslant b\}$ if and only if, $\phi(z)$ $(z = x + iy)$ is an entire analytic function of n complex variables and for any $\varepsilon > 0$, $\lambda > 0$, there exists $C_{\varepsilon,\lambda} > 0$, such that*

$$|\phi(z)| \leqslant C_{\varepsilon,\lambda}(1 + |x|)^{-\lambda} e^{(b+\varepsilon)|y|}, \quad x, y \in \mathbb{R}^n.$$

(2) *$f \in \mathscr{S}'$ and $\operatorname{supp} \widehat{f} \subset \{\xi : |\xi| \leqslant b\}$ if and only if, $f(z)$ $(z = x + iy)$ is an entire analytic function of n complex variables and for some $\lambda \in \mathbb{R}$ and for any $\varepsilon > 0$, there exists $C_\varepsilon > 0$ such that*

$$|f(z)| \leqslant C_\varepsilon(1 + |x|)^{\lambda} e^{(b+\varepsilon)|y|}, \quad x, y \in \mathbb{R}^n.$$

Proposition 1.7 is a fundamental result in the framework of the Fourier analysis, which says that if the Fourier transform of a tempered distribution f has a compact support, then it is an analytic function.

1.2 Fourier multiplier on L^p

We denote by $L^p := L^p(\mathbb{R}^n)$ the Lebesgue space, $1 \leqslant p \leqslant \infty$, the norm on L^p is written as $\|\cdot\|_p$. Let us consider the linear heat equation in L^p,

$$u_t - \Delta u = 0, \quad u|_{t=0} = u_0. \tag{1.3}$$

It is easy to see that $u = \mathscr{F}^{-1}e^{-t|\xi|^2}\mathscr{F}u_0 =: H(t)u_0$ is the solution of (1.3). For any $u_0 \in L^p$, we want to know if $H(t)u_0$ also belongs to L^p. More precisely, does

$$\|\mathscr{F}^{-1}e^{-t|\xi|^2}\mathscr{F}u_0\|_p \lesssim \|u_0\|_p, \quad 1 \leqslant p \leqslant \infty \tag{1.4}$$

hold for all $u_0 \in L^p$? The answer is affirmative (see Chapter 2). One can generalize (1.4) to the following notion.

Definition 1.1. Let $\rho \in \mathscr{S}'$. If there exists $C > 0$ such that[2]

$$\|\mathscr{F}^{-1}\rho\mathscr{F}f\|_p \leqslant C\|f\|_p, \quad \forall\, f \in \mathscr{S}, \tag{1.5}$$

then ρ is said to be a multiplier on L^p. The set of all multipliers on L^p is denoted by M_p , for which the norm is given by

$$\|\rho\|_{M_p} = \sup\{\|\mathscr{F}^{-1}\rho\mathscr{F}f\|_p : f \in \mathscr{S}, \|f\|_p = 1\}. \tag{1.6}$$

If $\rho \in M_p$, $1 \leqslant p < \infty$, since $\mathscr{S}$ is dense in L^p, one can extend $\mathscr{F}^{-1}\rho\mathscr{F}$ into a bounded operator on L^p and the norm of the extension operator is preserved. The extension operator is still written as $\mathscr{F}^{-1}\rho\mathscr{F}$ or $(\mathscr{F}^{-1}\rho)*$.

The multiplier theory is a fundamental tool in the harmonic analysis, and very useful in the definition of various function spaces. In what follows we discuss some basic properties of the multiplier space M_p.

Proposition 1.8. *Let $1 \leqslant p \leqslant q \leqslant 2$. Then we have*

(1) $M_p = M_{p'}$, $\quad \|\cdot\|_{M_p} = \|\cdot\|_{M_{p'}}$;
(2) $M_p \subset M_q$, $\quad \|\cdot\|_{M_q} \leqslant \|\cdot\|_{M_p}$;
(3) $M_2 = L^\infty$, $\quad \|\cdot\|_{M_2} = \|\cdot\|_\infty$;
(4) $M_1 = \{\rho \in \mathscr{S}' : \mathscr{F}^{-1}\rho$ *is a bounded measure*$\}$, $\|\rho\|_{M_1}$ *equals the total variation of* $\mathscr{F}^{-1}\rho$.

Proof. First, we prove (1). Let $\rho \in M_p$. For any $f,\ g \in \mathscr{S}$, $\|f\|_p = \|g\|_{p'} = 1$, we have[3]

$$|(\mathscr{F}^{-1}\rho\mathscr{F}f,\ g)| \leqslant \|\mathscr{F}^{-1}\rho\mathscr{F}f\|_p\|g\|_{p'} \leqslant \|\rho\|_{M_p}. \tag{1.7}$$

Noticing that $(\mathscr{F}^{-1}\rho\mathscr{F}f,\ g) = (\mathscr{F}^{-1}\rho * f * \widetilde{\overline{g}})(0)$, we see that (1.7) implies $\mathscr{F}^{-1}\rho\mathscr{F}g \in L^{p'}$ and $\rho \in M_{p'}$ with $\|\rho\|_{M_{p'}} \leqslant \|\rho\|_{M_p}$. It follows that (1) holds true.

[2]Notice that $\mathscr{F}^{-1}\rho\mathscr{F}f = (\mathscr{F}^{-1}\rho) * f$, in view of Proposition 1.6, we see that $\mathscr{F}^{-1}\rho\mathscr{F}f$ is a smooth function for any $f \in \mathscr{S}$.

[3]$(f,\ g) = \int f(x)\overline{g}(x)dx$.

Secondly, we show the result of (2). Due to $1 \leqslant p \leqslant q \leqslant 2$, we see that $p' \geqslant 2$. So, one can choose $\theta \in [0,1]$ satisfying $1/q = (1-\theta)/p + \theta/p'$. If $\rho \in M_p$, in view of (1) we have $\rho \in M_{p'}$. Namely,

$$\mathscr{F}^{-1}\rho\mathscr{F} : L^p \to L^p; \ \ \mathscr{F}^{-1}\rho\mathscr{F} : L^{p'} \to L^{p'} \tag{1.8}$$

are both bounded operators. It follows from Riesz-Thorin's interpolation theorem (see Appendix) that $\mathscr{F}^{-1}\rho\mathscr{F} : L^q \to L^q$ is bounded and

$$\|\rho\|_{M_q} \leqslant \|\rho\|^{1-\theta}_{M_p}\|\rho\|^{\theta}_{M_{p'}} = \|\rho\|_{M_p}. \tag{1.9}$$

Thirdly, we prove (3). If $\rho \in M_2$, by Plancherel's identity,

$$\|\mathscr{F}^{-1}\rho\mathscr{F}f\|_2 = \|\rho\mathscr{F}f\|_2 \leqslant \|\rho\|_\infty\|f\|_2. \tag{1.10}$$

So, $\|\rho\|_{M_2} \leqslant \|\rho\|_\infty$. Conversely, for any $\varepsilon > 0$, one can take a non-zero measurable closed subset E of $\mathbb{R}^n$, satisfying $|\rho(\xi)| \geqslant \|\rho\|_\infty - \varepsilon$ in E. Let $f \in L^2$ satisfy $\operatorname{supp}\mathscr{F}f \subset E$. It follows that $\|\rho\|_{M_2} \geqslant \|\rho\|_\infty - \varepsilon$.

Finally, we show that (4) holds. By the translation property of $\mathscr{F}^{-1}\rho\mathscr{F}$, we easily see that $\rho \in M_\infty$ if and only if

$$|(\mathscr{F}^{-1}\rho * f)(0)| = (2\pi)^{n/2}|(\mathscr{F}^{-1}\rho\mathscr{F}f)(0)| \leqslant C\|f\|_\infty, \quad \forall\, f \in \mathscr{S}. \tag{1.11}$$

Noticing that $\mathscr{S}$ is dense in $C_0(\mathbb{R}^n)$, from (1.11) and Proposition 1.6 one sees that $\mathscr{F}^{-1}\rho$ is a continuous functional on $C_0(\mathbb{R}^n)^4$. From the construction of $C_0(\mathbb{R}^n)^*$, one has the result. □

Proposition 1.9. *M_p is a Banach algebra.*

Proof. Obviously, $\|\cdot\|_{M_p}$ is a norm. By (3) in Proposition 1.8, one sees that $M_p \subset L^\infty$. If $\{\rho_k\}$ is a Cauchy sequence in M_p, so does in L^∞. We may assume, without loss of generality that it converges to ρ in L^∞. Since $L^\infty \subset \mathscr{S}'$, we conclude that for any $f \in \mathscr{S}$, $\mathscr{F}^{-1}\rho_k\mathscr{F}f \to \mathscr{F}^{-1}\rho\mathscr{F}f$ according to the strong topology on $\mathscr{S}'$. Recalling that $\mathscr{F}^{-1}\rho_k\mathscr{F}f$ is a Cauchy sequence in $L^p \subset \mathscr{S}'$, it has a limit ponit g. By the uniqueness of the limit in $\mathscr{S}'$ one obtains that $g = \mathscr{F}^{-1}\rho\mathscr{F}f$. So, $\|\rho_k - \rho\|_{M_p} \to 0$ as $k \to \infty$. It follows that M_p is a Banach space. Let $\rho_1, \rho_2 \in M_p$. For any $f \in \mathscr{S}$, we have

$$\|\mathscr{F}^{-1}\rho_1\rho_2\mathscr{F}f\|_p \leqslant \|\rho_1\|_{M_p}\|\mathscr{F}^{-1}\rho_2\mathscr{F}f\|_p \leqslant \|\rho_1\|_{M_p}\|\rho_2\|_{M_p}\|f\|_p, \tag{1.12}$$

which implies that $\rho_1\rho_2 \in M_p$ and

$$\|\rho_1\rho_2\|_{M_p} \leqslant \|\rho_1\|_{M_p}\|\rho_2\|_{M_p}. \tag{1.13}$$

Hence, M_p is a Banach algebra. □

[4]We denote by $C_0(\mathbb{R}^n)$ the space of all continuous functions $f(x)$ that vanish as $x \to \infty$.

In order to emphasize that M_p depends on $\mathbb{R}^n$, we write $M_p = M_p(\mathbb{R}^n)$. The following proposition indicates that $M_p(\mathbb{R}^n)$ is isometrically invariant under affine transforms[5] of $\mathbb{R}^n$.

Proposition 1.10. *Let $a: \mathbb{R}^n \to \mathbb{R}^m$ $(n \geqslant m)$ be a surjective affine transform, and $\rho \in M_p(\mathbb{R}^m)$. Then*

$$\|\rho(a(\cdot))\|_{M_p(\mathbb{R}^n)} = \|\rho\|_{M_p(\mathbb{R}^m)}. \tag{1.14}$$

In particular,

$$\|\rho(c\,\cdot)\|_{M_p(\mathbb{R}^n)} = \|\rho\|_{M_p(\mathbb{R}^n)}, \quad c \neq 0; \tag{1.15}$$

$$\|\rho(\langle x, \cdot\rangle)\|_{M_p(\mathbb{R}^n)} = \|\rho\|_{M_p(\mathbb{R})}, \quad x \neq 0, \tag{1.16}$$

where $\langle x, \xi\rangle = \sum_{i=1}^n x_i\xi_i$.

Proof. It suffices to consider the case that $a: \mathbb{R}^n \to \mathbb{R}^m$ is a linear transform. Make the coordinate transformation

$$\eta_i = a_i(\xi),\ 1 \leqslant i \leqslant m; \quad \eta_j = \xi_j,\ m+1 \leqslant j \leqslant n, \tag{1.17}$$

which is written as $\eta = A^{-1}\xi$ or $\xi = A\eta$. Let A_* be the transposed matrix of A. It is easy to see, for any $f \in \mathscr{S}$, that

$$\mathscr{F}^{-1}\rho(a(\xi))\mathscr{F}f = \left[\mathscr{F}^{-1}\rho(\eta_1, ..., \eta_m)\mathscr{F}f(A_*^{-1}\cdot)\right](A_*\cdot). \tag{1.18}$$

It follows from $\rho \in M_p(\mathbb{R}^m)$ that for any $f \in \mathscr{S}$

$$\begin{aligned}\|\mathscr{F}^{-1}\rho(a(\cdot))\mathscr{F}f\|_{L^p(\mathbb{R}^n)} &= |A|^{-1}\|\mathscr{F}^{-1}\rho\mathscr{F}f(A_*^{-1}\cdot)\|_{L^p(\mathbb{R}^n)}\\ &\leqslant \|\rho\|_{M_p(\mathbb{R}^m)}\|f\|_{L^p(\mathbb{R}^n)}.\end{aligned} \tag{1.19}$$

Thus, we have

$$\|\rho(a(\cdot))\|_{M_p(\mathbb{R}^n)} \leqslant \|\rho\|_{M_p(\mathbb{R}^m)}. \tag{1.20}$$

Taking $f(A_*^{-1}\cdot) = f_1(x_1, ..., x_m)f_2(x_{m+1}, ..., x_n)$, one can conclude that the inverse inequality of (1.20) also holds. □

In the following we give some criteria how to determine the multiplier in L^p.

Proposition 1.11 (Bernstein multiplier theorem). *Let $L > n/2$ be an integer, $\partial_{x_i}^\alpha \rho \in L^2$, $i = 1, ..., n$ and $0 \leqslant \alpha \leqslant L$. Then we have $\rho \in M_p$, $1 \leqslant p \leqslant \infty$ and*

$$\|\rho\|_{M_p} \lesssim \|\rho\|_2^{1-n/2L}\left(\sum_{i=1}^n \|\partial_{x_i}^L \rho\|_2\right)^{n/2L}. \tag{1.21}$$

[5]An affine transform of $\mathbb{R}^n$ is a map $F: \mathbb{R}^n \to \mathbb{R}^n$ of the form $F(\mathbf{p}) = A\mathbf{p} + \mathbf{q}$ for all $\mathbf{p} \in \mathbb{R}^n$, where A is a linear transform of $\mathbb{R}^n$.

Proof. By Proposition 1.8, it suffices to consider the case $p = 1$. Obviously,

$$\|\rho\|_{M_1} \leqslant \int_{\mathbb{R}^n} |\mathscr{F}^{-1}\rho(x)|dx. \tag{1.22}$$

For any $t > 0$,

$$\int_{|x|<t} |\mathscr{F}^{-1}\rho(x)|dx \lesssim t^{n/2}\|\rho\|_2. \tag{1.23}$$

Denote $J(x) = \sum_{i=1}^n |x_i|^L$. One has that

$$\begin{aligned}\int_{|x|>t} |\mathscr{F}^{-1}\rho(x)|dx &= \int_{|x|>t} J(x)^{-1}J(x)\left|\mathscr{F}^{-1}\rho(x)\right| dx \\ &\lesssim t^{n/2-L}\sum_{i=1}^n \left\|\partial^L_{x_i}\rho\right\|_2.\end{aligned} \tag{1.24}$$

Taking t such that $\|\rho\|_2 = t^{-L}\sum_{i=1}^n \|\partial^L_{x_i}\rho\|_2$, from (1.22)–(1.24) we have the result, as desired. □

Proposition 1.12 (Mihlin multiplier theorem). *Let $L > n/2$ be an integer and $\rho \in L^\infty$ satisfy*

$$|\xi|^{|\alpha|}|D^\alpha\rho(\xi)| \leqslant A, \quad |\alpha| \leqslant L, \quad \xi \in \mathbb{R}^n \setminus \{0\}. \tag{1.25}$$

Then $\rho \in M_p$ for any $1 < p < \infty$ and

$$\|\rho\|_{M_p} \leqslant C_p A, \tag{1.26}$$

where C_p may depend on p.

Proof. See [13]. □

1.3 Dyadic decomposition, Besov and Triebel spaces

Ingenious ideas are usually beautiful and resonating ideas, the dyadic decomposition is one of them.

We consider the decomposition on $\mathbb{R}^n$,

$$\mathcal{R}_0 = \{\xi : |\xi| < 1\}, \quad \mathcal{R}_k = \{\xi : 2^{k-1} \leqslant |\xi| < 2^k\}, \quad k \in \mathbb{N}. \tag{1.27}$$

It is easy to see that $\{\mathcal{R}_k\}_{k=0}^\infty$ is a pairwise disjoint sequence and $\mathbb{R}^n = \cup_{k=0}^\infty \mathcal{R}_k$, namely, $\{\mathcal{R}_k\}_{k=0}^\infty$ constitutes a decomposition of $\mathbb{R}^n$. According to this decomposition, we can roughly define the Littlewood-Paley decomposition operators in the following way

$$\triangle_k \sim \mathscr{F}^{-1}\chi_{\mathcal{R}_k}\mathscr{F}, \; k \in \mathbb{N} \cup \{0\}, \tag{1.28}$$

where $\chi_{\mathcal{R}_k}$ denotes the characteristic function on $\mathcal{R}_k$. The dyadic decomposition operator is a frequency-localized operator, which is one of the most delicate and clever ideas in harmonic analysis. If $\triangle_k$ takes the form as in (1.28), it has an advantage that $\triangle_k$'s ($k \geqslant 0$) orthogonalize each other, but the characteristic function in (1.28) has no smoothness and it is hard to use. In order to make multiplier calculations, we need to adopt the smooth version of (1.28).

Let $\psi: \mathbb{R}^n \to [0,1]$ be a smooth radial cut-off function, say

$$\psi(\xi) = \begin{cases} 1, & |\xi| \leqslant 1, \\ \text{smooth}, & 1 < |\xi| < 2, \\ 0, & |\xi| \geqslant 2. \end{cases} \tag{1.29}$$

Denote

$$\varphi(\xi) = \psi(\xi) - \psi(2\xi), \tag{1.30}$$

and we introduce the function sequence $\{\varphi_k\}_{k=0}^{\infty}$:

$$\begin{cases} \varphi_k(\xi) = \varphi(2^{-k}\xi), \quad k \in \mathbb{N}, \\ \varphi_0(\xi) = 1 - \sum_{k=1}^{\infty} \varphi_k(\xi) = \psi(\xi). \end{cases} \tag{1.31}$$

Since $\operatorname{supp}\varphi \subset \{\xi : \ 2^{-1} \leqslant |\xi| \leqslant 2\}$, we easily see that $\operatorname{supp}\varphi_k \subset \{\xi : 2^{k-1} \leqslant |\xi| \leqslant 2^{k+1}\}$, $k \in \mathbb{N}$, $\operatorname{supp}\varphi_0 \subset \{\xi : \ |\xi| \leqslant 2\}$. Define

$$\triangle_k = \mathscr{F}^{-1}\varphi_k\mathscr{F}, \quad k \in \mathbb{N} \cup \{0\}, \tag{1.32}$$

$\{\triangle_k\}_{k=0}^{\infty}$ is said to be the Littlewood-Paley (or dyadic) decomposition operator. Formally, we find that[6]

$$\sum_{k=0}^{\infty} \triangle_k = I. \tag{1.33}$$

Combining the dyadic decomposition operator with the function spaces[7] $\ell^q(L^p)$ and $L^p(\ell^q)$, we can introduce Besov spaces $B^s_{p,q}$ and Triebel-Lizorkin spaces $F^s_{p,q}$, respectively. Let

$$-\infty < s < \infty, \quad 1 \leqslant p, q \leqslant \infty. \tag{1.34}$$

[6]Actually, this decomposition works just for any locally integrable function which has some decay at the infinity, and one usually has all the convergence properties of the summation that one needs. In many applications, one can make the *a priori* assumption that f is Schwartz, in which case the convergence is uniform. However, if the function does not decay, this formula fails. For instance, if $f \equiv 1$, then all $\triangle_k f$'s vanish because $\triangle_k 1 = \varphi_k(0) = \varphi(0) = 0$.

[7]$\|(a_k)\|_{\ell^q(L^p)} := \left(\sum_k \|a_k(x)\|_p^q\right)^{1/q}, \quad \|(a_k)\|_{L^p(\ell^q)} := \left\|(\sum_k |a_k(x)|^q)^{1/q}\right\|_p.$

We define

$$B^s_{p,q} = \left\{ f \in \mathscr{S}'(\mathbb{R}^n) : \|f\|_{B^s_{p,q}} < \infty \right\}, \tag{1.35}$$

$$\|f\|_{B^s_{p,q}} := \left(\sum_{k=0}^{\infty} 2^{ksq} \|\triangle_k f\|_p^q \right)^{1/q}, \tag{1.36}$$

$B^s_{p,q}$ is said to be the Besov space[8]. Assume that

$$-\infty < s < \infty, \quad 1 \leqslant p < \infty, \quad 1 \leqslant q \leqslant \infty, \tag{1.37}$$

we define the following

$$F^s_{p,q} = \left\{ f \in \mathscr{S}'(\mathbb{R}^n) : \|f\|_{F^s_{p,q}} < \infty \right\}, \tag{1.38}$$

$$\|f\|_{F^s_{p,q}} := \left\| \left(\sum_{k=0}^{\infty} 2^{ksq} |\triangle_k f|^q \right)^{1/q} \right\|_p, \tag{1.39}$$

$F^s_{p,q}$ is said to be the Triebel-Lizorkin space[9]. Note that we need replace the ℓ^q-norm by the ℓ^∞-norm in the above definition if $q = \infty$.

Besov and Triebel-Lizorkin spaces had been formulated during 1960s–1980s, which have been widely applied in recent years. Roughly speaking, s is the spatial regularity index; $\triangle_k f$ is the frequency localization of f at the frequency $|\xi| \sim 2^k$; the norms of the function sequence $\{2^{sk}|\triangle_k f|\}$ in $\ell^q(L^p)$ and $L^p(\ell^q)$ generate Besov and Triebel norms of f, respectively.

The dyadic decomposition in the frequency space goes back to the equivalent norm on $L^p(\mathbb{R}^n)$

$$\|f\|_p \sim \|f\|_{F^0_{p,2}}, \quad 1 < p < \infty. \tag{1.40}$$

(1.40) is the well known Littlewood-Paley square function theorem (see Appendix).

If there is no explanation, we will assume that conditions (1.34) and (1.37) are satisfied for Besov and Triebel spaces $B^s_{p,q}$ and $F^s_{p,q}$, respectively. For simplicity, we will use $X^s_{p,q}$ to denote $B^s_{p,q}$ or $F^s_{p,q}$.[10]

Proposition 1.13. *Let $X^s_{p,q} = B^s_{p,q}$ ($X^s_{p,q} = F^s_{p,q}$). There hold the following inclusions.*

---

[8] In the definition of $B^s_{p,q}$, one can also consider the case $0 < p \wedge q < 1$, see Triebel [224].

[9] $F^s_{\infty,q}$ can be found in Triebel [224].

[10] When X appears in the both sides of an inclusion, X should be the same, namely $X = B$ in both sides, or $X = F$ in both sides.

(1) *Let* $q_1 \leqslant q_2$. *Then*

$$X^s_{p,q_1} \subset X^s_{p,q_2}. \tag{1.41}$$

(2) *Let* $\varepsilon > 0$ *and* $1 \leqslant q_1, q_2 \leqslant \infty$. *Then*

$$X^{s+\varepsilon}_{p,q_1} \subset X^s_{p,q_2}. \tag{1.42}$$

(3) *Let* $p < \infty$. *Then*

$$B^s_{p,p\wedge q} \subset F^s_{p,q} \subset B^s_{p,p\vee q}, \tag{1.43}$$

where $p \wedge q = \min(p,q)$ *and* $p \vee q = \max(p,q)$.

Proof. Since $\ell^p \subset \ell^{p+a}$, $a \geqslant 0$, we can easily get the result of (1). Let us observe that

$$\left(\sum_{k=0}^{\infty} 2^{skq_2}|a_k|^{q_2}\right)^{1/q_2} \lesssim \sup_{k\geqslant 0} 2^{(s+\varepsilon)k}|a_k|. \tag{1.44}$$

Taking $a_k = \|\triangle_k f\|_p$ or $a_k = |\triangle_k f|$, we can show that (2) holds with the help of (1). Finally, we prove (3). Denote $b_k = 2^{sk}\triangle_k f$. We divide the proof into the following two cases.

Case 1. $q \leqslant p$. In view of $\ell^q \subset \ell^p$ and Minkowski's inequality,[11] we have

$$\|b_k\|_{\ell^p(L^p)} \leqslant \|b_k\|_{L^p(\ell^q)} \leqslant \|b_k\|_{\ell^q(L^p)}, \tag{1.45}$$

which implies the result.

Case 2. $q \geqslant p$. Analogous to Case 1, we can use Minkowski's inequality and $\ell^p \subset \ell^q$ to get the result, as desired. □

Proposition 1.14. *Let* $X^s_{p,q}$ *be* $B^s_{p,q}$ *or* $F^s_{p,q}$. *Then*

(1) $X^s_{p,q}$ *is a Banach space;*
(2) $\mathscr{S}(\mathbb{R}^n) \subset X^s_{p,q} \subset \mathscr{S}'(\mathbb{R}^n)$;
(3) *If* $1 \leqslant p, q < \infty$, *then* $\mathscr{S}(\mathbb{R}^n)$ *is dense in* $X^s_{p,q}$.

Proof. Since $\ell^q(L^p)$ and $L^p(\ell^q)$ are normed spaces, one can conclude that $X^s_{p,q}$ is a normed space. In order to show the result of (1), it suffices to prove that $X^s_{p,q}$ is complete, whose proof will be left to the end. The proof of (2) is separated into the following four steps.

Step 1. We show that $\mathscr{S} \subset B^s_{p,\infty}$. In fact, for sufficiently large $L, M, N \in \mathbb{N}$,

$$\|f\|_{B^s_{p,\infty}} = \sup_{k\geqslant 0} 2^{sk}\|\triangle_k f\|_p$$

[11] Minkowski's inequalities read
i) $\|\sum_{j=0}^{\infty} f_j\|_p \leqslant \sum_{j=0}^{\infty}\|f_j\|_p$, for any $p \in [1,\infty]$;
ii) $\sum_{j=0}^{\infty}\|f_j\|_p \leqslant \|\sum_{j=0}^{\infty} f_j\|_p$, for any $p \in (0,1)$ and $f_j \geqslant 0$.

$$\begin{aligned}
&\lesssim \sup_{k\geqslant 0} 2^{sk}\|(1+|x|^2)^L \triangle_k f\|_\infty \\
&\lesssim \sup_{k\geqslant 0} 2^{sk}\|(I-\triangle)^L \varphi_k \mathscr{F} f\|_1 \\
&\lesssim \|(1+|\cdot|^2)^M (I-\triangle)^L \mathscr{F} f\|_\infty \\
&\lesssim p_N(\mathscr{F} f). \qquad (1.46)
\end{aligned}$$

Since $\mathscr{F} : \mathscr{S} \to \mathscr{S}$ is a continuous map, from (1.46) it follows the result, as desired.

Step 2. We prove that $\mathscr{S} \subset X^s_{p,q}$. Since $\mathscr{S} \subset B^{s+\varepsilon}_{p,\infty}$, by Proposition 1.13, we see that $B^{s+\varepsilon}_{p,\infty} \subset B^s_{p,p\wedge q} \subset B^s_{p,q} \cap F^s_{p,q}$. Thus, we have the desired result.

Step 3. We show that $B^s_{p,\infty} \subset \mathscr{S}'$. For convenience, we denote $\varphi_{-1} \equiv 0$. From the construction of φ_k, one has that $\varphi_k \varphi_{k+\ell} \equiv 0$ for $\ell \neq -1, 0, 1$. For any $f \in B^s_{p,\infty}$, $\psi \in \mathscr{S}$, we can take $N \in \mathbb{N}$ sufficiently large,

$$\begin{aligned}
|\langle f, \psi\rangle| &\leqslant \sum_{k=0}^{\infty} \sum_{\ell=-1}^{1} |\langle \triangle_k f, \ \mathscr{F}\varphi_{k+\ell}\mathscr{F}^{-1}\psi\rangle| \\
&\lesssim \sum_{k=0}^{\infty} \sum_{\ell=-1}^{1} \|\triangle_k f\|_p \|\triangle_{k+\ell}\psi\|_{p'} \\
&\lesssim \|f\|_{B^s_{p,\infty}} \sum_{k=0}^{\infty} 2^{-sk} \|\triangle_k \psi\|_{p'} \\
&\lesssim \|f\|_{B^s_{p,\infty}} \|\psi\|_{B^{-s+\varepsilon}_{p',\infty}} \\
&\lesssim \|f\|_{B^s_{p,\infty}} p_N(\mathscr{F}\psi). \qquad (1.47)
\end{aligned}$$

For any bounded set B in $\mathscr{S}$, we have $p_N(\mathscr{F}\psi) \lesssim 1$ for $\psi \in B$. The result follows.

Step 4. Similarly as in Step 2, we can show that $X^s_{p,q} \subset \mathscr{S}'$ and the details of the proof are omitted.

Finally, we prove the completeness of $B^s_{p,q}$. In an analogous way we can get that $F^s_{p,q}$ is also complete. Assume that $\{f_\ell\}_{\ell=1}^\infty$ is a Cauchy sequence in $B^s_{p,q}$. By (2) we see that it is also a Cauchy sequence in $\mathscr{S}'$. Since $\mathscr{S}'$ is a complete and locally convex topological linear space, there exists an $f \in \mathscr{S}'$ such that $f_\ell \to f$ in the sense of the strongly topology of $\mathscr{S}'$. On the other hand, that $\{f_\ell\}_{\ell=1}^\infty$ is a Cauchy sequence in $B^s_{p,q}$ implies that $\{\triangle_k f_\ell\}_{\ell=1}^\infty$ is a Cauchy sequence in L^p. By the completeness of L^p, there exists a $g_k \in L^p$ such that

$$\|\triangle_k f_\ell - g_k\|_p \to 0, \quad \text{as } \ell \to \infty. \qquad (1.48)$$

Due to $\triangle_k f_\ell \to \triangle_k f$ in $\mathscr{S}'$ as $\ell \to \infty$ and $L^p \subset \mathscr{S}'$, we immediately have $g_k = \triangle_k f$. Hence, (1.48) implies that

$$\|\triangle_k(f_\ell - f)\|_p \to 0, \quad \text{as } \ell \to \infty. \tag{1.49}$$

(1.49) and Fatou's lemma yield $\|f_\ell - f\|_{B^s_{p,q}} \to 0$ as $\ell \to \infty$. □

1.4 Embeddings on $X^s_{p,q}$

It is known that there is no inclusion between $L^p(\mathbb{R}^n)$ and $L^q(\mathbb{R}^n)$ $(p \neq q)$ and we can not compare $\|\cdot\|_p$ and $\|\cdot\|_q$. However, if we localize f in a compact subset Ω of $\mathbb{R}^n$ in the frequency space, then we have

$$\|\mathscr{F}^{-1}\chi_\Omega \mathscr{F} f\|_q \lesssim \|\mathscr{F}^{-1}\chi_\Omega \mathscr{F} f\|_p, \quad p \leqslant q. \tag{1.50}$$

Combining this fact with the dyadic decomposition operator, we can obtain the inclusions between Besov spaces in a simple way. This is a great advantage of the frequency localized techniques.

Let Ω be a compact subset of $\mathbb{R}^n$. Denote

$$\mathscr{S}_\Omega = \{f \in \mathscr{S} : \ \operatorname{supp} \mathscr{F} f \subset \Omega\}. \tag{1.51}$$

Proposition 1.15. *Let $1 \leqslant p \leqslant q \leqslant \infty$. Then we have*

$$\|f\|_q \lesssim \|f\|_p, \quad \forall\, f \in \mathscr{S}_\Omega. \tag{1.52}$$

Proof. Let $\psi \in \mathscr{S}$ satisfy $\mathscr{F}\psi(\xi) = 1$, $\forall\, \xi \in \Omega$. In view of $f \in \mathscr{S}_\Omega$ we see that $\mathscr{F} f = \mathscr{F} f \cdot \mathscr{F}\psi$. It follows that

$$f(x) = C \int_{\mathbb{R}^n} \psi(x-y) f(y) dy. \tag{1.53}$$

Applied Hölder's inequality,

$$\|f\|_\infty \lesssim \|\psi\|_{p'} \|f\|_p \lesssim \|f\|_p, \tag{1.54}$$

which implies that for $q \geqslant p$,

$$\|f\|_q \leqslant \|f\|_\infty^{1-p/q} \|f\|_p^{p/q} \lesssim \|f\|_p, \tag{1.55}$$

the result follows. □

Denote

$$L^p_\Omega = \{f \in L^p : \ \operatorname{supp} \mathscr{F} f \subset \Omega\}. \tag{1.56}$$

Using (1.51) and standard approximate techniques, one obtains that

Proposition 1.16. *Let $1 \leqslant p \leqslant q \leqslant \infty$. Then*

$$\|f\|_q \lesssim \|f\|_p, \quad \forall\, f \in L^p_\Omega. \tag{1.57}$$

Corollary 1.1. *Let $1 \leqslant p \leqslant q \leqslant \infty$, $B_\lambda = \{\xi : |\xi| \leqslant \lambda\}$. Then*

$$\|f\|_q \lesssim \lambda^{n(1/p-1/q)}\|f\|_p, \quad \forall\, f \in L^p_{B_\lambda}. \tag{1.58}$$

Proof. Due to

$$\|f\|_q = \lambda^{-n/q}\|f(\cdot/\lambda)\|_q, \tag{1.59}$$

$$\mathscr{F}(f(\cdot/\lambda)) = \lambda^n(\mathscr{F}f)(\lambda\cdot), \tag{1.60}$$

we see, for any $f \in L^p_{B_\lambda}$, that $f(\cdot/\lambda) \in L^p_{B_1}$. In view of (1.59), (1.60) and Proposition 1.16 we obtain the result, as desired. □

Let φ_k be defined by (1.31). Since $\operatorname{supp}\varphi_k \subset B_{2^{k+1}}$, for any $1 \leqslant p_1 \leqslant p_2 \leqslant \infty$, we have, from (1.58), that Bernstein's inequality holds

$$\|\triangle_k f\|_{p_2} \lesssim 2^{n(1/p_1-1/p_2)k}\|\triangle_k f\|_{p_1}. \tag{1.61}$$

Assume that s_1 and s_2 satisfy

$$s_1 - \frac{n}{p_1} = s_2 - \frac{n}{p_2}. \tag{1.62}$$

By (1.61) and (1.62), we immediately have

$$2^{s_2k}\|\triangle_k f\|_{p_2} \lesssim 2^{s_1k}\|\triangle_k f\|_{p_1}. \tag{1.63}$$

Taking the ℓ^r-norm in both sides of (1.63), one has that

$$\|f\|_{B^{s_2}_{p_2,r}} \lesssim \|f\|_{B^{s_1}_{p_1,r}}. \tag{1.64}$$

Thus, we have shown that

Theorem 1.1. *Let $1 \leqslant p_1 \leqslant p_2 \leqslant \infty$, $1 \leqslant r \leqslant \infty$ satisfy $s_1 - n/p_1 = s_2 - n/p_2$. Then*

$$B^{s_1}_{p_1,r} \subset B^{s_2}_{p_2,r}. \tag{1.65}$$

Considering the inclusions on $F^s_{p,q}$, we have a similar result which is slightly better than that of (1.65).

Theorem 1.2. *Let $1 \leqslant p_1 < p_2 < \infty$, $1 \leqslant q, r \leqslant \infty$, $-\infty < s_2 < s_1 < \infty$ satisfy $s_1 - n/p_1 = s_2 - n/p_2$. Then*

$$F^{s_1}_{p_1,q} \subset F^{s_2}_{p_2,r}. \tag{1.66}$$

Proof. By Proposition 1.13, we need to show

$$F^{s_1}_{p_1,\infty} \subset F^{s_2}_{p_2,1}. \tag{1.67}$$

We may assume that $\|f\|_{F^{s_1}_{p_1,\infty}} = 1$. Let us recall an equivalent norm on L^p:

$$\|g\|_p^p = p\int_0^\infty t^{p-1}|\{x : |g(x)| > t\}|dt, \tag{1.68}$$

where $|E|$ denotes the measure of the set E. So,

$$\begin{aligned}\|f\|^{p_2}_{F^{s_2}_{p_2,1}} \sim & \int_0^A t^{p_2-1}\left|\left\{x : \sum_{k=0}^\infty 2^{ks_2}|(\triangle_k f)(x)| > t\right\}\right| dt \\ & + \int_A^\infty t^{p_2-1}\left|\left\{x : \sum_{k=0}^\infty 2^{ks_2}|(\triangle_k f)(x)| > t\right\}\right| dt \\ =: & I + II,\end{aligned} \tag{1.69}$$

where $A \gg 1$ is a constant which can be chosen as below. It is easy to see that

$$\sum_{k=K+1}^\infty 2^{ks_2}|\triangle_k f| \lesssim 2^{K(s_2-s_1)} \sup_{k\geqslant 0} 2^{ks_1}|\triangle_k f|. \tag{1.70}$$

Applying (1.70) we can get the estimate of I ($K = -1$),

$$\begin{aligned} I \lesssim & \int_0^A t^{p_2-1}\left|\left\{x : \sup_{k\geqslant 0} 2^{ks_1}|(\triangle_k f)(x)| > ct\right\}\right| dt \\ \lesssim & \int_0^{cA} \tau^{p_1-1}\left|\left\{x : \sup_{k\geqslant 0} 2^{ks_1}|(\triangle_k f)(x)| > \tau\right\}\right| d\tau \\ \lesssim & 1. \end{aligned} \tag{1.71}$$

Now we estimate II. By Corollary 1.1,

$$\|\triangle_k f\|_\infty \lesssim 2^{kn/p_1}\|\triangle_k f\|_{p_1} \lesssim 2^{k(n/p_1-s_1)}\|f\|_{F^{s_1}_{p_1,\infty}}. \tag{1.72}$$

Hence, for $K \in \mathbb{N}\cup\{0\}$,

$$\sum_{k=0}^K 2^{ks_2}|\triangle_k f| \lesssim \sum_{k=0}^K 2^{k(s_2-s_1+n/p_1)} \lesssim 2^{Kn/p_2}. \tag{1.73}$$

Taking K to be the largest natural number satisfying $C2^{Kn/p_2} \leqslant t/2$, we have $2^K \sim t^{p_2/n}$. It is easy to see that such a K exists if $t \geqslant A \gg 1$. For $t \geqslant A$ and $\sum_{k=0}^\infty 2^{ks_2}|(\triangle_k f)(x)| \geqslant t$, from (1.70) and (1.73) we get

$$C2^{K(s_2-s_1)} \sup_{k\geqslant 0} 2^{ks_1}|\triangle_k f| \geqslant \sum_{k=K+1}^\infty 2^{ks_2}|\triangle_k f| > t/2. \tag{1.74}$$

Combining (1.68) and (1.74), we obtain that

$$\begin{aligned} II &\lesssim \int_A^\infty t^{p_2-1} \left| \left\{ x : \sup_{k\geqslant 0} 2^{ks_1} |(\triangle_k f)(x)| > ct^{p_2/p_1} \right\} \right| dt \\ &\lesssim \int_{A'}^\infty \tau^{p_1-1} \left| \left\{ x : \sup_{k\geqslant 0} 2^{ks_1} |(\triangle_k f)(x)| > \tau \right\} \right| d\tau \\ &\lesssim 1. \end{aligned} \tag{1.75}$$

The estimates of I and II imply the conclusion. □

Proposition 1.17. *Let $1 \leqslant p < \infty$, $s > n/p$ and $1 \leqslant q \leqslant \infty$. Let $X^s_{p,q}$ be $B^s_{p,q}$ or $F^s_{p,q}$. Then*

$$X^s_{p,q} \subset B^0_{\infty,1} \subset L^\infty. \tag{1.76}$$

Proof. Using the dyadic decomposition and (1.61), we have

$$\begin{aligned} \|u\|_\infty &\leqslant \sum_{k=0}^\infty \|\triangle_k u\|_\infty \lesssim \sum_{k=0}^\infty 2^{kn/p} \|\triangle_k u\|_p \\ &\lesssim \left(\sum_{k=0}^\infty 2^{k(n/p-s)} \right) \|u\|_{B^s_{p,\infty}} \leqslant \|u\|_{X^s_{p,q}}, \end{aligned} \tag{1.77}$$

which is the result, as desired. □

1.5 Differential-difference norm on $X^s_{p,q}$

The earliest version of Besov spaces was defined by an integration with a differential-difference form, not by dyadic decomposition operators. In this section we consider the equivalent differential-difference norm, which is quite convenient when we make nonlinear estimates in the study of nonlinear evolution equations.

Denote

$$\triangle_h^m f(x) = \sum_{k=0}^m C_m^k (-1)^{m-k} f(x+kh), \tag{1.78}$$

$$\omega_p^m(t,f) = \sup_{|h|\leqslant t} \| \triangle_h^m f\|_p. \tag{1.79}$$

We have

Proposition 1.18. *Let $s > 0$, $m, N \in \mathbb{N}$ and $m + N > s$, $0 \leqslant N < s$; $1 \leqslant p, q \leqslant \infty$. Then*

$$\|f\|_{B^s_{p,q}} \sim \|f\|_p + \sum_{j=1}^n \left(\int_0^\infty \left(t^{N-s} \omega_p^m(t, \partial^N_{x_j} f) \right)^q \frac{dt}{t} \right)^{1/q}. \tag{1.80}$$

Proof. Since $\omega^m_p(t,f)$ is an increasing function of $t>0$, it suffices to show that

$$\|f\|_{B^s_{p,q}} \sim \|f\|_p + \sum_{j=1}^n \left(\sum_{i=-\infty}^{\infty} \left(2^{i(s-N)}\omega^m_p(2^{-i},\ \partial^N_{x_j} f)\right)^q \right)^{1/q}. \quad (1.81)$$

Let $f \in B^s_{p,q}$. We write $\rho_h(\xi) = e^{ih\xi} - 1$. One has that

$$\| \triangle^m_h \partial^N_{x_j} f\|_p = \|\mathscr{F}^{-1}\rho^m_h \mathscr{F}\partial^N_{x_j} f\|_p \leqslant \sum_{k=0}^{\infty} \|\mathscr{F}^{-1}\rho^m_h \varphi_k \mathscr{F}\partial^N_{x_j} f\|_p. \quad (1.82)$$

In the following we show that[12]

$$\|\mathscr{F}^{-1}\rho^m_h \varphi_k \mathscr{F}\partial^N_{x_j} f\|_p \lesssim \min(1, |h|^m 2^{km}) 2^{Nk} \|\triangle_k f\|_p. \quad (1.83)$$

Noticing the support set of φ_k ($\varphi_{-1} = 0$),

$$\|\mathscr{F}^{-1}\rho^m_h \varphi_k \mathscr{F}\partial^N_{x_j} f\|_p \lesssim \sum_{\ell=-1}^{1} \|\mathscr{F}^{-1}\rho^m_h \varphi_{k+\ell} \mathscr{F}\triangle_k\partial^N_{x_j} f\|_p. \quad (1.84)$$

By the definition of ρ_h we have $\rho_h \in M_p$. From the fact that M_p is a Banach algebra, it follows that $\rho^m_h \in M_p$. In view of (1.15) in Proposition 1.10, we get $\varphi_k \in M_p$. Hence, $\rho^m_h\varphi_k \in M_p$. By (1.84),

$$\|\mathscr{F}^{-1}\rho^m_h \varphi_k \mathscr{F}\partial^N_{x_j} f\|_p \lesssim \|\triangle_k \partial^N_{x_j} f\|_p. \quad (1.85)$$

Observing the following identity

$$\rho^m_h \varphi_k = \frac{\rho^m_h}{\langle h, \xi\rangle^m} \left\langle \frac{h}{|h|}, \frac{\xi}{|\xi|} \right\rangle^m (|h||\xi|)^m \varphi_k \quad (1.86)$$

and noticing that $\|\rho_h/\langle h,\xi\rangle\|_{M_p(\mathbb{R}^n)} = \|(e^{i\xi}-1)/\xi\|_{M_p(\mathbb{R})} < \infty$, it follows that for all $k \geqslant 1$,

$$\|\rho^m_h \varphi_k\|_{M_p} \lesssim 2^{km}|h|^m. \quad (1.87)$$

Obviously, (1.87) also holds for $k = 0$. Hence, in view of (1.84), we have

$$\|\mathscr{F}^{-1}\rho^m_h \varphi_k \mathscr{F}\partial^N_{x_j} f\|_p \lesssim 2^{km}|h|^m \|\triangle_k \partial^N_{x_j} f\|_p. \quad (1.88)$$

Applying the technique as in (1.84), we have from (1.15) that

$$\|\partial^N_{x_j} \triangle_k f\|_p \lesssim 2^{Nk}\|\triangle_k f\|_p. \quad (1.89)$$

Collecting (1.85), (1.88) and (1.89), we get (1.83). Inserting (1.83) into (1.82), one has that

$$2^{i(s-N)}\omega^m_p(2^{-i}, \partial^N_{x_j} f) \lesssim \sum_{k=0}^{\infty} 2^{(i-k)(s-N)}(1 \wedge 2^{(k-i)m}) 2^{sk}\|\triangle_k f\|_p. \quad (1.90)$$

[12] Notice that $\triangle_h$ and $\triangle_k$ are different operators.

By (1.90) and Young's inequality, we obtain

$$\sum_{j=1}^{n}\left(\sum_{i=-\infty}^{\infty}\left(2^{i(s-N)}\omega_p^m(2^{-i},\ \partial_{x_j}^N f)\right)^q\right)^{1/q} \lesssim \|f\|_{B^s_{p,q}}. \tag{1.91}$$

Clearly, $\|f\|_p \lesssim \|f\|_{B^s_{p,q}}$. Thus, the right hand side of (1.80) can be controlled by its left hand side.

Next, we show the remaining part of (1.80). Let $\rho_{jk}(\xi) = e^{i2^{-k}\xi_j} - 1$. It suffices to prove that

$$\|\triangle_k f\|_p \lesssim 2^{-Nk}\sum_{j=1}^{n}\|\mathscr{F}^{-1}\rho_{jk}^m \mathscr{F}\partial_{x_j}^N f\|_p, \quad k \geqslant 1. \tag{1.92}$$

In fact, if (1.92) holds, in view of the definition of $\omega_p^m(t, f)$, we have

$$\|\mathscr{F}^{-1}\rho_{jk}^m \mathscr{F}\partial_{x_j}^N f\|_p \lesssim \omega_p^m(2^{-k},\ \partial_{x_j}^N f), \quad k \geqslant 1. \tag{1.93}$$

It follows from (1.93) and the Besov norm that ($\varphi_0 \in M_p$),

$$\|f\|_{B^s_{p,q}} \lesssim \|f\|_p + \sum_{j=1}^{n}\left(\sum_{k=1}^{\infty}\left(2^{k(s-N)}\omega_p^m(2^{-k},\ \partial_{x_j}^N f)\right)^q\right)^{1/q}, \tag{1.94}$$

which implies that the left hand side of (1.80) can be controlled by its right hand side.

Finally, we prove (1.92). We need the following lemma

Lemma 1.1. *There exist smooth* χ_j $(j = 1, ..., n)$ *satisfying*

$$\sum_{j=1}^{n}\chi_j(\xi) = 1, \ \forall\, \xi \in \{\xi : 1/2 \leqslant |\xi| \leqslant 2\}; \tag{1.95}$$

$$\operatorname{supp}\chi_j \subset \{\xi = (\xi_1, ..., \xi_n) : \ |\xi_j| \geqslant 1/3\sqrt{n}\}. \tag{1.96}$$

Proof. We can choose that $\kappa \in \mathscr{S}(\mathbb{R})$, $\zeta \in \mathscr{S}(\mathbb{R}^{n-1})$ with $\operatorname{supp}\kappa = \{\xi \in \mathbb{R} : |\xi| \geqslant 1/3\sqrt{n}\}$, $\operatorname{supp}\zeta = \{\xi \in \mathbb{R}^{n-1} : |\xi| \leqslant 3\}$ and with positive values in the interiors of $\operatorname{supp}\kappa$ and $\operatorname{supp}\zeta$, respectively. We write $\xi^j = (\xi_1, ..., \xi_{j-1}, \xi_{j+1}, ..., \xi_n)$ and

$$\chi_j(\xi) = \begin{cases} \frac{\kappa(\xi_j)\zeta(\xi^j)}{\sum_{j=1}^n \kappa(\xi_j)\zeta(\xi^j)}, & \sum_{j=1}^n \kappa(\xi_j)\zeta(\xi^j) \neq 0, \\ 0, & \sum_{j=1}^n \kappa(\xi_j)\zeta(\xi^j) = 0. \end{cases} \tag{1.97}$$

It is easy to see that χ_j satisfies (1.95) and (1.96). □

Now we can finish the proof of (1.92). By Lemma 1.1, for any $k \geqslant 1$,

$$\|\triangle_k f\|_p \leqslant \sum_{j=1}^{n} \|\mathscr{F}^{-1}\rho_{jk}^{-m}\varphi_k\chi_j(2^{-k}\cdot)\xi_j^{-N}\mathscr{F}(\mathscr{F}^{-1}\rho_{jk}^{m}\mathscr{F}\partial_{x_j}^N f)\|_p. \quad (1.98)$$

By Proposition 1.11,

$$(e^{\mathrm{i}\xi_j} - 1)^{-m}\varphi\chi_j\xi_j^{-N} \in M_p, \quad (1.99)$$

where φ is as in (1.30). It follows from Propositions 1.10 and (1.98) that (1.92) holds. □

As a consequence of Proposition 1.18, we have

Corollary 1.2. *Let $s > 0$, $s \notin \mathbb{N}$. we denote by $[s]$ the integer part of s, $1 \leqslant p, q \leqslant \infty$. Then*

$$\|f\|_{B^s_{p,q}} \sim \|f\|_p + \sum_{j=1}^{n} \left(\int_0^\infty \left(t^{[s]-s} \sup_{|h| \leqslant t} \| \triangle_h \, \partial_{x_j}^{[s]} f\|_p \right)^q \frac{dt}{t} \right)^{1/q}, \quad (1.100)$$

where $\triangle_h := \triangle_h^1$.

It is very convenient to use (1.100) when we make nonlinear mapping estimates.

1.6 Homogeneous space $\dot{X}^s_{p,q}$

Let φ be as in (1.30). Analogous to (1.31), we write

$$\varphi_k(\xi) = \varphi(2^{-k}\xi), \quad k \in \mathbb{Z}. \quad (1.101)$$

It is easy to see that

$$\sum_{k \in \mathbb{Z}} \varphi_k(\xi) = 1, \quad \xi \in \mathbb{R}^n \setminus \{0\}. \quad (1.102)$$

According to (1.32), we introduce the homogeneous dyadic decomposition operators

$$\triangle_k = \mathscr{F}^{-1}\varphi_k\mathscr{F}, \quad k \in \mathbb{Z}. \quad (1.103)$$

Similar to $X^s_{p,q}$, using $\{\triangle_k\}_{k\in\mathbb{Z}}$ and function spaces $\ell^q(L^p)$ and $L^p(\ell^q)$, one can define $\dot{X}^s_{p,q}$. Notice that in (1.101)–(1.103), there is no restriction on $\xi = 0$ and so, one needs to modify the Schwartz space $\mathscr{S}$ and it dual $\mathscr{S}'$. Denote

$$\dot{\mathscr{S}}(\mathbb{R}^n) = \{f \in \mathscr{S}(\mathbb{R}^n) : (D^\alpha \widehat{f})(0) = 0, \ \forall \alpha\}. \quad (1.104)$$

As a subspace of $\mathscr{S}$, $\dot{\mathscr{S}} := \dot{\mathscr{S}}(\mathbb{R}^n)$ is equipped with the same topology as $\mathscr{S}$. We denote by $\dot{\mathscr{S}}' := \dot{\mathscr{S}}'(\mathbb{R}^n)$ the dual space of $\dot{\mathscr{S}}$. We now introduce the homogeneous space $\dot{X}^s_{p,q}$. Let

$$-\infty < s < \infty, \ 1 \leqslant p, q \leqslant \infty. \tag{1.105}$$

Denote

$$\dot{B}^s_{p,q} = \left\{ f \in \dot{\mathscr{S}}'(\mathbb{R}^n) : \|f\|_{\dot{B}^s_{p,q}} < \infty \right\}, \tag{1.106}$$

$$\|f\|_{\dot{B}^s_{p,q}} := \left(\sum_{k=-\infty}^{\infty} 2^{ksq} \|\triangle_k f\|_p^q \right)^{1/q}, \tag{1.107}$$

$\dot{B}^s_{p,q}$ is said to be a homogeneous Besov space. Let

$$-\infty < s < \infty, \quad 1 \leqslant p < \infty, \quad 1 \leqslant q \leqslant \infty. \tag{1.108}$$

Define

$$\dot{F}^s_{p,q} = \left\{ f \in \dot{\mathscr{S}}'(\mathbb{R}^n) : \|f\|_{\dot{F}^s_{p,q}} < \infty \right\}, \tag{1.109}$$

$$\|f\|_{\dot{F}^s_{p,q}} := \left\| \left(\sum_{k=-\infty}^{\infty} 2^{ksq} |\triangle_k f|^q \right)^{1/q} \right\|_p, \tag{1.110}$$

$\dot{F}^s_{p,q}$ is said to be a homogeneous Triebel-Lizorkin space.

If there is no confusion, we will write $\dot{X}^s_{p,q}$ to stand for $\dot{B}^s_{p,q}$ or $\dot{F}^s_{p,q}$. Using the dilation, one has that

$$\|f(2^\ell \cdot)\|_{\dot{X}^s_{p,q}} = 2^{\ell(n-s/p)} \|f\|_{\dot{X}^s_{p,q}}, \tag{1.111}$$

from which one can easily understand why we call $\dot{X}^s_{p,q}$ as a homogeneous space. $\dot{X}^s_{p,q}$ and $X^s_{p,q}$ are quite similar. In the following we only state some results on $\dot{X}^s_{p,q}$, whose proofs follow an analogous way as the nonhomogeneous case. If there is no explanation, we will always assume that conditions (1.105) and (1.108) are satisfied for the spaces $\dot{B}^s_{p,q}$ and $\dot{F}^s_{p,q}$, respectively.

Proposition 1.19. *Let $\dot{X}^s_{p,q} = \dot{B}^s_{p,q}$ (or $\dot{X}^s_{p,q} = \dot{F}^s_{p,q}$). Then*

(1) *$\dot{X}^s_{p,q}$ is a Banach space;*
(2) *$\dot{\mathscr{S}} \subset \dot{X}^s_{p,q} \subset \dot{\mathscr{S}}'$;*
(3) *If $1 \leqslant p, q < \infty$, then $\dot{\mathscr{S}}(\mathbb{R}^n)$ is dense in $\dot{X}^s_{p,q}$;*
(4) *If $q_1 \leqslant q_2$, then $\dot{X}^s_{p,q_1} \subset \dot{X}^s_{p,q_2}$;*
(5) *Let $1 \leqslant p < \infty$. Then $\dot{B}^s_{p,p\wedge q} \subset \dot{F}^s_{p,q} \subset \dot{B}^s_{p,p\vee q}$.*

Theorem 1.3. *Let* $-\infty < s_2 < s_1 < \infty$, $s_1 - n/p_1 = s_2 - n/p_2$ *and* $1 \leqslant r, q \leqslant \infty$

$$\dot{B}^{s_1}_{p_1,r} \subset \dot{B}^{s_2}_{p_2,r}, \quad \dot{F}^{s_1}_{p_1,r} \subset \dot{F}^{s_2}_{p_2,q}.$$

In the sequel, we will frequently use Theorem 1.3, where the first inclusion can be shown in the same way as in Besov space, the second one is slightly different from the Triebel-Lizorkin space, see Appendix.

Theorem 1.4. *Let* $s > 0$, $m, N \in \mathbb{N}$, $N < s, N+m > s$ *and* $1 \leqslant p, q \leqslant \infty$. *Then we have*

$$\|f\|_{\dot{B}^s_{p,q}} \sim \sum_{j=1}^{n} \left(\int_0^\infty t^{q(N-s)} \omega_p^m (t, \partial^N_{x_j} f)^q \frac{dt}{t} \right)^{1/q}.$$

In particular, if $s > 0$ *and* $s \notin \mathbb{N}$, *then*

$$\|f\|_{\dot{B}^s_{p,q}} \sim \sum_{j=1}^{n} \left(\int_0^\infty t^{q([s]-s)} \sup_{|h| \leqslant t} \| \Delta_h \, \partial^{[s]}_{x_j} f\|_p^q \frac{dt}{t} \right)^{1/q}.$$

The following result is a straightforward consequence of Proposition 1.18 and Theorem 1.4, which indicates the relation between homogeneous and nonhomogeneous spaces.

Proposition 1.20. *Let* $s > 0$ *and* $1 \leqslant p, q \leqslant \infty$. *Then we have*

$$B^s_{p,q} = L^p \cap \dot{B}^s_{p,q}.$$

The following is an interpolation inequality in Besov spaces, which are very useful in nonlinear estimates, see [79; 89].

Proposition 1.21 (Convexity Hölder's inequality). *Let* $1 \leqslant p_i, q_i \leqslant \infty$, $0 \leqslant \theta_i \leqslant 1$, $\sigma_i, \sigma \in \mathbb{R}$ $(i = 1, \ldots, N)$, $\sum_{i=1}^N \theta_i = 1$, $\sigma = \sum_{i=1}^N \theta_i \sigma_i$, $1/p = \sum_{i=1}^N \theta_i/p_i$, $1/q = \sum_{i=1}^N \theta_i/q_i$. *Then* $\cap_{i=1}^N \dot{B}^{\sigma_i}_{p_i,q_i} \subset \dot{B}^\sigma_{p,q}$ *and for any* $v \in \cap_{i=1}^N \dot{B}^{\sigma_i}_{p_i,q_i}$,

$$\|v\|_{\dot{B}^\sigma_{p,q}} \leqslant \prod_{i=1}^N \|v\|^{\theta_i}_{\dot{B}^{\sigma_i}_{p_i,q_i}}.$$

This estimate also holds if one substitutes $\dot{B}^\sigma_{p,q}$ *by* $\dot{F}^\sigma_{p,q}$ $(p, p_i \neq \infty)$.

Proof. Applying Hölder's inequality on L^p and ℓ^q respectively, one can easily deduces the result:

$$\|v\|_{\dot{B}^\sigma_{p,q}} = \left(\sum_{k \in \mathbb{Z}} 2^{\sigma k q} \|\Delta_k v\|_p^q \right)^{1/q}$$

$$\leqslant \left(\sum_{k\in\mathbb{Z}} 2^{\sigma kq} \prod_{i=1}^{N} \|\triangle_k v\|_{p_i}^{\theta_i q} \right)^{1/q}$$

$$= \left(\sum_{k\in\mathbb{Z}} \prod_{i=1}^{N} (2^{k\sigma_i\theta_i} \|\triangle_k v\|_{p_i}^{\theta_i})^q \right)^{1/q}$$

$$\leqslant \prod_{i=1}^{N} \|v\|_{\dot{B}^{\sigma_i}_{p_i,q_i}}^{\theta_i}.$$

The case $\dot{F}^{\sigma}_{p,q}$ is similar. □

1.7 Bessel (Riesz) potential spaces H^s_p ($\dot{H}^s_p$)

Recall that $J_s = (I-\Delta)^{s/2}$ and $I_s = (-\Delta)^{s/2}$ are said to be the Bessel and the Riesz potentials, respectively. Assume that

$$1 < p < \infty, \quad -\infty < s < \infty. \tag{1.112}$$

Denote

$$H^s_p = \left\{ f \in \mathscr{S}' : \|f\|_{H^s_p} := \|J_s f\|_p < \infty \right\}; \tag{1.113}$$

$$\dot{H}^s_p = \left\{ f \in \dot{\mathscr{S}}' : \|f\|_{\dot{H}^s_p} := \|I_s f\|_p < \infty \right\}. \tag{1.114}$$

H^s_p and $\dot{H}^s_p$ are said to be Bessel and Riesz potential spaces, respectively. In the following we give the equivalent norms on H^s_p and $\dot{H}^s_p$:

Theorem 1.5 (Littlewood-Paley square function theorem). *Let s and p satisfy* (1.112). *Then*

$$H^s_p = F^s_{p,2}, \quad \dot{H}^s_p = \dot{F}^s_{p,2} \tag{1.115}$$

with equivalent norms.

Theorem 1.5 can be found in Stein [202] and Triebel [224].

Theorem 1.6. *Let $-\infty < s_2 \leqslant s_1 < \infty$ and $1 < p_1 \leqslant p_2 < \infty$ with $s_1 - n/p_1 = s_2 - n/p_2$. Then we have*

$$H^{s_1}_{p_1} \subset H^{s_2}_{p_2}, \quad \dot{H}^{s_1}_{p_1} \subset \dot{H}^{s_2}_{p_2}. \tag{1.116}$$

Proof. It is a consequence of Theorems 1.5, 1.2 and 1.3. □

Proposition 1.22. *Let s and p satisfy* (1.112). *We have*

(1) H_p^s *is a Banach space;*
(2) $\mathscr{S} \subset H_p^s \subset \mathscr{S}'$ *and* $\mathscr{S}(\mathbb{R}^n)$ *is dense in* H_p^s;
(3) $B_{p,p}^s \subset H_p^s \subset B_{p,2}^s$ $(1 < p \leqslant 2)$, $B_{p,2}^s \subset H_p^s \subset B_{p,p}^s$ $(2 \leqslant p < \infty)$;
(4) $H_p^{s+\varepsilon} \subset H_p^s$ $(\varepsilon > 0)$;
(5) $H_p^s \subset L^\infty$ $(s > n/p)$.

Remark 1.1. The results in (1)–(3) of Proposition 1.22 also hold for Riesz potential spaces. The proof of Proposition 1.22 follows from Theorem 1.5 and the corresponding results of $F_{p,2}^s$.

Proposition 1.23. *Let* $s > 0$ *and* $1 < p < \infty$. *We have* $H_p^s = L^p \cap \dot{H}_p^s$. *Moreover, if* s *is an integer, then*

$$\|f\|_{\dot{H}_p^s} \sim \sum_{|\alpha|=s} \|D^\alpha f\|_p.$$

In the following we establish a modified Hölder inequality, as far as the author have seen, Pecher [190] gave the first result on the modified Hölder inequality. The next two results were obtained in Wang [234] by following Pecher [190].

Proposition 1.24 (Modified Hölder inequality I). *Let* $1 \leqslant p < \infty$ *and* $1 < p_i < \infty$. *Assume that* α_i *is a multi-index,* $|\alpha_i| \leqslant s_i$, $\rho_i > 0$, *and*

$$a_i = \rho_i \left(\frac{1}{p_i} - \frac{s_i - |\alpha_i|}{n} \right), \quad i = 0, 1, ..., N+1. \tag{1.117}$$

If $a_i > 0$ *and* $\sum_{i=0}^{N+1} a_i = 1/p$, *then we have*

$$\left\| \prod_{i=0}^{N+1} |D^{\alpha_i} u_i|^{\rho_i} \right\|_p \leqslant C \prod_{i=0}^{N+1} \|u_i\|_{\dot{H}_{p_i}^{s_i}}^{\rho_i}. \tag{1.118}$$

Substituted $\dot{H}_{p_i}^{s_i}$ *by* $H_{p_i}^{s_i}$ *in* (1.118), *the conclusion also holds.*

Proof. Let $q_i = 1/a_i$. Since $1/p = \sum_{i=0}^{N+1} 1/q_i$, we have, from Hölder's inequality, that

$$\left\| \prod_{i=0}^{N+1} |D^{\alpha_i} u_i|^{\rho_i} \right\|_p \leqslant \prod_{i=0}^{N+1} \|D^{\alpha_i} u_i\|_{\rho_i q_i}^{\rho_i} \lesssim \prod_{i=0}^{N+1} \|u_i\|_{\dot{H}_{\rho_i q_i}^{|\alpha_i|}}^{\rho_i}. \tag{1.119}$$

Noticing that $1/\rho_i q_i - |\alpha_i|/n = 1/p_i - s_i/n$, and using Theorem 1.6, we get $\dot{H}_{p_i}^{s_i} \subset \dot{H}_{\rho_i q_i}^{|\alpha_i|}$. By (1.119), we have the desired result. □

Proposition 1.25 (Modified Hölder's inequality II). *Let* $1 \leqslant p < \infty$, $1 < p_i < \infty$, $|\alpha_i| \leqslant s_i$, $\rho_i > 0$, *and* a_i *be as in* (1.117). *We have*

(1) *If each* $a_i \neq 0$, *and* $\sum_{a_i>0} a_i = 1/p$, *then*

$$\Big\| \prod_{i=0}^{N+1} |D^{\alpha_i} u_i|^{\rho_i} \Big\|_p \leqslant C \prod_{i=0}^{N+1} \|u_i\|^{\rho_i}_{H^{s_i}_{p_i}}. \tag{1.120}$$

(2) *If* $\sum_{a_i>0} a_i < 1/p \leqslant \sum_{i=1}^{N} \rho_i/p_i$, *then* (1.120) *also holds.*

Proof. First, we prove (1). We may assume $a_0, ..., a_K > 0$ and $a_{K+1}, ..., a_{N+1} < 0$. By Hölder's inequality,

$$\Big\| \prod_{i=0}^{N+1} |D^{\alpha_i} u_i|^{\rho_i} \Big\|_p \lesssim \prod_{i=0}^{K} \|D^{\alpha_i} u_i\|^{\rho_i}_{\rho_i q_i} \prod_{i=K+1}^{N+1} \|D^{\alpha_i} u_i\|^{\rho_i}_{\infty}. \tag{1.121}$$

By Proposition 1.24,

$$\prod_{i=0}^{K} \|D^{\alpha_i} u_i\|^{\rho_i}_{\rho_i q_i} \lesssim \prod_{i=0}^{K} \|u_i\|^{\rho_i}_{\dot{H}^{s_i}_{p_i}}. \tag{1.122}$$

Since $a_i < 0$ $(i = K+1, ..., N+1)$, in view of (5) of Proposition 1.22, we have

$$\prod_{i=K+1}^{N+1} \|D^{\alpha_i} u_i\|^{\rho_i}_{\infty} \lesssim \prod_{i=K+1}^{N+1} \|D^{\alpha_i} u_i\|^{\rho_i}_{H^{s_i-|\alpha_i|}_{p_i}} \lesssim \prod_{i=K+1}^{N+1} \|u_i\|^{\rho_i}_{H^{s_i}_{p_i}}. \tag{1.123}$$

Collecting (1.121)–(1.123), we obtain the result.

Next, we prove (2). We can assume that $a_0, ..., a_K > 0$, $a_{K+1} = ... = a_J = 0$ and $a_{J+1}, ..., a_{N+1} < 0$.

Case 1. $1/p \leqslant \sum_{i=0}^{K} \rho_i/p_i$. Take $q_{K+1}, ..., q_N \gg 1$ (for any $i > K$ and $q_i > p_i/\rho_i$) satisfying

$$\sum_{i=0}^{K} a_i + \sum_{i=K+1}^{N+1} \frac{1}{q_i} < \frac{1}{p} \leqslant \sum_{i=0}^{K} \frac{\rho_i}{p_i}. \tag{1.124}$$

Thus, we can choose appropriate q_i $(i = 0, ..., K)$ verifying $p_i/\rho_i \leqslant q_i \leqslant 1/a_i$ and $1/p = \sum_{i=0}^{N+1} 1/q_i$. By Proposition 1.25, we obtain

$$\Big\| \prod_{i=0}^{N+1} |D^{\alpha_i} u_i|^{\rho_i} \Big\|_p \leqslant C \prod_{i=0}^{N+1} \|u_i\|^{\rho_i}_{H^{|\alpha_i|}_{\rho_i q_i}}. \tag{1.125}$$

Since $1/q_i\rho_i \geqslant 1/p_i - (s - |\alpha_i|)/n$ and $p_i \leqslant q_i\rho_i$, we see that $H^{s_i}_{p_i} \subset H^{|\alpha_i|}_{\rho_i q_i}$. From (1.125) we have the result.

Case 2. $\sum_{i=0}^{K} \rho_i/p_i < 1/p \leqslant \sum_{i=0}^{N+1} \rho_i/p_i$. Let $q_i = p_i/\rho_i$ for $i = 0, ..., K$. For $i = K+1, ..., N+1$, we can find a $q_i \in [p_i/\rho_i, \infty)$ satisfying $1/p = \sum_{i=0}^{N+1} 1/q_i$. Analogous to the above discussions, we can obtain our result. □

1.8 Fractional Gagliardo-Nirenberg inequalities

We consider the Gagliardo-Nirenberg (GN) inequality in Besov and Triebel spaces. Using the dyadic decomposition, we obtain a very simple proof for the GN inequality. On the other hand, the GN inequality with fractional order derivatives is of interest for its own sake and has an independent significance. The results in this section are obtained in [99], earlier results in this topic can be found in [23; 162; 183; 233; 242].

1.8.1 *GN inequality in* $\dot{B}^s_{p,q}$

Lemma 1.2. *Let* $1 \leqslant p, p_0, p_1, q, q_0, q_1 \leqslant \infty$, $s, s_0, s_1 \in \mathbb{R}$, $0 \leqslant \theta \leqslant 1$. *Suppose that the following conditions hold:*

$$\frac{n}{p} - s = (1-\theta)\left(\frac{n}{p_0} - s_0\right) + \theta\left(\frac{n}{p_1} - s_1\right), \tag{1.126}$$

$$s \leqslant (1-\theta)s_0 + \theta s_1, \tag{1.127}$$

$$\frac{1}{q} \leqslant \frac{1-\theta}{q_0} + \frac{\theta}{q_1}. \tag{1.128}$$

Then the fractional GN inequality of the following type

$$\|u\|_{\dot{B}^s_{p,q}} \lesssim \|u\|^{1-\theta}_{\dot{B}^{s_0}_{p_0,q_0}} \|u\|^{\theta}_{\dot{B}^{s_1}_{p_1,q_1}} \tag{1.129}$$

holds for all $u \in \dot{B}^{s_0}_{p_0,q_0} \cap \dot{B}^{s_1}_{p_1,q_1}$.

Proof. First, we consider the case $1/q \leqslant (1-\theta)/q_0 + \theta/q_1$. By (1.127), we have

$$\frac{1}{p} - \frac{1-\theta}{p_0} - \frac{\theta}{p_1} = \frac{s}{n} - (1-\theta)\frac{s_0}{n} - \theta\frac{s_1}{n} := -\eta \leqslant 0. \tag{1.130}$$

Take p^* and s^* satisfying

$$\frac{1}{p^*} = \frac{1}{p} + \eta, \quad s^* = s + n\eta.$$

Applying the convexity Hölder inequality, we have

$$\|f\|_{\dot{B}^{s^*}_{p^*,q}} \leqslant \|f\|^{1-\theta}_{\dot{B}^{s_0}_{p_0,q_0}} \|f\|^{\theta}_{\dot{B}^{s_1}_{p_1,q_1}}. \tag{1.131}$$

Using the inclusion $\dot{B}^{s^*}_{p^*,q} \subset \dot{B}^s_{p,q}$, we get the conclusion. $\square$

The result below clarify how the third indices q in $\dot{B}^s_{p,q}$ contribute the validity of the GN inequalities.

Theorem 1.7. *Let $1 \leqslant q < \infty$, $1 \leqslant p$, p_0, $p_1 \leqslant \infty$, $0 < \theta < 1$, $s, s_0, s_1 \in \mathbb{R}$. Then the fractional GN inequality of the following type*

$$\|u\|_{\dot{B}^s_{p,q}} \lesssim \|u\|^{1-\theta}_{\dot{B}^{s_0}_{p_0,\infty}} \|u\|^{\theta}_{\dot{B}^{s_1}_{p_1,\infty}} \tag{1.132}$$

holds if and only if

$$\frac{n}{p} - s = (1-\theta)\left(\frac{n}{p_0} - s_0\right) + \theta\left(\frac{n}{p_1} - s_1\right), \tag{1.133}$$

$$s_0 - \frac{n}{p_0} \neq s_1 - \frac{n}{p_1}, \tag{1.134}$$

$$s \leqslant (1-\theta)s_0 + \theta s_1, \tag{1.135}$$

$$p_0 = p_1 \quad \text{if} \quad s = (1-\theta)s_0 + \theta s_1. \tag{1.136}$$

Proof. (Sufficiency) We can assume that $s_0 = 0$ and the case $s_0 \neq 0$ can be shown by a similar way.

Step 1. We consider the case $p \geqslant p_0 \vee p_1$. By definition[13],

$$\|u\|_{\dot{B}^s_{p,q}} = \left(\sum_{N \text{ dyadic}} N^{sq}\|\triangle_N u\|^q_p\right)^{1/q}. \tag{1.137}$$

From (1.135), it follows that

$$\theta\left(\frac{n}{p} - \frac{n}{p_1} + s_1 - s\right) = (1-\theta)\left(s + \frac{n}{p_0} - \frac{n}{p}\right). \tag{1.138}$$

Since $0 < \theta < 1$, (1.134) implies that $(\frac{n}{p} - \frac{n}{p_1} + s_1 - s)(s + \frac{n}{p_0} - \frac{n}{p}) > 0$.

Case 1. We consider the case

$$s_1 - s + \frac{n}{p} - \frac{n}{p_1} > 0, \quad s + \frac{n}{p_0} - \frac{n}{p} > 0. \tag{1.139}$$

Using the inclusion $\dot{B}^s_{p,r_1} \subset \dot{B}^s_{p,r_2}$ for any $r_1 \leqslant r_2$, it suffices to consider the case $q < 1/2$, $q^{-1} \in \mathbb{N}$. For brevity, we write $K := q^{-1}$.

$$\|u\|_{\dot{B}^s_{p,q}} \leqslant \sum_{N_1 \geqslant \ldots \geqslant N_K} (N_1^s \ldots N_K^s \|\triangle_{N_1} u\|_p \ldots \|\triangle_{N_K} u\|_p)^{q^2} \times (N_1^s \ldots N_K^s \|\triangle_{N_1} u\|_p \ldots \|\triangle_{N_K} u\|_p)^{q(1-q)}. \tag{1.140}$$

[13] Here $\triangle_N$ is different from the notation as in Section 1.6, which is identical with $\triangle_{\log_2^N}$ as in (1.103).

In view of Bernstein's inequality,

$$\|\triangle_N u\|_p \leqslant N^{\frac{n}{p_0}-\frac{n}{p}}\|\triangle_N u\|_{p_0}, \quad \|\triangle_N u\|_p \leqslant N^{\frac{n}{p_1}-\frac{n}{p}}\|\triangle_N u\|_{p_1}. \tag{1.141}$$

We can choose $a \in (0,1], k \geqslant 1$ satisfying $\theta K = k-1+a$. Hence,

$$\begin{aligned}
&\|\triangle_{N_1}u\|_p \ldots \|\triangle_{N_K}u\|_p \\
&= (\|\triangle_{N_1}u\|_p \ldots \|\triangle_{N_{k-1}}u\|_p\|\triangle_{N_k}u\|_p^a)(\|\triangle_{N_k}u\|_p^{1-a}\|\triangle_{N_{k+1}}u\|_p \ldots \|\triangle_{N_K}u\|_p) \\
&\lesssim N_k^{(1-a)(\frac{n}{p_0}-\frac{n}{p})} N_{k+1}^{\frac{n}{p_0}-\frac{n}{p}} \ldots N_K^{\frac{n}{p_0}-\frac{n}{p}} \|\triangle_{N_k}u\|_{p_0}^{1-a}\|\triangle_{N_{k+1}}u\|_{p_0} \ldots \|\triangle_{N_K}u\|_{p_0} \\
&\times N_1^{\frac{n}{p_1}-\frac{n}{p}} \ldots N_{k-1}^{\frac{n}{p_1}-\frac{n}{p}} N_k^{a(\frac{n}{p_1}-\frac{n}{p})} \|\triangle_{N_1}u\|_{p_1} \ldots \|\triangle_{N_{k-1}}u\|_{p_1}\|\triangle_{N_k}u\|_{p_1}^a.
\end{aligned} \tag{1.142}$$

Inserting (1.142) into (1.140), we have

$$\begin{aligned}
\|u\|_{\dot{B}^s_{p,q}} \lesssim & \sum_{N_1 \geqslant \ldots \geqslant N_K} (N_1^s \ldots N_K^s \|\triangle_{N_1}u\|_p \ldots \|\triangle_{N_K}u\|_p)^{q^2} \\
& \times \Lambda(N_1, ..., N_K)\|u\|_{\dot{B}^{s_1}_{p_1,\infty}}^{q(1-q)\theta K}\|u\|_{\dot{B}^0_{p_0,\infty}}^{(1-\theta)Kq(1-q)},
\end{aligned} \tag{1.143}$$

where

$$\begin{aligned}
\Lambda(N_1, \ldots N_K) = \Big(& N_1^{-\frac{n}{p}+\frac{n}{p_1}-s_1+s} \ldots N_{k-1}^{-\frac{n}{p}+\frac{n}{p_1}-s_1+s} N_k^{a(-\frac{n}{p}+\frac{n}{p_1}-s_1+s)} \\
& \times N_k^{(1-a)(-\frac{n}{p}+\frac{n}{p_0}+s)} N_{k+1}^{-\frac{n}{p}+\frac{n}{p_0}+s} \ldots N_K^{-\frac{n}{p}+\frac{n}{p_0}+s}\Big)^{q(1-q)}.
\end{aligned} \tag{1.144}$$

By (1.143), we have

$$\|u\|_{\dot{B}^s_{p,q}} \lesssim \sum_{N_1 \geqslant \ldots \geqslant N_K} \Lambda(N_1, \ldots N_K) \sum_{i=1}^K (N_i^s \|\triangle_{N_i}u\|_p)^q \tag{1.145}$$

$$\times \|u\|_{\dot{B}^{s_1}_{p_1,\infty}}^{(1-q)\theta}\|u\|_{\dot{B}^0_{p_0,\infty}}^{(1-\theta)(1-q)}. \tag{1.146}$$

So, it suffices to prove

$$\sum_{N_1 \geqslant \ldots \geqslant N_K} \Lambda(N_1, \ldots N_K) \sum_{i=1}^K (N_i^s \|\triangle_{N_i}u\|_p)^q \lesssim \|u\|^q_{\dot{B}^s_{p,q}}. \tag{1.147}$$

In fact, (1.144)–(1.147) imply the result. Finally, we prove (1.147). Applying the condition (1.139), we have

$$\begin{aligned}
&\sum_{N_1 \geqslant \ldots \geqslant N_K} \Lambda(N_1, \ldots N_K)(N_k^s\|\triangle_{N_k}u\|_p)^q \\
&\lesssim \sum_{N_{k-1} \geqslant N_k} \Big(N_{k-1}^{(k-1)(s-s_1+\frac{n}{p_1}-\frac{n}{p})} N_k^{(K-k+1-a)(s+\frac{n}{p_0}-\frac{n}{p})+a(s-s_1+\frac{n}{p_1}-\frac{n}{p})}\Big)^{q(1-q)}
\end{aligned}$$

$$\times N_k^{sq}\|\Delta_{N_k}u\|_p^q$$
$$\lesssim \sum_{N_{k-1}\geqslant N_k}\left(\frac{N_{k-1}}{N_k}\right)^{(k-1)(s-s_1+\frac{n}{p_1}-\frac{n}{p})q(1-q)} N_k^{sq}\|\Delta_{N_k}u\|_p^q$$
$$\lesssim \|u\|^q_{\dot B^s_{p,q}}. \tag{1.148}$$

Case 2. We consider the case

$$s_1 - s + \frac{n}{p} - \frac{n}{p_1} < 0, \quad s + \frac{n}{p_0} - \frac{n}{p} < 0. \tag{1.149}$$

Substituting the summation $\sum_{N_1\geqslant\ldots\geqslant N_K}$ by $\sum_{N_1\leqslant\ldots\leqslant N_K}$ in (1.140) and repeating the procedure as in Case 1, we can get the result, as desired.

Up to now, we have shown the results for the following two cases: (i) $s = (1-\theta)s_0 + \theta s_1$ and $p_0 = p_1$; (ii) $s < (1-\theta)s_0 + \theta s_1$ and $p \geqslant p_0 \vee p_1$.

Step 2. We consider the case $p < p_0 \vee p_1$ and $s < (1-\theta)s_0 + \theta s_1$. Due to $\theta \in (0,1)$ and $1/p \leqslant (1-\theta)/p_0 + \theta/p_1$, we see that $p_0 \neq p_1$ and $p_0 \wedge p_1 < p < p_0 \vee p_1$. Let $0 < \varepsilon \ll 1$. In view of the result as in Step 1, we see that

$$\|f\|_{\dot B^s_{p,q}} \lesssim \|f\|^{1/2}_{\dot B^{s-\varepsilon}_{p,\infty}}\|f\|^{1/2}_{\dot B^{s+\varepsilon}_{p,\infty}}. \tag{1.150}$$

Since $s_0 - n/p_0 \neq s_1 - n/p_1$, we can assume that $s_0 - n/p_0 < s_1 - n/p_1$. It follows that $1/p - s/n \in (1/p_0 - s_0/n,\ 1/p_1 - s_1/n)$. Hence, for sufficiently small $\varepsilon > 0$,

$$\frac{1}{p} - \frac{s\pm\varepsilon}{n} \in \left(\frac{1}{p_0} - \frac{s_0}{n},\ \frac{1}{p_1} - \frac{s_1}{n}\right).$$

It follows that there exist $\theta_\pm \in (0,1)$ satisfying

$$\frac{1}{p} - \frac{s\pm\varepsilon}{n} = (1-\theta_\pm)\left(\frac{1}{p_0} - \frac{s_0}{n}\right) + \theta_\pm\left(\frac{1}{p_1} - \frac{s_1}{n}\right).$$

Due to $\lim_{\varepsilon\to 0}\theta_\pm = \theta$, we see that for sufficiently small $\varepsilon > 0$,

$$s \pm \varepsilon \leqslant (1-\theta_\pm)s_0 + \theta_\pm s_1.$$

Therefore, by Lemma 1.2, we have

$$\|f\|_{\dot B^{s-\varepsilon}_{p,\infty}} \lesssim \|f\|^{1-\theta_-}_{\dot B^{s_0}_{p_0,\infty}}\|f\|^{\theta_-}_{\dot B^{s_1}_{p_1,\infty}}, \tag{1.151}$$
$$\|f\|_{\dot B^{s+\varepsilon}_{p,\infty}} \lesssim \|f\|^{1-\theta_+}_{\dot B^{s_0}_{p_0,\infty}}\|f\|^{\theta_+}_{\dot B^{s_1}_{p_1,\infty}}. \tag{1.152}$$

We easily see that $\theta = (\theta_+ + \theta_-)/2$. Inserting (1.151) and (1.152) into (1.150), we have the result, as desired.

We omit the proof of the necessity, one can refer to [99] for details. □

For the most general case, we have the following

Theorem 1.8. *Let* $0 < p, p_0, p_1, q, q_0, q_1 \leqslant \infty$, $s, s_0, s_1 \in \mathbb{R}$, $0 \leqslant \theta \leqslant 1$. *Then the fractional GN inequality of the following type*

$$\|u\|_{\dot{B}^s_{p,q}} \lesssim \|u\|^{1-\theta}_{\dot{B}^{s_0}_{p_0,q_0}} \|u\|^{\theta}_{\dot{B}^{s_1}_{p_1,q_1}} \tag{1.153}$$

holds for all $u \in \dot{B}^{s_0}_{p_0,q_0} \cap \dot{B}^{s_1}_{p_1,q_1}$ *if and only if*

$$\frac{n}{p} - s = (1-\theta)\left(\frac{n}{p_0} - s_0\right) + \theta\left(\frac{n}{p_1} - s_1\right), \tag{1.154}$$

$$s \leqslant (1-\theta)s_0 + \theta s_1, \tag{1.155}$$

$$\frac{1}{q} \leqslant \frac{1-\theta}{q_0} + \frac{\theta}{q_1}, \quad \text{if } p_0 \neq p_1 \text{ and } s = (1-\theta)s_0 + \theta s_1, \tag{1.156}$$

$$s_0 \neq s_1 \quad \text{or} \quad \frac{1}{q} \leqslant \frac{1-\theta}{q_0} + \frac{\theta}{q_1}, \quad \text{if } p_0 = p_1 \text{ and } s = (1-\theta)s_0 + \theta s_1, \tag{1.157}$$

$$s_0 - \frac{n}{p_0} \neq s - \frac{n}{p} \quad \text{or} \quad \frac{1}{q} \leqslant \frac{1-\theta}{q_0} + \frac{\theta}{q_1}, \quad \text{if } s < (1-\theta)s_0 + \theta s_1. \tag{1.158}$$

Proof. See [99]. □

1.8.2 *GN inequality in* $\dot{F}^s_{p,q}$

In homogeneous Triebel-Lizorkin spaces $\dot{F}^s_{p,q}$, we have the following (cf. [99]):

Theorem 1.9. *Let* $1 \leqslant p, p_i, q < \infty$, $s, s_0, s_1 \in \mathbb{R}$, $0 < \theta < 1$. *Then the fractional GN inequality of the following type*

$$\|u\|_{\dot{F}^s_{p,q}} \lesssim \|u\|^{1-\theta}_{\dot{F}^{s_0}_{p_0,\infty}} \|u\|^{\theta}_{\dot{F}^{s_1}_{p_1,\infty}} \tag{1.159}$$

holds if and only if

$$\frac{n}{p} - s = (1-\theta)\left(\frac{n}{p_0} - s_0\right) + \theta\left(\frac{n}{p_1} - s_1\right), \tag{1.160}$$

$$s \leqslant (1-\theta)s_0 + \theta s_1, \tag{1.161}$$

$$s_0 \neq s_1 \quad \text{if} \quad s = (1-\theta)s_0 + \theta s_1. \tag{1.162}$$

Proof. We only prove the sufficiency, which is a reformulation of Oru's [183] (see also [23]) and the proof of the necessity can be found in [99]. First, we consider the case $s < (1-\theta)s_0 + \theta s_1$. We can take sufficiently small $\varepsilon > 0$ satisfying

$$s \leqslant (1-\theta)s_0^* + \theta s_1^*, \quad s_0^* := s_0 - \varepsilon,\ s_1^* := s_1 - \varepsilon.$$

Since $\varepsilon \ll 1$, we can assume that

$$\frac{1}{p_0^*} := \frac{1}{p_0} - \frac{\varepsilon}{n} > 0, \quad \frac{1}{p_1^*} := \frac{1}{p_1} - \frac{\varepsilon}{n} > 0.$$

Hence,

$$\frac{n}{p} - s = (1-\theta)\left(\frac{n}{p_0^*} - s_0^*\right) + \theta\left(\frac{n}{p_1^*} - s_1^*\right), \tag{1.163}$$

which implies that

$$\frac{1}{p} - \frac{1-\theta}{p_0^*} - \frac{\theta}{p_1^*} = \frac{s}{n} - (1-\theta)\frac{s_0^*}{n} - \theta\frac{s_1^*}{n} := -\eta \leqslant 0. \tag{1.164}$$

Putting

$$\frac{1}{p^*} = \frac{1}{p} + \eta, \quad s^* = s + n\eta, \tag{1.165}$$

we see that

$$\frac{1}{p^*} = \frac{1-\theta}{p_0^*} + \frac{\theta}{p_1^*}, \quad s^* = (1-\theta)s_0^* + \theta s_1^*. \tag{1.166}$$

Using Hölder's inequality, in an analogous way as in Besov spaces, we have

$$\|f\|_{\dot F^{s^*}_{p^*,q}} \lesssim \|f\|^{1-\theta}_{\dot F^{s_0^*}_{p_0^*,q}} \|f\|^{\theta}_{\dot F^{s_1^*}_{p_1^*,q}}.$$

Recalling the inclusions

$$\dot F^{s_0}_{p_0,\infty} \subset F^{s_0^*}_{p_0^*,q}, \quad \dot F^{s_1}_{p_1,\infty} \subset F^{s_1^*}_{p_1^*,q}$$

we immediately get the conclusion.

Next, we consider the case $s = (1-\theta)s_0 + \theta s_1$ and $s_0 \neq s_1$. In this case we easily see that $1/p = (1-\theta)/p_0 + \theta/p_1$. The result follows from Lemma C.1. □

The following is the GN inequality with fractional derivatives.

Corollary 1.3. *Let* $1 < p, p_0, p_1 < \infty$, s, $s_1 \in \mathbb{R}$, $0 \leqslant \theta \leqslant 1$. *Then the fractional GN inequality of the following type*

$$\|u\|_{\dot H^s_p} \lesssim \|u\|^{1-\theta}_{L^{p_0}} \|u\|^{\theta}_{\dot H^{s_1}_{p_1}} \tag{1.167}$$

holds if and only if

$$\frac{n}{p} - s = (1-\theta)\frac{n}{p_0} + \theta\left(\frac{n}{p_1} - s_1\right), \quad s \leqslant \theta s_1. \tag{1.168}$$

As the end of this chapter, we state some recent results on the generalizations of Besov's and Triebel-Lizorkin's spaces.

Remark 1.2.

(1) In [5], the Q_α space was introduced for which the norm is given by

$$\|f\|_{Q_\alpha(\mathbb{R}^n)} = \sup_I \left\{ \int_I \int_I \frac{|f(x) - f(y)|^2}{|x-y|^{n+2\alpha}} dx dy \right\}^{1/2} < \infty,$$

where I ranges over all cubes in $\mathbb{R}^n$. Xiao [253] considered the well posedness of the Navier-Stokes equation in Q type spaces.

(2) Yang and Yuan [254] gave a unified way to handle the Q_α and Triebel-Lizorkin spaces. Indeed, for any $p \in (1, \infty)$, $q \in (1, \infty]$, $\tau, s \in \mathbb{R}$, they introduced the space $\dot{F}^{s,\tau}_{p,q}$ in the following way

$$\|f\|_{\dot{F}^{s,\tau}_{p,q}} = \sup_{P\ dyadic} \frac{1}{|P|^\tau} \left\{ \int_P \left[\sum_{j=-\log_2^{l(P)}}^{\infty} (2^{sj} |\Delta_j f(x)|)^q \right]^{p/q} dx \right\}^{1/p},$$

where $l(P)$ denotes the side length of the cube P. They showed that $Q_\alpha = \dot{F}^{\alpha,1/2-\alpha/n}_{2,2}$. In a similar way as $\dot{F}^{s,\tau}_{p,q}$, Yang and Yuan [255] introduced Besov type space $\dot{B}^{s,\tau}_{p,q}$. Some recent progress on $B^{s,\tau}_{p,q}$ and $F^{s,\tau}_{p,q}$ spaces can be found in [257].

(3) Function space like $\dot{F}^s_{\infty,q}$ are not considered in this chapter. It is of importance in the case $s = 0$ and $q = 2$, which is equivalent to BMO space introduced by F. John and L. Nirenberg in 1961. An interesting generalization of BMO type on homogeneous-type spaces or on measurable subsets of such spaces was given by X. T. Duong and L. X. Yan [69]. Their spaces are defined by certain maximal functions associated with generalized approximation of the identity, which coincide with the classical BMO spaces of F. John and L. Nirenberg in some special cases.

Chapter 2

Navier-Stokes equation

To be without some of the things you want is an indispensable part of happiness. —— B. Russell

We consider the initial value problem for a nonlinear evolution equation, which is said to be locally well-posed in L^p if for any initial value $u|_{t=0} \in L^p$, there exists a unique solution in $C(0,T;L^p)$ for some $T > 0$, moreover, the solution map $u|_{t=0} \to u$ is at least a continuous mapping from L^p to $C(0,T;L^p)$. Furthermore, if $T = \infty$, then we say that it is globally well-posed.

The Navier-Stokes (NS) equation is a fundamental equation in the theory of fluid mechanics. However, the existence for its global smooth solutions in three spatial dimensions has been open for many years.

In two spatial dimensions, the global well-posedness for the NS equation was established by Ladyzhenskaya [154], and Kato [119] gave an alternate proof based on the semi-group method and he also obtained the local well posedness of solutions in L^n for $n \geqslant 3$.

We will use harmonic analysis techniques to study the NS equation. First, we consider the $L^r \to L^p$ estimate for the heat semi-group $H(t) = e^{t\Delta}$ and establish the time-space estimates in mixed Lebesgue spaces $L_t^q L_x^p$ for the solutions of the linear heat equation, those estimates are similar to the Strichartz estimates for the dispersive equations. Next, applying the time-space estimates we show the local well-posedness in L^n for the NS equation in a very simple way. In 2D case, in view of the *a priori* estimate in L^2, we immediately obtain the global well-posedness of the NS equation in L^2. The method here is also useful for the other nonlinear parabolic equations, such as the semi-linear heat equation, the Ginzburg-Landau equation and so on.

Finally, the regularity behavior for the NS equation will be investigated by applying the frequency-uniform decomposition techniques.

2.1 Introduction

2.1.1 *Model, energy structure*

We study the Cauchy problem for the (incompressible) Navier-Stokes (NS) equation

$$u_t - \Delta u + (u\cdot\nabla)u + \nabla p = 0, \quad \mathrm{div}\, u = 0, \quad u(0,x) = u_0(x), \tag{2.1}$$

where $\Delta = \sum_{i=1}^n \partial^2_{x_i}$, $\nabla = (\partial_{x_1}, ..., \partial_{x_n})$, $\mathrm{div}\, u = \partial_{x_1} u_1 + ... + \partial_{x_n} u_n$, $u = (u_1, ..., u_n)$ and p are real-valued unknown functions of $(t,x) \in \mathbb{R}_+ \times \mathbb{R}^n$, $u_0 = (u_0^1, ..., u_0^n)$ denotes the initial value of u at $t = 0$.

In view of its own structure, one can easily deduce that the smooth solution of (2.1) satisfies the following conservation law:

$$\frac{1}{2}\|u(t)\|_2^2 + \int_0^t \|\nabla u(s)\|_2^2 ds = \frac{1}{2}\|u_0\|_2^2, \tag{2.2}$$

where $\|u\|_2^2 := \sum_{i=1}^n \|u_i\|_2^2$, $\|\nabla u\|_2^2 := \sum_{i,j=1}^n \|\partial_{x_j} u_i\|_2^2$ for $u = (u_1, ..., u_n)$. Indeed, multiplying u in (2.1) and then integrating by part, we have (2.2). According to (2.2), it is natural to ask what happens if $u_0 \in L^2(\mathbb{R}^n)$.

Using the compactness method, we can obtain the existence of the weak solutions if $u_0 \in L^2(\mathbb{R}^n)$, however, the uniqueness, the persistence of the regularity of the solution and the continuity of the solution map are hard to obtain, see [159].

2.1.2 *Equivalent form of NS*

Let (u, p) be a smooth solution of the NS equation. Taking the divergence of the first equation in (2.1) and noticing that $\mathrm{div}\, u = 0$, we immediately obtain that

$$\Delta p + \mathrm{div}[(u\cdot\nabla)u] = 0. \tag{2.3}$$

It follows that $\nabla p = (-\Delta)^{-1}\nabla \mathrm{div}[(u\cdot\nabla)u]$. For convenience, we write

$$\mathbb{P} = I + (-\Delta)^{-1}\nabla\mathrm{div}. \tag{2.4}$$

Solving p from (2.3) and inserting it into (2.1), we get

$$u_t - \Delta u + \mathbb{P}\left[(u\cdot\nabla)u\right] = 0, \quad u(0,x) = u_0(x). \tag{2.5}$$

From (2.5), we see that the NS equation belongs to a nonlinear parabolic equation.

2.1.3 *Critical spaces*

If u solves Eq. (2.5), so does $u_\lambda = \lambda u(\lambda^2 t, \lambda x)$ ($\forall\ \lambda > 0$) with initial data $u_\lambda(0,x) = \lambda u_0(\lambda x)$. We say that $X := X(\mathbb{R}^n)$ is a critical space for the NS equation, if the norm of $u_\lambda(0,x)$ in X is invariant for all $\lambda > 0$. Taking notice of

$$\|u_\lambda(0,\cdot)\|_{L^r(\mathbb{R}^n)} = \lambda\|u_0(\lambda\ \cdot)\|_{L^r(\mathbb{R}^n)} = \lambda^{1-n/r}\|u_0\|_{L^r(\mathbb{R}^n)}, \tag{2.6}$$

we see that $r = n$ is the unique index so that the norm of $u_\lambda(0,x)$ in $L^r(\mathbb{R}^n)$ is invariant for all $\lambda > 0$. So, L^n is a critical space of the NS equation.

Using the same way as in the above, we can verify that $\dot{H}^{n/2-1}(\mathbb{R}^n)$ is also the unique space so that the norm of $u_\lambda(0,x)$ in $\dot{H}^s(\mathbb{R}^n)$ is invariant. $\dot{H}^{n/2-1}$ is another critical space in all $\dot{H}^s$.

On the other hand, noticing that $\dot{H}^{n/2-1} \subset L^n$ is a sharp inclusion, one can easily understand $\dot{H}^{n/2-1}$ to be a critical space.

We can calculate that $\|u_\lambda(0,\cdot)\|_{\dot{B}^{-1}_{\infty,\infty}} \sim \|u_0\|_{\dot{B}^{-1}_{\infty,\infty}}$, from this point of view, we see that $\dot{B}^{-1}_{\infty,\infty}$ is the largest critical space.

2.2 Time-space estimates for the heat semi-group

We consider the Cauchy problem for the heat equation:

$$u_t - \Delta u = f, \quad u(0,x) = u_0(x).$$

Taking the Fourier transform, we get

$$\widehat{u}_t + |\xi|^2\widehat{u} = \widehat{f}, \quad \widehat{u}(0) = \widehat{u}_0.$$

Solving the ordinary differential equation, we obtain that

$$u(t) = H(t)u_0 + \int_0^t H(t-\tau)f(\tau,\cdot)d\tau, \tag{2.7}$$

where $H(t) = e^{t\Delta} = \mathscr{F}^{-1}e^{-t|\xi|^2}\mathscr{F}$.

2.2.1 $L^r \to L^p$ *estimate for the heat semi-group*

In $H(t)$, $e^{-t|\xi|^2}$ as an exponential decay function, which corresponds to the dissipation for $H(t)$, determines that the semi-group $H(t)$ has very good properties. Let us start with an $L^r \to L^p$ estimate of $H(t)$, which can be found in [193].

Proposition 2.1. *Let* $1 \leqslant r \leqslant p \leqslant \infty$. *Then*

$$\|\nabla^k H(t)f\|_p \lesssim t^{-\frac{k}{2}-\frac{n}{2}(\frac{1}{r}-\frac{1}{p})}\|f\|_r, \quad k = 0,1,\ t > 0. \tag{2.8}$$

Proof. First, we show that $H(t) : L^r \to L^r$. Using Young's inequality, one has that

$$\|H(t)f\|_r \leqslant \|\mathscr{F}^{-1}e^{-t|\xi|^2}\|_1\|f\|_r \lesssim \|f\|_r. \tag{2.9}$$

Next, we consider $\|H(t)\|_{L^r \to L^\infty}$. Again, by Young's inequality,

$$\|H(t)f\|_\infty \leqslant \|\mathscr{F}^{-1}e^{-t|\xi|^2}\|_{r'}\|f\|_r \lesssim t^{-\frac{n}{2r}}\|f\|_r. \tag{2.10}$$

Now, for any $p \geqslant r$, using Hölder's inequality, (2.9) and (2.10), we have

$$\|H(t)f\|_p \leqslant \|H(t)f\|_\infty^{1-r/p}\|H(t)f\|_r^{r/p} \lesssim t^{-\frac{n}{2}(\frac{1}{r}-\frac{1}{p})}\|f\|_r. \tag{2.11}$$

We estimate $\|\partial_{x_1}H(t)f\|_r$. By Young's inequality,

$$\|\partial_{x_1}H(t)f\|_r \leqslant \|\mathscr{F}^{-1}(\xi_1 e^{-t|\xi|^2})\|_1\|f\|_r \lesssim t^{-\frac{1}{2}}\|f\|_r. \tag{2.12}$$

So,

$$\|\nabla H(t)f\|_r \lesssim t^{-\frac{1}{2}}\|f\|_r. \tag{2.13}$$

Combining (2.11) with (2.13), we immediately have

$$\begin{aligned}\|\nabla H(t)f\|_p = \|H(t/2)\nabla H(t/2)f\|_p &\lesssim t^{-\frac{n}{2}(\frac{1}{r}-\frac{1}{p})}\|\nabla H(t/2)f\|_r \\ &\lesssim t^{-\frac{1}{2}-\frac{n}{2}(\frac{1}{r}-\frac{1}{p})}\|f\|_r. \end{aligned} \tag{2.14}$$

This is the result, as desired. □

As a generalization, we can show that for any $s \geqslant 0$ and $1 < r \leqslant p \leqslant \infty$,

$$\|(-\Delta)^{s/2}H(t)f\|_p \lesssim t^{-\frac{s}{2}-\frac{n}{2}(\frac{1}{r}-\frac{1}{p})}\|f\|_r. \tag{2.15}$$

2.2.2 *Time-space estimates for the heat semi-group*

We study the heat semi-group in the mixed space $L_t^\gamma L_x^p$ and the technique is to extensively use the dyadic decomposition together with the exponential decay of $e^{-t|\xi|^2}$. On the heat semi-group, the frequency localization idea goes back to Chemin [29] (see also [31]). The techniques used in this section are also adapted to more general semi-groups, say $e^{-t(1+ia)|\xi|^\theta}$ with $0 < \theta < \infty$, see [240]. The following result is due to [240].

Proposition 2.2. Let $a \geqslant 0$, $1 \leqslant r \leqslant p \leqslant \infty$, $0 < \lambda \leqslant \infty$ and $2/\gamma = a + n(1/r - 1/p)$. Then we have

$$\|H(t)f\|_{L^\gamma(\mathbb{R}_+;\dot{B}^0_{p,\lambda})} \leqslant C\|f\|_{\dot{B}^{-a}_{r,\lambda\wedge\gamma}}. \tag{2.16}$$

Proof. Noticed that $e^{-t|\xi|^2}$ is exponentially decaying, in order to extensively use this fact, the frequency localization techniques will be applied. We have

$$\|\Delta_j H(t) f\|_r \leqslant \|\varphi_j e^{-t|\xi|^2}\|_{M_r} \|f\|_r. \tag{2.17}$$

Using the multiplier criteria and noticing that $\varphi_j = \varphi(2^{-j}\xi)$, $\mathrm{supp}\varphi \subset \{\xi : 1/2 \leqslant |\xi| \leqslant 2\}$, one has that

$$\|\varphi_j e^{-t|\xi|^2}\|_{M_r} = \|\varphi e^{-t2^{2j}|\xi|^2}\|_{M_r} \leqslant \|\varphi e^{-t2^{2j}|\xi|^2}\|_{H^L}, \tag{2.18}$$

where $L > n/2$. Due to $\sup_{x>0} x^k/e^x \leqslant C$, it is easy to see that

$$\|\varphi e^{-t2^{2j}|\xi|^2}\|_{H^L} \lesssim e^{-ct2^{2j}}. \tag{2.19}$$

So, by (2.17)–(2.19) we get

$$\|\Delta_j H(t) f\|_r \lesssim e^{-ct2^{2j}} \|f\|_r. \tag{2.20}$$

Since $\Delta_j = \Delta_j(\sum_{\ell=0,\pm1} \Delta_{j+\ell})$, (2.20) implies that

$$\|\Delta_j H(t) f\|_r \lesssim e^{-ct2^{2j}} \|\Delta_j f\|_r. \tag{2.21}$$

Taking the ℓ^λ-norm in (2.21), we obtain that

$$\|H(t)f\|_{\dot{B}^0_{r,\lambda}} \lesssim \left(\sum_j e^{-ct2^{2j}} \|\Delta_j f\|_r^\lambda \right)^{1/\lambda}. \tag{2.22}$$

In what follows the proof is separated into two cases. The first case is that $\gamma \geqslant \lambda$. Taking the norm in $L^\gamma_t(\mathbb{R}_+)$ on (2.22) and using Minkowski's inequality, we have

$$\begin{aligned} \|H(t)f\|_{L^\gamma(\mathbb{R}_+, \dot{B}^0_{r,\lambda})} &\lesssim \left\| \sum_{k=-\infty}^{\infty} e^{-ct\lambda 2^{2k}} \|\Delta_k f\|_r^\lambda \right\|_{L_t^{\gamma/\lambda}}^{1/\lambda} \\ &\lesssim \left(\sum_{k=-\infty}^{\infty} \left\| e^{-ct\lambda 2^{2k}} \right\|_{L_t^{\gamma/\lambda}} \|\Delta_k f\|_r^\lambda \right)^{1/\lambda}. \end{aligned} \tag{2.23}$$

Noticing that

$$\left\| e^{-ct\lambda 2^{2k}} \right\|_{L_t^{\gamma/\lambda}} \lesssim 2^{-\lambda 2k/\gamma}, \tag{2.24}$$

from (2.23) and (2.24) we get the consequence in the case $p = r$ and $\gamma \geqslant \lambda$. As $r < p$, applying the inclusion $\dot{B}^{-2/\gamma+n(1/r-1/p)}_{r,\lambda} \subset \dot{B}^{-2/\gamma}_{p,\lambda}$, we can obtain the result.

Secondly, we consider the case $\gamma < \lambda$. By (2.22),

$$\int_{\mathbb{R}_+} \|H(t)f\|^\gamma_{\dot{B}^0_{p,\lambda}} dt \lesssim \sum_j 2^{-2j} \|\Delta_j f\|_p^\gamma. \tag{2.25}$$

We get the result in the case $p = r$ and $\gamma < \lambda$. If $r < p$, in an analogous way to the first case one can prove the conclusion. □

In fact, the above estimates (2.15) and (2.16) can be developed to the case $0 < r \leqslant p \leqslant \infty$ and we have no restriction on $\lambda, \gamma > 0$. We have a little bit of surprise about this fact, see [240] for details. Some earlier results related to (2.16) were due to Weissler [250] and Giga [77], where their proof is based on Marcinkiewicz' interpolation theorem.

Corollary 2.1. *Let $2/\gamma(p) = n(1/2 - 1/p)$. For any $2 \leqslant p < \infty$, we have*

$$\|\nabla H(t)f\|_{L^2(\mathbb{R}_+;L^2)} \lesssim \|f\|_{L^2}, \tag{2.26}$$

$$\|H(t)f\|_{L^{\gamma(p)}(\mathbb{R}_+;L^p)} \lesssim \|f\|_{L^2}. \tag{2.27}$$

Proof. Taking $\lambda = r = 2$ in Proposition 2.2 and noticing that $\dot{B}^0_{p,2} \subset L^p$, one immediately has the consequence. □

For simplicity, we write

$$(\mathscr{A}f)(t,x) := \int_0^t H(t-\tau)f(\tau,x)d\tau. \tag{2.28}$$

In what follows we consider the estimate of $\mathscr{A}f$. By Proposition 2.1,

$$\|\nabla^k \mathscr{A}f\|_p \lesssim \int_0^t (t-\tau)^{-\frac{k}{2}-\frac{n}{2}(\frac{1}{r}-\frac{1}{p})}\|f(\tau)\|_r \, d\tau, \quad k=0,1. \tag{2.29}$$

Applying the Hardy-Littlewood-Sobolev inequality, we obtain that

Proposition 2.3. *Let $1 \leqslant r \leqslant p \leqslant \infty$ and $1 < \gamma, \gamma_1 < \infty$ satisfy*

$$\frac{1}{\gamma} = \frac{1}{\gamma_1} + \frac{k}{2} + \frac{n}{2}\left(\frac{1}{r} - \frac{1}{p}\right) - 1, \quad \frac{k}{2} + \frac{n}{2}\left(\frac{1}{r} - \frac{1}{p}\right) < 1, \quad k = 0,1. \tag{2.30}$$

Then we have

$$\|\nabla^k \mathscr{A}f\|_{L^\gamma(\mathbb{R}_+;L^p)} \lesssim \|f\|_{L^{\gamma_1}(\mathbb{R}_+;L^r)}. \tag{2.31}$$

Proposition 2.3 can not handle the case $\gamma = \infty$, but we have (see [240])

Proposition 2.4. *Let $1 \leqslant r \leqslant \infty$, $1 \leqslant q' \leqslant \lambda \leqslant \infty$*[1]*. Then*

$$\|\mathscr{A}f\|_{L^\infty(\mathbb{R}_+,\dot{B}^0_{r,\lambda})} \lesssim \|f\|_{L^{q'}(\mathbb{R}_+,\dot{B}^{-2/q}_{r,\lambda})}. \tag{2.32}$$

Proof. Using (2.21) and Young's inequality,

$$\begin{aligned}\|\triangle_k \mathscr{A}f\|_r &\lesssim \int_0^t e^{-c(t-\tau)2^{2k}}\|\triangle_k f(\tau)\|_r d\tau \\ &\lesssim 2^{-2k/q}\|\triangle_k f\|_{L^{q'}(\mathbb{R}_+,L^r)}.\end{aligned} \tag{2.33}$$

[1] p' stands for the conjugate number of p, i.e. $1/p + 1/p' = 1$.

Taking the ℓ^λ-norm in (2.33) and using Minkowski's inequality, we get

$$\|\mathscr{A} f\|_{\dot{B}^0_{r,\lambda}} \lesssim \left\{ \sum_{k=-\infty}^{\infty} \left(2^{-2k/q} \|\Delta_k f\|_{L^{q'}(\mathbb{R}_+,L^r)} \right)^\lambda \right\}^{1/\lambda}$$
$$\lesssim \|f\|_{L^{q'}(\mathbb{R}_+,\dot{B}^{-2/q}_{r,\lambda})}, \tag{2.34}$$

which implies the result, as desired. □

Corollary 2.2. *Let $2/\gamma(p) = n(1/2 - 1/p)$. For any $2 \leqslant p < \infty$, $2/\gamma(p) < 1$, we have*

$$\|\nabla \mathscr{A} f\|_{L^2(\mathbb{R}_+;L^2)} \lesssim \|f\|_{L^{\gamma(p)'}(\mathbb{R}_+,L^{p'})}, \tag{2.35}$$
$$\|\mathscr{A} f\|_{L^\infty(\mathbb{R}_+,L^2)\cap L^{\gamma(p)}(\mathbb{R}_+;L^p)} \lesssim \|f\|_{L^{\gamma(p)'}(\mathbb{R}_+,L^{p'})}. \tag{2.36}$$

Clearly, all of the conclusions hold if we substitute the time interval $\mathbb{R}_+$ by $[0,T]$.

2.3 Global well-posedness in L^2 of NS in 2D

For any Banach function space $X := X(\mathbb{R}^n)$ and $u = (u_1, ..., u_n) \in X^n$, we write $\|u\|_X^2 = \sum_i \|u_i\|_X^2$ and $\|\nabla u\|_X^2 = \sum_{i,j} \|\partial_{x_i} u_j\|_X^2$. For any vector function $u(t) = (u_1(t), ..., u_n(t))$ defined in I and valued in X^n, we write

$$\|u\|_{L^q(I,X^n)} = \left(\int_I \|u(t)\|_X^q dt \right)^{1/q}. \tag{2.37}$$

If there is no confusion, we will write $L^q(I,X) := L^q(I,X^n)$. We denote by $[X]_0^n$ the completion of the set

$$\{u \in \mathscr{S}^n : \operatorname{div} u = 0\} \quad \text{in} \quad X^n$$

and by $L^q(I,[X]_0^n)$ the space of all of the vector functions $u(t) = (u_1(t), ..., u_n(t))$ in $L^q(I,X^n)$ such that $u(t) \in [X]_0^n$ for a.e. $t \in I$. It is easy to verify that

$$[L^p]_0^n = \{u \in [L^p]^n : \operatorname{div} u = 0\},$$
$$L^q(I,[X]_0^n) = \{u \in L^q(I,[X]^n) : \operatorname{div} u(t) = 0, \ a.e.\ t \in I\},$$

where "div" is in the sense of tempered distributions.

Taking $n = 2$ and $p = 4$ in Corollaries 2.1 and 2.2, we have

$$\|\nabla H(t) u_0\|_{L^2(0,T;\ L^2)} \lesssim \|u_0\|_{L^2}, \tag{2.38}$$

$$\|H(t)u_0\|_{L^4(0,T;\ L^4)} \lesssim \|u_0\|_{L^2}, \tag{2.39}$$

$$\|\nabla \mathscr{A} f\|_{L^2(0,T;\ L^2)} \lesssim \|f\|_{L^{4/3}(0,T;\ L^{4/3})}, \tag{2.40}$$

$$\|\mathscr{A} f\|_{L^\infty(0,T;\ L^2)\cap L^4(0,T;\ L^4)} \lesssim \|f\|_{L^{4/3}(0,T;\ L^{4/3})}. \tag{2.41}$$

We write $L^p_{x,t\in[0,T]} = L^p(0,T;\ L^p)$. Define the metric space:

$$\mathfrak{D} = \left\{u : \|u\|_{L^4_{x,t\in[0,T]}} + \|\nabla u\|_{L^2_{x,t\in[0,T]}} \leqslant M\right\}, \tag{2.42}$$

$$d(u,v) = \|u-v\|_{L^4_{x,t\in[0,T]}} + \|\nabla(u-v)\|_{L^2_{x,t\in[0,T]}}. \tag{2.43}$$

Assume that $u_0 \in [L^2]^2_0$. We consider the mapping:

$$\mathfrak{M} : u(t) \to H(t)u_0 + \mathscr{A}\mathbb{P}[(u\cdot\nabla)u]. \tag{2.44}$$

Now we fix an $M > 0$ such that $CM = 1/2$, where C is the largest constant appeared in the following inequalities. We show that $\mathfrak{M} : (\mathfrak{D}, d) \to (\mathfrak{D}, d)$ is a contraction mapping. In fact, for any $u \in \mathfrak{D}$,

$$\|\mathfrak{M}u\|_{L^4_{x,t\in[0,T]}} \leqslant \|H(t)u_0\|_{L^4_{x,t\in[0,T]}} + C\|\mathbb{P}[(u\cdot\nabla)u]\|_{L^{4/3}_{x,t\in[0,T]}}, \tag{2.45}$$

$$\|\nabla\mathfrak{M}u\|_{L^2_{x,t\in[0,T]}} \leqslant \|\nabla H(t)u_0\|_{L^2_{x,t\in[0,T]}} + C\|\mathbb{P}[(u\cdot\nabla)u]\|_{L^{4/3}_{x,t\in[0,T]}}. \tag{2.46}$$

In view of Mihlin's multiplier theorem, $\mathbb{P} : L^{4/3} \to L^{4/3}$. So, it follows from Hölder's inequality that,

$$\|\mathbb{P}[(u\cdot\nabla)u]\|_{L^{4/3}_{x,t\in[0,T]}} \lesssim \|u\|_{L^4_{x,t\in[0,T]}} \|\nabla u\|_{L^2_{x,t\in[0,T]}} \leqslant M^2. \tag{2.47}$$

By (2.38) and (2.39), there exists a $T > 0$ such that

$$\|\nabla H(t)u_0\|_{L^2(0,T;\ L^2)} + \|H(t)u_0\|_{L^4(0,T;\ L^4)} \leqslant M/2. \tag{2.48}$$

So,

$$\|\mathfrak{M}u\|_{L^4_{x,t\in[0,T]}} \leqslant M/2 + CM^2 \leqslant M, \tag{2.49}$$

$$\|\nabla\mathfrak{M}u\|_{L^2_{x,t\in[0,T]}} \leqslant M/2 + CM^2 \leqslant M. \tag{2.50}$$

Similarly,

$$d(\mathfrak{M}u,\ \mathfrak{M}v) \leqslant \frac{1}{2}d(u,v). \tag{2.51}$$

In view of Banach's contraction mapping principle, there exists a $u \in \mathfrak{D}$ satisfying

$$u(t) = H(t)u_0 + \mathscr{A}\mathbb{P}[(u\cdot\nabla)u]. \tag{2.52}$$

In view of the regularity theory, we see that the solution is smooth in the domain $(0,T)\times\mathbb{R}^2$. Since $u_0 \in [L^2]^2_0$ and $\operatorname{div}\mathbb{P} = 0$, it follows from the integral equation that $\operatorname{div} u = 0$ holds in $(0,T) \times \mathbb{R}^n$. Again, by (2.41), one sees that $u \in C(0,T;\ [L^2]_0^2)$. The conservation (2.2) can be deduced by a straightforward calculation to the integral equation. It follows that $\|u(T)\|_2 \leqslant \|u_0\|_2$. According to the standard semi-group theory, one can extend the solution above step by step and finally finds a maximal time T_m. It suffices to show that $T_m = \infty$. If not, then $T_m < \infty$. T_m is maximal implies that

$$\|u\|_{C([0,T_m);\ [L^2]_0^2) \cap L^4_{x,t \in (0,T_m)}} + \|\nabla u\|_{L^2(0,T_m;\ L^2)} = \infty.$$

By the energy estimate,

$$\frac{1}{2}\|u(t)\|_2^2 + \int_0^t \|\nabla u(s)\|_2^2 ds = \frac{1}{2}\|u_0\|_2^2, \quad \forall\, t < T_m. \tag{2.53}$$

In view of the Gagliardo-Nirenberg inequality,

$$\|u\|_{L^4_{x,t\in[0,T]}} \lesssim \|u\|^{1/2}_{L^\infty(0,T;\ L^2)} \|\nabla u\|^{1/2}_{L^2(0,T;\ L^2)} \lesssim \|u_0\|_2, \quad \forall\, T < T_m. \tag{2.54}$$

A contradiction. We have shown

Theorem 2.1. *Let $u_0 \in [L^2]_0^2$. Then the NS equation* (2.5) *has a unique solution u satisfying*

$$u \in C(0,\infty;\ [L^2]_0^2) \cap L^2(0,\infty;\ [\dot{H}^1]_0^2), \tag{2.55}$$

and

$$\frac{1}{2}\|u(t)\|_2^2 + \int_0^t \|\nabla u(s)\|_2^2 ds = \frac{1}{2}\|u_0\|_2^2, \quad 0 < t < \infty. \tag{2.56}$$

The contraction mapping argument implies that the solution map is analytic and of course, is continuous and the persistence of the regularity of solutions is also easy to obtain by using the integral equation. So the NS equation is globally well-posed in L^2.

Finally, we point out that the method used here can be applied to some other kinds of nonlinear parabolic equations. For examples,

$$u_t - \Delta u + |\nabla u| u = 0, \quad u(0,x) = u_0(x)$$

is globally well posed in $L^2(\mathbb{R}^2)$; the Hamilton-Jacobi equation

$$u_t - \Delta u + |\nabla u|^{3/2} = 0, \quad u(0,x) = u_0(x)$$

is locally well-posed in $L^2(\mathbb{R}^2)$.

2.4 Well-posedness in L^n of NS in higher dimensions

For convenience, we still write $L^p_{x,t\in[0,T]} = L^p(0,T;\ L^p(\mathbb{R}^n))$ and use the same notations as in the previous section. By Corollaries 2.1 and 2.2, we have for $n \geqslant 3$,

$$\|\nabla H(t)u_0\|_{L^2(0,T;\,L^2)} \lesssim \|u_0\|_{L^2}, \tag{2.57}$$

$$\|H(t)u_0\|_{L^{2+4/n}_{x,t\in[0,T]}} \lesssim \|u_0\|_{L^2}, \tag{2.58}$$

$$\|\nabla \mathscr{A} f\|_{L^2(0,T;\ L^2)} \lesssim \|f\|_{L^{(2+4/n)'}_{x,t\in[0,T]}}, \tag{2.59}$$

$$\|\mathscr{A} f\|_{L^\infty(0,T;\ L^2)\cap L^{2+4/n}_{x,t\in[0,T]}} \lesssim \|f\|_{L^{(2+4/n)'}_{x,t\in[0,T]}}. \tag{2.60}$$

One may expect that the techniques in the previous section also work very well. Unfortunately,

$$\|(u\cdot\nabla)u\|_{L^{(2+4/n)'}_{x,t\in[0,T]}} \leqslant \|\nabla u\|_{L^2_{x,t\in[0,T]}} \|u\|_{L^{n+2}_{x,t\in[0,T]}}. \tag{2.61}$$

$L^{n+2}_{x,t\in[0,T]}$ is out of the control of the above linear estimates. So, we need the following

Corollary 2.3. *We have*

$$\|H(t)u_0\|_{L^{n+2}_{x,t\in[0,T]}} \lesssim \|u_0\|_n, \tag{2.62}$$

$$\|H(t)u_0\|_{L^\infty(0,T;\ L^n)} \lesssim \|u_0\|_n, \tag{2.63}$$

$$\|\nabla \mathscr{A} f\|_{L^{n+2}_{x,t\in[0,T]}} \lesssim \|f\|_{L^{(n+2)/2}_{x,t\in[0,T]}}, \tag{2.64}$$

$$\|\nabla \mathscr{A} f\|_{L^\infty(0,T;\ L^n)} \lesssim \|f\|_{L^{(2+n)/2}_{x,t\in[0,T]}}. \tag{2.65}$$

Proof. Taking $p = r = \lambda = n$ and $\gamma = 2+n$ in Proposition 2.2, we get

$$\|H(t)u_0\|_{L^{2+n}(\mathbb{R}_+;\dot B^{2/(2+n)}_{n,n})} \lesssim \|u_0\|_{\dot B^0_{n,n}} \lesssim \|u_0\|_n. \tag{2.66}$$

Using the inclusions $\dot B^{2/(2+n)}_{n,n} = \dot F^{2/(2+n)}_{n,n} \subset \dot F^0_{n+2,2} = L^{n+2}$, we see that (2.66) implies that (2.62) holds. Obviously, we have (2.63).

(2.64) is a straightforward consequence of Proposition 2.3. Taking $r = \lambda = (n+2)/2$ and $q' = (n+2)/2$ in Proposition 2.4, we get

$$\|\mathscr{A} f\|_{L^\infty(\mathbb{R}_+;\,\dot B^{2n/(2+n)}_{(n+2)/2,\,(n+2)/2})} \lesssim \|f\|_{L^{(n+2)/2}(\mathbb{R}_+;\,\dot B^0_{(n+2)/2,\,(n+2)/2})}. \tag{2.67}$$

This implies

$$\|\nabla \mathscr{A} f\|_{L^\infty(\mathbb{R}_+;\,\dot B^{(n-2)/(2+n)}_{(n+2)/2,\,(n+2)/2})} \lesssim \|f\|_{L^{(n+2)/2}_{x,t\in\mathbb{R}_+}}. \tag{2.68}$$

Using the embedding $\dot B^{(n-2)/(2+n)}_{(n+2)/2,\,(n+2)/2} = \dot F^{(n-2)/(2+n)}_{(n+2)/2,\,(n+2)/2} \subset \dot F^0_{n,2} = L^n$, we see that (2.68) implies that (2.65) holds. $\square$

The above linear estimates can derive the local well-posedness and the global well-posedness with small data in L^n for $n \geqslant 3$. We will use the same notations as in the case $n = 2$. We have

Theorem 2.2. *Let $u_0 \in [L^n]_0^n$. Then there exists a $T_m > 0$ such that the NS equation (2.5) has a unique solution u satisfying*

$$u \in C([0, T_m);\ [L^n]_0^n) \cap L^{2+n}_{\rm loc}(0, T_m;\ [L^{2+n}]_0^n). \tag{2.69}$$

If $T_m < \infty$, then we have $\|u\|_{L^{2+n}(0,T_m;\ L^{2+n})} = \infty$. If $\|u_0\|_n$ is sufficiently small, then $T_m = \infty$. Moreover, if $u_0 \in [L^2]_0^n$, then $u \in C(0, T_m; [L^2]_0^n)$, $\partial_{x_i} u \in L^2(0, T_m;\ [L^2]_0^n)$, and

$$\frac{1}{2}\|u(t)\|_2^2 + \int_0^t \|\nabla u(s)\|_2^2 ds = \frac{1}{2}\|u_0\|_2^2, \quad 0 < t < T_m. \tag{2.70}$$

Proof. Put

$$\mathfrak{D} = \left\{u : \|u\|_{L^{2+n}_{x,t\in[0,T]}} \leqslant \delta,\ \|u\|_{L^\infty([0,T];L^n)} \leqslant 2C\|u_0\|_n\right\}, \tag{2.71}$$

$$d(u,v) = \|u - v\|_{L^{2+n}_{x,t\in[0,T]}}. \tag{2.72}$$

We consider the mapping:

$$\mathfrak{M} : u(t) \to H(t)u_0 + \mathscr{A}\mathbb{P}\,\mathrm{div}\,(u \otimes u), \tag{2.73}$$

by Corollary 2.3, we have

$$\begin{aligned} \|\mathfrak{M}u\|_{L^{2+n}_{x,t\in[0,T]}} &\lesssim \|H(t)u_0\|_{L^{2+n}_{x,t\in[0,T]}} + \|u \otimes u\|_{L^{(2+n)/2}_{x,t\in[0,T]}} \\ &\lesssim \|H(t)u_0\|_{L^{2+n}_{x,t\in[0,T]}} + \|u\|^2_{L^{2+n}_{x,t\in[0,T]}} \\ &\lesssim \|H(t)u_0\|_{L^{2+n}_{x,t\in[0,T]}} + \delta^2, \end{aligned} \tag{2.74}$$

$$\begin{aligned} \|\mathfrak{M}u\|_{L^\infty(0,T;\ L^n)} &\lesssim \|u_0\|_n + \|u \otimes u\|_{L^{(2+n)/2}_{x,t\in[0,T]}} \\ &\lesssim \|u_0\|_n + \delta^2. \end{aligned} \tag{2.75}$$

If $C\delta \leqslant 1/4$, we can show that $\mathfrak{M}$ is a contraction mapping from $\mathfrak{D}$ into itself. So, there exists a u satisfying

$$u(t) = H(t)u_0 + \mathscr{A}\mathbb{P}\nabla \cdot (u \otimes u). \tag{2.76}$$

By a standard argument, we see that u is unique in $L^{2+n}(0, T;\ [L^{2+n}]_0^n)$. Moreover, one can extend the solution step by step and find a maximal T_m such that $u \in C([0, T_m);\ [L^n]_0^n) \cap L^{2+n}_{\rm loc}(0, T_m; [L^{2+n}]_0^n)$.

If $u_0 \in [L^2]_0^n$, then we can use (2.57)–(2.61) to obtain $u \in C(0, T;\ [L^2]_0^n)$ and $\partial_{x_i} u \in L^2(0, T; [L^2]_0^n)$ in a similar way as in the 2D case. According to the regularity theory (see [67] for instance), we see that u is infinitely smooth in the domain $(0, T) \times \mathbb{R}^n$. So, a straightforward computation will lead to (2.2). Moreover, if $\|u_0\|_n$ is small enough, we can take $T = \infty$ in (2.71) and (2.72). □

The result of Theorem 2.2 has a long history. Weissler [251] gives a detailed L^p-theory in half-space for local solutions. Kato [119] used a different way to obtain the well posedness result in L^n by constructing the resolution norm like $\sup_t t^\theta \|u(t)\|_p$. The idea only taking $L^{n+2}_{x,t}$ as a resolution space was hidden in some references, see for instance [67]. Here the proof is to take $L^{n+2}_{x,t\in[0,T]} \cap C([0,T]; L^n)$ as a resolution space which enables us to get the local well posedness of solutions in L^n, where (2.65) plays a crucial role. Similar bilinear estimate to (2.65) in 3D was shown in [70] in a different way:

$$\|\mathscr{A}\mathbb{P}\,\mathrm{div}\,(u \otimes v)\|_{L^\infty([0,T];L^3)} \lesssim \|u\|_{L^5_{x,t}} \|v\|_{L^5_{x,t}}. \tag{2.77}$$

In 3D case, Kenig and Koch [123] obtained the result of Theorem 2.2 by applying (2.77).

2.5 Regularity of solutions for NS

We will show that, for any $t > 0$, the solution of the NS equation is infinitely differentiable and in fact, is really analytic.

2.5.1 *Gevrey class and function space* $E^s_{2,1}$

First, we consider the Gevrey classes. Let $s > 0$, denote

$$G_s(\mathbb{R}^n) = \left\{ f \in C^\infty(\mathbb{R}^n) : \exists \rho, M > 0 \ s.t. \ \|f\|_{\dot{H}^m} \leqslant M \left(\frac{m!}{\rho^m}\right)^s \ \forall m \in \mathbb{Z}_+ \right\}. \tag{2.78}$$

$G_s(\mathbb{R}^n)$ is said to be a Gevrey s-class.

Our aim is to show that the solution of the NS equation belongs to the Gevrey 1-class. In order to show this fact, we will use Wiener's decomposition of $\mathbb{R}^n$. In Chapter 6, we will continuously use Wiener's decomposition to study the nonlinear dispersive equations. Let Q_α be the unit cube with the center at $\alpha \in \mathbb{Z}^n$, i.e., $Q_\alpha = \alpha + Q_0, Q_0 = \{x = (x_1, ..., x_n) : -1/2 \leqslant x_i < 1/2\}$. It is easy to see that

$$\mathbb{R}^n = \cup_{\alpha\in\mathbb{Z}^n} Q_\alpha, \quad Q_\alpha \cap Q_\beta = \varnothing, \quad \alpha \neq \beta.$$

So, $\{Q_\alpha\}$ constitutes a decomposition of $\mathbb{R}^n$, which is said to be the uniform decomposition (or Wiener's decomposition) of $\mathbb{R}^n$. Comparing it with the dyadic decomposition, we see that it is more delicate than the dyadic decomposition of $\mathbb{R}^n$. Putting the uniform decomposition into the frequency space and combining it with the $\ell^1(L^2)$ space, we have the following

$$E^s_{2,1} = \left\{ f \in \mathscr{S}'(\mathbb{R}^n) : \|f\|_{E^s_{2,1}} = \sum_{k\in\mathbb{Z}^n} 2^{s|k|} \left\| \mathscr{F}^{-1}\chi_{Q_k}\widehat{f}\right\|_2 < \infty \right\}. \tag{2.79}$$

Such a kind of space and its generalized version $E^s_{p,q}$ were studied in [247], which is a kind of generalized modulation spaces [86]. It is easy to see that $E^s_{2,1}$ is a Banach space, which has an exponential regularity weight. From PDE's point of view, $E^s_{2,1}$ has infinite smoothness, which is different from any Besov spaces and any classical modulation spaces. The classical modulation space $M^s_{2,1}$ takes the following norm:

$$M^s_{2,1} = \left\{ f \in \mathscr{S}'(\mathbb{R}^n) : \|f\|_{M^s_{2,1}} = \sum_{k\in\mathbb{Z}^n} \langle k\rangle^s \left\| \mathscr{F}^{-1}\chi_{Q_k}\widehat{f}\right\|_2 < \infty \right\}, \tag{2.80}$$

where $\langle k\rangle = 1 + |k|$. $M^s_{2,1}$ and its generalizations $M^s_{p,q}$ have finite smoothness, which are similar to Besov spaces (see Chapter 6). $E^s_{2,1}$ is of importance in our regularity argument for the parabolic equation, whose infinite regularity can be seen in the following (cf. [247]):

Proposition 2.5. *We have*

$$G_1 = \bigcup_{s>0} E^s_{2,1}. \tag{2.81}$$

Proof. First, we show that for any $s > 0$, $E^s_{2,1}$ is a subset of G_1. Indeed, for any $f \in E^s_{2,1}$ and $0 < c \ll 1$,

$$\begin{aligned}
\|f\|_{\dot H^m} &\lesssim \|\nabla^m f\|_{E^0_{2,1}} \\
&\lesssim \sum_{k\in\mathbb{Z}^n} \|\chi_{Q_k}|\xi|^m \widehat{f}\,\|_2 \\
&\lesssim \sum_{k\in\mathbb{Z}^n} (|k| + \sqrt{n/2})^m \|\chi_{Q_k}\widehat{f}\,\|_2 \\
&\lesssim \frac{m!}{(cs)^m} \sum_{k\in\mathbb{Z}^n} \frac{(cs)^m(|k|+C)^m}{m!\cdot 2^{s|k|}} 2^{s|k|} \|\chi_{Q_k}\widehat{f}\,\|_2 \\
&\lesssim \frac{m!}{(2s)^m} \|f\|_{E^s_{2,1}}.
\end{aligned}$$

On the other hand, let $f \in G_1$ and $L > n/2$. Using Taylor's expansion and Hölder's inequality,

$$\|f\|_{E^s_{2,1}} = \sum_{k\in\mathbb{Z}^n} 2^{s|k|} \|\chi_{Q_k}\widehat{f}\,\|_2$$

$$\leqslant \|f\|_2 + \sum_{m=0}^{\infty} \frac{(s\ln 2)^m}{m!} \sum_{k\in\mathbb{Z}^n,\ |k|\gg 1} |k|^m \|\chi_{Q_k}\widehat{f}\,\|_2$$

$$\lesssim \|f\|_2 + \sum_{m=0}^{\infty} \frac{(s\ln 2)^m}{m!} \sum_{k\in\mathbb{Z}^n,\ |k|\gg 1} C^m |k|^{-L} \|\chi_{Q_k}\widehat{\nabla^{m+L} f}\,\|_2$$

$$\lesssim \|f\|_2 + \sum_{m=0}^{\infty} \frac{(Cs)^m}{m!} \|f\|_{\dot{H}^{L+m}}$$

$$\lesssim \|f\|_2 + \sum_{m=0}^{\infty} \frac{(Cs)^m}{m!} \cdot \frac{(m+L)!}{\rho^m}.$$

Choosing $0 < s \ll 1$, we see that $\sum_{m=0}^{\infty}(Cs/\rho)^m (m+L)^L$ is a convergent series. Thus, $f \in E^s_{2,1}$. □

2.5.2 *Estimates of heat semi-group in* $E^s_{2,1}$

Using Proposition 2.5, we want to get the solution of the NS equation $u \in G_1$. It suffices to show that u belongs to $E^s_{2,1}$ $(s > 0)$. Hence, one needs to estimate $\|H(t)u_0\|_{E^s_{2,1}}$ and $\|\mathscr{A} f\|_{E^s_{2,1}}$.

Proposition 2.6. *Let* $H(t) = e^{t\Delta}$. *There exists a* $c > 0$ *such that for* $j = 0, 1$ *and* $T \leqslant 1$,

$$\|\nabla^j H(t)u_0\|_{E^{ct}_{2,1}} \lesssim t^{-j/2}\|u_0\|_{E^0_{2,1}}, \quad t \leqslant T, \tag{2.82}$$

$$\sup_{t\in[0,T]} \left\| \nabla^j \int_0^t H(t-\tau) f(\tau) d\tau \right\|_{E^{ct}_{2,1}} \lesssim (T + T^{1-j/2}) \sup_{\tau\in[0,T]} \|f(\tau)\|_{E^{c\tau}_{2,1}}. \tag{2.83}$$

Proof. For convenience, we will use the notation

$$\square_k = \mathscr{F}^{-1}\chi_{Q_k}\mathscr{F}. \tag{2.84}$$

By Plancherel's identity, for $|k| \geqslant 1$,

$$\|\square_k H(t)u_0\|_2 = \|\chi_{Q_k}(\xi)e^{-t|\xi|^2}\mathscr{F}u_0\|_2 \leqslant e^{-ct|k|^2}\|\square_k u_0\|_2. \tag{2.85}$$

If $k = 0$, we see that (2.85) also holds for $t \leqslant 1$. Taking the summation over all $k \in \mathbb{Z}^n$, we can get (2.82) hold for $j = 0$.

We now prove (2.82) for $j = 1$. If $k = 0$ and $t \leqslant 1$,

$$\|\nabla\square_k H(t)u_0\|_2 \lesssim \|\chi_{Q_k}(\xi)e^{-t|\xi|^2}\mathscr{F}u_0\|_2 \leqslant \|\square_k u_0\|_2. \tag{2.86}$$

For $|k| \geqslant 1$,

$$\|\nabla\square_k H(t)u_0\|_2 \lesssim \||\xi|\chi_{Q_k}(\xi)e^{-t|\xi|^2}\mathscr{F}u_0\|_2 \leqslant t^{-1/2}e^{-ct|k|}\|\square_k u_0\|_2. \tag{2.87}$$

Combining (2.86) with (2.87), we have

$$\|\nabla\Box_k H(t)u_0\|_2 \lesssim t^{-1/2}e^{-ct|k|}\|\Box_k u_0\|_2. \tag{2.88}$$

This proves (2.82) for $j=1$. For $t\in[0,T]$ with $T\leqslant 1$,

$$\left\|\Box_k\nabla\int_0^t H(t-\tau)f(\tau)d\tau\right\|_2 \lesssim \int_0^t (t-\tau)^{-1/2}e^{-c(t-\tau)|k|}\|\Box_k f(\tau)\|_2 d\tau, \tag{2.89}$$

which implies that (2.83) holds in the case $j=1$. Another case $j=0$ is similar to the case $j=1$. □

2.5.3 *Bilinear estimates in* $E^s_{2,1}$

By Proposition 2.6, our goal is to show that the solution of the NS equation belongs to $E^{ct}_{2,1}$, it is necessary to estimate $\|u\otimes u\|_{E^{ct}_{2,1}}$. We have

Proposition 2.7. $E^\lambda_{2,1}$ is a Banach algebra. More precisely, we hace

$$\|uv\|_{E^\lambda_{2,1}} \leqslant C2^{C\lambda}\|u\|_{E^\lambda_{2,1}}\|v\|_{E^\lambda_{2,1}}, \tag{2.90}$$

where C is independent of $\lambda>0$ and $u,v\in E^\lambda_{2,1}$.

Proof. Noticing that for $k=(k_1,...,k_n)\in\mathbb{Z}^n$ with $|k|=|k_1|+...+|k_n|$, we have

$$\|uv\|_{E^\lambda_{2,1}} = \sum_{k\in\mathbb{Z}^n} 2^{\lambda|k|}\|\Box_k(uv)\|_2. \tag{2.91}$$

It is easy to see that

$$uv = \sum_{i,j\in\mathbb{Z}^n}(\Box_i u)(\Box_j v), \tag{2.92}$$

which implies that

$$\Box_k(uv) = \sum_{i,j\in\mathbb{Z}^n}\Box_k(\Box_i u\,\Box_j v). \tag{2.93}$$

We have $\mathscr{F}(\Box_i u\,\Box_j v) = (\chi_{Q_i}\widehat{u})*(\chi_{Q_j}\widehat{v})$ and

$$\mathrm{supp}(\chi_{Q_i}\widehat{u})*(\chi_{Q_j}\widehat{v})\subset\Omega := \{\xi:\ |\xi-i-j|\leqslant 2\sqrt{n}\}. \tag{2.94}$$

So,

$$\Box_k(\Box_i u\,\Box_j v)=0,\quad |k-i-j|\geqslant 3\sqrt{n}. \tag{2.95}$$

It follows that

$$\|\Box_k(\Box_i u\,\Box_j v)\|_2 = \|\Box_k(\Box_i u\,\Box_j v)\|_2\,\chi_{(|k-i-j|\leqslant 3\sqrt{n})}. \tag{2.96}$$

In view of Plancherel's identity and Hölder's inequality,

$$\begin{aligned}\|\Box_k(\Box_i u\,\Box_j v)\|_2 &\leqslant \|\Box_i u\,\Box_j v\|_2\,\chi_{(|k-i-j|\leqslant 3\sqrt{n})}\\&\leqslant \|\Box_i u\|_\infty\|\Box_j v\|_2\,\chi_{(|k-i-j|\leqslant 3\sqrt{n})}\\&\lesssim \|\Box_i u\|_2\|\Box_j v\|_2\,\chi_{(|k-i-j|\leqslant 3\sqrt{n})}.\end{aligned}\tag{2.97}$$

Hence,

$$\begin{aligned}\|uv\|_{E^\lambda_{2,1}} &\leqslant \sum_{k\in\mathbb{Z}^n}2^{\lambda|k|}\sum_{i,j\in\mathbb{Z}^n}\|\Box_k(\Box_i u\,\Box_j v)\|_2\\&\lesssim \sum_{k\in\mathbb{Z}^n}\sum_{i,j\in\mathbb{Z}^n}2^{\lambda|k|}\|\Box_i u\|_2\|\Box_j v\|_2\,\chi_{(|k-i-j|\leqslant 3\sqrt{n})}\\&\lesssim 2^{C\lambda}\sum_{i,j\in\mathbb{Z}^n}2^{\lambda(|i|+|j|)}\|\Box_i u\|_2\|\Box_j v\|_2.\end{aligned}\tag{2.98}$$

The result follows. □

2.5.4 *Gevrey regularity of NS equation*

Once we get that the solution of the NS equation is in $E^{ct}_{2,1}$, then it belongs to the Gevrey 1-class. Let

$$\mathcal{D}=\{u:\sup_{0\leqslant t\leqslant t_0}\|u(t)\|_{E^{ct}_{2,1}}\leqslant M\},\tag{2.99}$$

$$d(u,v)=\sup_{0\leqslant t\leqslant t_0}\|u(t)-v(t)\|_{E^{ct}_{2,1}}.\tag{2.100}$$

Considering the mapping

$$\mathcal{T}:u(t)\to H(t)u_0-\int_0^t H(t-\tau)[\mathbb{P}\nabla\cdot(u\otimes u)](\tau)d\tau,\tag{2.101}$$

we show that for $u_0\in E^0_{2,1}$, there exists a $t_0>0$ such that $\mathcal{T}:(\mathcal{D},d)\to(\mathcal{D},d)$ is a contraction mapping. For convenience, we denote

$$|\!|\!|u|\!|\!|=\sup_{0\leqslant t\leqslant t_0}\|u(t)\|_{E^{ct}_{2,1}}.\tag{2.102}$$

Let $u\in\mathcal{D}$. By Proposition 2.6 and 2.7,

$$\begin{aligned}|\!|\!|\mathcal{T}u|\!|\!| &\lesssim \|u_0\|_{E^0_{2,1}}+t_0|\!|\!|u|\!|\!|+t_0^{1/2}|\!|\!|(u\otimes u)|\!|\!|\\&\lesssim \|u_0\|_{E^0_{2,1}}+t_0^{1/2}|\!|\!|u|\!|\!|^2.\end{aligned}\tag{2.103}$$

Take $M=2C\|u_0\|_{E^0_{2,1}}$ and assume $t_0<1$. If $Ct_0^{1/2}M\leqslant 1/4$, then $\mathcal{T}u\in\mathcal{D}$. Similarly, for any $u,v\in\mathcal{D}$,

$$|\!|\!|\mathcal{T}u-\mathcal{T}v|\!|\!|\leqslant\frac{1}{2}|\!|\!|u-v|\!|\!|.\tag{2.104}$$

So, there exists a $u \in \mathcal{D}$ satisfying $\mathcal{T}u = u$.

Theorem 2.3. *Let $u_0 \in [E^0_{2,1}]^n_0$. Then there exists a $T_m > 0$, such that the NS equation (2.5) has a unique solution $u \in C\left([0, T_m); [E^0_{2,1}]^n_0\right)$. Moreover, there exists a $t_0 \in (0, T_m)$ satisfying*

$$u(t) \in [E^{c(t\wedge t_0)}_{2,1}]^n_0, \quad t \in [0, T_m). \tag{2.105}$$

If $T_m < \infty$, then $\sup_{0\leqslant t<T_m} \|u(t)\|_{E^0_{2,1}} = \infty$. Moreover, $u \in C\left(0, T_m; [L^2]^n_0\right)$, $\partial_{x_i} u \in L^2\left(0, T_m; [L^2]^n_0\right)$, and

$$\frac{1}{2}\|u(t)\|_2^2 + \frac{1}{2}\int_0^t \|\nabla u(s)\|_2^2 ds = \frac{1}{2}\|u_0\|_2^2, \quad 0 < t < T_m. \tag{2.106}$$

If $\|u_0\|_n$ is sufficiently small, then $T_m = \infty$.

Remark 2.1.

(1) By Theorem 2.3, we can get that the solution of the NS equation belongs to $C^\infty((0, T_m) \times \mathbb{R}^n)$.
(2) Theorem 2.3 implies that the solution of the NS equation belongs to the Gevrey 1-class and so, is really analytic.
(3) Theorem 2.3 also describes the disappearing process of the regularity of the solutions to the NS equation.

Remark 2.2. The Gevrey regularity for the evolution equations is of importance for its own sake. The Gevrey regularity of the weak solutions for a class of linear and semi-linear Fokker-Planck equations

$$(\partial_t + v \cdot \nabla_x - \Delta_v)u = F(t, x, v, u, \nabla_v u)$$

was recently studied by Chen, Li and Xu [35], see also [37] and references therein for a class of the linear model of spatially inhomogeneous Boltzmann equations without an angular cutoff.

Remark 2.3. As the end of this chapter, we state some recent progress on NS equation without proofs.

(1) Let us mention the result by Koch and Tartaru [151] where the global solutions for NS in 3D are obtained with the small data in the space BMO^{-1} with the norm:

$$\|u\|_{\dot{B}^{-1}_{\infty,\infty}} + \sup_{x\in\mathbb{R}^3,\ R>0} R^{-3/2}\left(\int_{[0,R^2]\times\{y:|x-y|\leqslant R\}} |e^{t\Delta}u(y)|dydt\right)^{1/2},$$

see also Chemin and Gallagher's generalizations in [30; 31; 32] for a class of large data.

(2) Recently, Escauriaza, Seregin and Šverák [70] showed that any "Leray-Hopf" weak solution in 3D which remains bounded in $L^3(\mathbb{R}^3)$ cannot develop a singularity in finite time. Their proof used a blow-up procedure and reduction to a backwards uniqueness question for the heat equation, and was then completed using Carleman-type inequalities and the theory of unique continuation. Kenig and Koch [123] gave an alternative proof by substituting L^3 with $\dot{H}^{1/2}$. Dong and Du [68] generalized their results in higher spatial dimensions $n \geqslant 3$.

(3) Noticing that $L^3 \subset B^{-1}_{\infty,\infty}$ in 3D is a sharp embedding, for any solution u of the NS equation in $C([0,T^*); L^3)$, we see that $u \in C([0,T^*); B^{-1}_{\infty,\infty})$. May [163] prove that if $T^* < \infty$, then there exists a constant $c > 0$ independent of the solution of NS equation such that

$$\lim \sup_{t \to T^*} \|u(t) - \omega\|_{B^{-1}_{\infty,\infty}} \geqslant c$$

for all $\omega \in \mathscr{S}$. For the Leray-Hopf weak solution, Cheskidov and Shvydkoy [39] obtained similar result.

(4) Recently, the Cauchy problem for the 3D anisotropic Navier-Stokes equation

$$u_t - (\partial^2_{x_1} + \partial^2_{x_2})u + u \cdot \nabla u + \nabla p = 0, \quad \nabla \cdot u = 0, \quad u|_{t=0} = u_0$$

is considered in [187; 33] (see also references therein). These equations come from meteorology models. More recent results based on the harmonic analysis method can be found in [155].

Chapter 3

Strichartz estimates for linear dispersive equations

First, let us imagine the large-time-decaying phenomena in a dispersive system... .

We begin to study the dispersive equations. Roughly speaking, the dispersive equation takes the following form

$$\partial_t u - \mathrm{i}P(D)u = F, \tag{3.1}$$

where $u(t,x)$ is an unknown function, $\mathrm{i} = \sqrt{-1}$, $\widehat{P(D)u} = P(\xi)\widehat{u}$, and $P(\xi)$ is a real-valued function, which is said to be the dispersion relation of (3.1)[1]. If F is a nonlinear function of u, then (3.1) is said to be a nonlinear dispersive equation.

In this chapter we study the time-space estimates for the solutions of the linear dispersive equation in mixed Lebesgue spaces $L^q_t L^p_x$ (or more general spaces $L^q(0,T;B^s_{p,r})$), so called the Strichartz inequalities, which is a starting point to the study of nonlinear dispersive equations. For instant, we consider the Schrödinger equation[2]

$$\mathrm{i}u_t + \Delta u = f, \quad u(0,x) = u_0(x), \tag{3.2}$$

where $\Delta = \sum_{i=1}^n \partial^2_{x_i}$, $u(t,x)$ is a complex valued function of $(t,x) \in \mathbb{R}\times\mathbb{R}^n$, u_0 denotes the initial value at $t=0$. f is a known complex function of $(t,x) \in \mathbb{R}\times\mathbb{R}^n$.

We will use its integral form. Taking the Fourier transform to (3.2), we get

$$\mathrm{i}\widehat{u}_t - |\xi|^2\widehat{u} = \widehat{f}, \quad \widehat{u}(0) = \widehat{u}_0.$$

[1] We emphasize that $P(\cdot)$ must be a real function.

[2] Erwin Schrödinger (1887-1961) was an Austrian theoretical physicist who achieved fame for his fundamental contributions to quantum mechanics, especially the Schrödinger equation, for which he received the Nobel Prize in 1933.

Solving the ordinary differential equation, we obtain that

$$u(t) = S(t)u_0 - \mathrm{i}\int_0^t S(t-\tau)f(\tau,\cdot)d\tau, \tag{3.3}$$

where $S(t) = e^{\mathrm{i}t\Delta} := \mathscr{F}^{-1}e^{-\mathrm{i}t|\xi|^2}\mathscr{F}$. By Plancherel's identity, we see that $\|S(t)u_0\|_2 = \|u_0\|_2$, which means that $S(t)u_0$ is invariant in L^2 for any $t \in \mathbb{R}$. However, we can show that $S(t)$ satisfies the following Strichartz estimates:

$$\|S(t)u_0\|_{L^{\gamma(r)}(\mathbb{R},L^r)} \lesssim \|u_0\|_2,$$
$$\left\|\int_0^t S(t-\tau)f(\tau,\cdot)d\tau\right\|_{L^{\gamma(r)}(\mathbb{R},L^r)} \lesssim \|f\|_{L^{\gamma(\rho)'}(\mathbb{R},L^{\rho'})},$$

where $2 \leqslant r,\rho \leqslant \infty$, $2/\gamma(\cdot) = n(1/2 - 1/\cdot) \in [0,1)$, and p' is the conjugate number of p. Such kinds of estimates are time-decaying versions, which are of importance in the study of nonlinear dispersive equations. In the next chapter, we will further indicate why Strichartz inequalities are useful.

3.1 $L^{p'} \to L^p$ estimates for the dispersive semi-group

Let us consider the Schrödinger semi-group $e^{\mathrm{i}t\Delta} := \mathscr{F}^{-1}e^{\mathrm{i}t|\xi|^2}\mathscr{F}$. Since

$$e^{\mathrm{i}t\Delta}u_0 = ct^{-n/2}\int_{\mathbb{R}^n} e^{\frac{|x-y|^2}{4\mathrm{i}t}}u_0(y)dy, \tag{3.4}$$

we immediately have the $L^1 \to L^\infty$ decay

$$\|e^{\mathrm{i}t\Delta}u_0\|_\infty \lesssim t^{-n/2}\|u_0\|_1.$$

Taken noticing of $\|e^{\mathrm{i}t\Delta}u_0\|_2 = \|u_0\|_2$, an interpolation yields

$$\|e^{\mathrm{i}t\Delta}u_0\|_p \lesssim t^{-n/2}\|u_0\|_{p'}, \quad p \geqslant 2,\ 1/p + 1/p' = 1.$$

This is the fundamental $L^{p'} \to L^p$ estimate, or the time decay estimate for the Schrödinger semi-group. For the general semi-group, it is not expected to have an analytic expression as in (3.4), we need to look for other ways to get the $L^{p'} \to L^p$ estimates.

In this section we study the $L^{p'} \to L^p$ estimates for a class of dispersive semi-groups $U(t) := \mathscr{F}^{-1}\exp(\mathrm{i}tP(\xi))\mathscr{F}$ and use two different ways to consider its $L^{p'} \to L^p$ decay. The first method is very effective for homogeneous functions $P(\cdot)$, the second method can deal with nonhomogeneous radial functions $P(\cdot)$.

First, we consider the decay estimates for the semi-group $U_m(t) := \mathscr{F}^{-1}\exp(\mathrm{i}t|\xi|^m)\mathscr{F}$. Some earlier decay estimates on $U_m(t)$ were obtained by Pecher [191]. Following Pecher's proof, we can get an $L^{p'} \to L^p$ decay estimate with smooth effects in the case $m > 2$. Some nontrivial generalized dispersive $L^q - L^p$ estimates can be founded in Cui [52; 53], Sugimoto [207; 208] (see also [195]).

We need the following lemma, which is due to Littman [161].

Lemma 3.1. *Let $v \in C_0^\infty(\mathbb{R}^n)$, $\mathrm{supp}\, v = \Omega$. $P(\xi) : \mathbb{R}^n \to \mathbb{R}$ is an infinitely smooth function on Ω. For any $\xi \in \Omega$, the rank of $(\partial^2 P(\xi)/\partial\xi_i\partial\xi_j)_{i,j=1}^n$ is at least $\rho > 0$, Then there exists a $K \in \mathbb{N}$ such that for any $\lambda \in \mathbb{R}$,*

$$\|\mathscr{F}^{-1} e^{\mathrm{i}\lambda P(\xi)} v\|_\infty \lesssim (1+|\lambda|)^{-\rho/2} \sum_{|\alpha|\leqslant K} \|D^\alpha v\|_\infty. \tag{3.5}$$

Now we derive the decay estimates of $U_m(t)$ in homogeneous Besov spaces. Let $\{\triangle_k\}_{k\in\mathbb{Z}}$ be as in Sec. 1.6. In view of Young's inequality (see Appendix), one has that

$$\|\mathscr{F}^{-1} e^{\mathrm{i}t|\xi|^m} \mathscr{F} \triangle_k f\|_\infty \leqslant \|\mathscr{F}^{-1} e^{\mathrm{i}t|\xi|^m} \varphi(2^{-k}\xi)\|_\infty \|f\|_1. \tag{3.6}$$

Now we estimate $\|\mathscr{F}^{-1} e^{\mathrm{i}t|\xi|^m} \varphi(2^{-k}\xi)\|_\infty$. Applying Proposition 1.4 and Lemma 3.1, we have

$$\begin{aligned} &\|\mathscr{F}^{-1} e^{\mathrm{i}t|\xi|^m} \varphi(2^{-k}\xi)\|_\infty \\ &= 2^{kn} \|\mathscr{F}^{-1} e^{\mathrm{i}t2^{km}|\xi|^m} \varphi(\xi)\|_\infty \lesssim t^{-\rho/2} 2^{k(n-m\rho/2)}, \end{aligned} \tag{3.7}$$

where ρ denotes the rank of $(\partial^2|\xi|^m/\partial\xi_i\partial\xi_j)_{i,j=1}^n$ on the support of φ ($\mathrm{supp}\varphi \subset \{\xi : |\xi| \in (2^{-1}, 2)\}$). It is easy to see that

$$\rho = \begin{cases} n-1, & m = 1, \\ n, & m \geqslant 2. \end{cases} \tag{3.8}$$

So,

$$\|\mathscr{F}^{-1} e^{\mathrm{i}t|\xi|^m} \mathscr{F} \triangle_k f\|_\infty \lesssim t^{-\rho/2} 2^{k(n-m\rho/2)} \|f\|_1. \tag{3.9}$$

Obviously, by Plancherel's identity,

$$\|\mathscr{F}^{-1} e^{\mathrm{i}t|\xi|^m} \mathscr{F} \triangle_k f\|_2 \leqslant \|f\|_2. \tag{3.10}$$

In view of Riesz-Thorin's interpolation theorem, (3.9) and (3.10) imply that for any $2 \leqslant p \leqslant \infty$, $1/p + 1/p' = 1$,

$$2^{-k(2n-m\rho)(1/2-1/p)} \|U_m(t)\triangle_k f\|_p \lesssim t^{-\rho(1/2-1/p)} \|f\|_{p'}. \tag{3.11}$$

For convenience, we write

$$2\sigma(m,p) := (2n - m\rho)(1/2 - 1/p). \tag{3.12}$$

Substituting f by $\sum_{\ell=-1}^{1} \triangle_{k+\ell} f$ in (3.11), we get

$$2^{-2\sigma(m,p)k}\|U_m(t)\triangle_k f\|_p \lesssim t^{-\rho(1/2-1/p)} \sum_{\ell=-1}^{1} \|\triangle_{k+\ell} f\|_{p'}. \tag{3.13}$$

Taking the ℓ^q-norm in (3.13), we have

$$\|U_m(t)f\|_{\dot{B}^{-2\sigma(m,p)}_{p,q}} \lesssim t^{-\rho(1/2-1/p)}\|f\|_{\dot{B}^0_{p',q}}, \; 2 \leqslant p \leqslant \infty. \tag{3.14}$$

Using the inclusion $\dot{B}^s_{p,2} \subset \dot{F}^s_{p,2} = \dot{H}^s_p$ and $\dot{B}^s_{p',2} \supset \dot{F}^s_{p',2} = \dot{H}^s_{p'}$, we have

$$\|U_m(t)f\|_{\dot{H}^{-2\sigma(m,p)}_p} \lesssim t^{-\rho(1/2-1/p)}\|f\|_{p'}, \; 2 \leqslant p < \infty. \tag{3.15}$$

Taking $m = 1, 2$, we immediately have

Proposition 3.1. *Let* $n \geqslant 2$, $W(t) = \mathscr{F}^{-1}e^{\mathrm{i}t|\xi|}\mathscr{F}$, $2 \leqslant p < \infty$, $1 \leqslant q \leqslant \infty$, *and* $1/p + 1/p' = 1$. *Then*

$$\|W(t)f\|_{\dot{B}^{-(n+1)(1/2-1/p)}_{p,q}} \lesssim t^{-(n-1)(1/2-1/p)}\|f\|_{\dot{B}^0_{p',q}}, \tag{3.16}$$

$$\|W(t)f\|_{\dot{H}^{-(n+1)(1/2-1/p)}_p} \lesssim t^{-(n-1)(1/2-1/p)}\|f\|_{p'}. \tag{3.17}$$

Proposition 3.2. *Let* $n \geqslant 1$, $S(t) = \mathscr{F}^{-1}e^{\mathrm{i}t|\xi|^2}\mathscr{F}$, $2 \leqslant p < \infty$, $1 \leqslant q \leqslant \infty$, *and* $1/p + 1/p' = 1$. *Then we have*

$$\|S(t)f\|_{\dot{B}^0_{p,q}} \lesssim t^{-n(1/2-1/p)}\|f\|_{\dot{B}^0_{p',q}}, \tag{3.18}$$

$$\|S(t)f\|_p \lesssim t^{-n(1/2-1/p)}\|f\|_{p'}. \tag{3.19}$$

In the higher order case $m > 2$, we have

Proposition 3.3. *Let* $n \geqslant 1$, $m > 2$, $2 \leqslant p < \infty$, $1 \leqslant q \leqslant \infty$, *and* $1/p + 1/p' = 1$. *Then we have*

$$\|U_m(t)f\|_{\dot{B}^{-2\sigma(m,p)}_{p,q}} \lesssim t^{-n(1/2-1/p)}\|f\|_{\dot{B}^0_{p',q}}, \tag{3.20}$$

$$\|U_m(t)f\|_{\dot{H}^{-2\sigma(m,p)}_p} \lesssim t^{-n(1/2-1/p)}\|f\|_{p'}. \tag{3.21}$$

Noticed that for $m > 2$, $2\sigma(m,p) = n(2-m)(1/2-1/p) < 0$, $U_m(t)$ gains some regularities in Proposition 3.3.

Below, we consider another decay estimate of $U_m(t)$ without the smoothness in the case $m > 2$. We have

Proposition 3.4. *Let* $n \geqslant 1$, $m > 2$, $2 \leqslant p < \infty$, *and* $1/p + 1/p' = 1$. *Then*

$$\|U_m(t)f\|_p \lesssim t^{-n(1/p'-1/p)/m}\|f\|_{p'}. \tag{3.22}$$

Proof. In view of the dilation property, it suffices to consider the case $t = 1$. Let $\{\varphi_k\}_{k=0}^{\infty}$ be as in Sec. 1.3. (3.13) implies that for any $k \geqslant 1$,

$$\|U_m(1)\triangle_k f\|_p \lesssim 2^{2\sigma(m,p)k} \sum_{\ell=-1}^{1} \|\triangle_{k+\ell} f\|_{p'}. \tag{3.23}$$

By $\operatorname{supp}\varphi_0 \subset \{\xi : |\xi| \leqslant 2\}$, we have

$$\begin{aligned}\|U_m(1)\triangle_0 f\|_p &\leqslant \sum_{\ell=0}^{1} \|U_m(1)\triangle_0\triangle_\ell f\|_p \\ &\lesssim \sum_{\ell=0}^{1} \|\varphi_0 \mathscr{F} \triangle_\ell f\|_{p'} \lesssim \sum_{\ell=0}^{1} \|\triangle_\ell f\|_{p'}. \end{aligned} \tag{3.24}$$

We write $\triangle_{-1} = 0$. It follows that (3.23) holds for all $k \geqslant 0$. Taking the ℓ^2 in (3.23), one has that

$$\|U_m(1)f\|_{B^0_{p,2}} \lesssim \|f\|_{B^0_{p',2}}. \tag{3.25}$$

Similar to (3.15), (3.25) implies that

$$\|U_m(1)f\|_p \lesssim \|f\|_{p'}. \tag{3.26}$$

In view of the dilation property, we obtain the result, as desired. □

Using (3.21), (3.22) and the convexity Hölder inequality (Proposition 1.21), we immediately have

Proposition 3.5. *Let $n \geqslant 1$, $m > 2$, $2 \leqslant p < \infty$, and $1/p + 1/p' = 1$. Then for any $\theta \in [0, 1]$,*

$$\|U_m(t)f\|_{\dot{H}_p^{-2\sigma(m,p)\theta}} \lesssim t^{-(n\theta+2n(1-\theta)/m)(1/2-1/p)} \|f\|_{p'}. \tag{3.27}$$

The above method is also valid for some other homogeneous non-radial functions, say $P(\xi) = \xi_1^4 \pm ... \pm \xi_k^4$, $k \leqslant n$. Noticing that, if $P(\xi)$ is not a homogeneous function, (3.7) can not be obtained by scaling, we need to look for another way to handle the nonhomogeneous case.

Our idea is to simplify $P(\cdot)$ as a radial function $P(\xi) := P(|\xi|)$, which is essentially reduced to one dimensional case. $P(|\xi|)$ can be separated into two parts, lower and higher frequency parts, which correspond to $|\xi| \lesssim 1$ and $|\xi| \gg 1$, respectively. We further assume that $P(\xi)$ has a different growth as $|\xi| \lesssim 1$ and $|\xi| \gg 1$, which is sufficient for many semi-groups.

In what follows, we always assume that $P : (0, \infty) \to \mathbb{R}$ is a smooth radial function satisfying

(H1) there exists an $m_1 > 0$, such that for any $\alpha \geqslant 2$, $\alpha \in \mathbb{N}$,

$$|P'(r)| \sim r^{m_1-1}, \quad |P^{(\alpha)}(r)| \lesssim r^{m_1-\alpha}, \quad r \geqslant 1;$$

(H2) there exists an $m_2 > 0$, such that for any $\alpha \geqslant 3$, $\alpha \in \mathbb{N}$,

$$|P'(r)| \sim r^{m_2-1}, \quad |P^{(\alpha)}(r)| \lesssim r^{m_2-\alpha}, \quad 0 < r < 1;$$

(H3) there exists an α_1 such that

$$|P''(r)| \sim r^{\alpha_1-2}, \quad r \geqslant 1;$$

(H4) there exists an α_2 such that

$$|P''(r)| \sim r^{\alpha_2-2}, \quad 0 < r < 1.$$

Remark 3.1. The following are some examples for which conditions (H1)–(H4) are satisfied: (1) $P(\xi) = \sqrt{1+|\xi|^2}$ (relevant to the Klein-Gordon semi-group); (2) $P(\xi) = |\xi|^4 + |\xi|^2$ (relevant to the fourth order Schrödinger semi-group); (3) $P(\xi) = \sqrt{1+|\xi|^4}$ (relevant to the beam semi-group).

Comparing $P(|\xi|) = |\xi|$ with $P(|\xi|) = \sqrt{1+|\xi|^2}$, although they have the same growth as $t \to \infty$, we easily see that the rank of $(\partial^2_{ij}\sqrt{1+|\xi|^2})_{n\times n}$ is n and $\sqrt{1+|\xi|^2}$ is better than $|\xi|$ at $\xi = 0$. So, we can expect that the Klein-Gordon semi-group has a better $L^{p'} \to L^p$ decay.

We first consider 1D case, which is easier to find the ideas. The following result is due to [94].

Proposition 3.6. *Let $n = 1$ and $U(t) = \mathscr{F}^{-1}e^{itP(|\xi|)}\mathscr{F}$. We have the following results.*

(a) *Let $\{\triangle_k\}_{k\in\mathbb{Z}}$ be defined in Sec. 1.6. Then for any $k \geqslant 0$,*

$$\|U(t)\triangle_k u_0\|_\infty \lesssim 2^k\|u_0\|_1.$$

Moreover, if P satisfies (H3), then

$$\|U(t)\triangle_k u_0\|_\infty \lesssim |t|^{-\theta/2}2^{k(1-\alpha_1\theta/2)}\|u_0\|_1, \quad 0 \leqslant \theta \leqslant 1.$$

(b) *Let $\{\triangle_k\}_{k\in\mathbb{Z}}$ be as in Sec. 1.6. Then for any $k < 0$, one has that*

$$\|U(t)\triangle_k u_0\|_\infty \lesssim 2^k\|u_0\|_1.$$

Moreover, if P satisfies (H4), then

$$\|U(t)\triangle_k u_0\|_\infty \lesssim |t|^{-\theta/2}2^{k(1-\alpha_2\theta/2)}\|u_0\|_1, \quad 0 \leqslant \theta \leqslant 1.$$

(c) *Let $\triangle_0$ be as in Sec. 1.3. Assume that P satisfies (H2) and (H4) with $m_2 = \alpha_2$, then*

$$\|U(t)\triangle_0 u_0\|_\infty \lesssim (1+|t|)^{-\theta}\|u_0\|_1, \quad 0 \leqslant \theta \leqslant \min\left(\frac{1}{m_2}, \frac{1}{2}\right).$$

We need the following Van der Corput lemma, see Appendix.

Lemma 3.2. *Let $\varphi \in C_0^\infty(\mathbb{R})$, and $P \in C^2(\mathbb{R})$ satisfy $|P''(\xi)| \geqslant \lambda > 0$ for all $\xi \in \operatorname{supp} \varphi$. Then*

$$\left| \int e^{iP(\xi)} \varphi(\xi) d\xi \right| \lesssim \lambda^{-1/2} (\|\varphi\|_\infty + \|\varphi'\|_1).$$

Proof. [Proof of Proposition 3.6] First, we prove (a). Using Young's inequality, one has that

$$\|U(t)\triangle_k u_0\|_\infty \lesssim \|J_k\|_\infty \|u_0\|_1,$$

where

$$J_k(x) = \mathscr{F}^{-1}(e^{itP(|\xi|)} \varphi(2^{-k}|\xi|))(2^{-k}x). \tag{3.28}$$

By (3.28), we have

$$\|J_k\|_\infty \lesssim 2^k. \tag{3.29}$$

Denote $P_1(\xi) = x\xi + tP(2^k|\xi|)$. We have $|P_1''(\xi)| \gtrsim |t|2^{k\alpha_1}$ for $\xi \in \operatorname{supp} \varphi$. Using Van der Corput lemma, we get

$$\|J_k\|_\infty \lesssim |t|^{-1/2} 2^{k(1-\alpha_1/2)}. \tag{3.30}$$

Making an interpolation between (3.29) and (3.30), we get that for any $0 \leqslant \theta \leqslant 1$,

$$\|J_k\|_\infty \lesssim |t|^{-\theta/2} 2^{k(1-\theta\alpha_1/2)}.$$

It follows that (a) holds.

The proof of (b) is analogous to (a) and the details will be omitted. We now prove (c). In view of the first conclusion in (b), we have

$$\|U(t)\triangle_0 u_0\|_\infty \lesssim \sum_{k<0} 2^k \|u_0\|_1 \lesssim \|u_0\|_1. \tag{3.31}$$

We first consider the case $m_2 < 2$. Since $\min(1/m_2, 1/2) = 1/2$, from (b) it follows that

$$\|U(t)\triangle_0 u_0\|_\infty \lesssim \sum_{k<0} |t|^{-1/2} 2^{k(1-m_2/2)} \|u_0\|_1 \lesssim |t|^{-1/2} \|u_0\|_1. \tag{3.32}$$

From (3.31) and (3.32), we can get the conclusions.

Next, we discuss the case $m_2 \geqslant 2$. It suffices to consider the case $m_2 \geqslant 2$ and $\theta = \min(1/m_2, 1/2) = 1/m_2$. A straightforward calculation yields

$$\frac{d}{d\xi}\left(\frac{1}{P'(2^k|\xi|)} \right) \lesssim 2^{-k(m_2-1)}, \quad \xi \in \operatorname{supp}\varphi. \tag{3.33}$$

So, if $|x| \leqslant 1$, then we have $|\partial_\xi^m(e^{ix\xi}\varphi(\xi))| \lesssim 1$. Integrating by part to $J_k(\cdot)$, we have

$$|J_k(x)| \lesssim |t|^{-1}2^{k(1-m_2)}.$$

If $|x| > 1$, let k_0 be the minimal integer so that $|x| \leqslant |t|2^{k_0 m_2}$, then $|x| \approx |t|2^{k_0 m_2}$. For $|k-k_0| > C \gg 1$, we have $|P_1'(\xi)| \geqslant c|t|2^{km_2}$. Integrating by part to $J_k(\cdot)$, we have

$$|J_k(x)| \lesssim |t|^{-1}2^{k(1-m_2)}.$$

For $|k-k_0| \leqslant C$, noticing that $|x| > 1$ and $m_2 \geqslant 2$, we have

$$|J_k(x)| \lesssim |t|^{-1/2}2^{k(1-m_2/2)} \lesssim |t|^{-1/2}\left(\frac{|x|}{|t|}\right)^{(1-m_2/2)/m_2} \lesssim |t|^{-1/m_2}.$$

So,

$$\begin{aligned}
|\sum_{k\leqslant 0} J_k(x)| &\lesssim \sum_{|k-k_0|\leqslant C} |J_k(x)| + \sum_{|k-k_0|\geqslant C} |J_k(x)| \\
&\lesssim \sum_{|k-k_0|\leqslant C} |t|^{-1/m_2} + \sum_{2^k<|t|^{-1/m_2}} 2^k + \sum_{2^k>|t|^{-1/m_2}} |t|^{-1}2^{k(1-m_2)} \\
&\lesssim |t|^{-1/m_2},
\end{aligned}$$

which finishes the proof of (c). □

In higher spatial dimensions, we have

Proposition 3.7. *Let $n \geqslant 2$. Denote $U(t) = \mathscr{F}^{-1}e^{itP(|\xi|)}\mathscr{F}$. We have the following decay estimates.*

(a) *Let $\{\triangle_k\}_{k\in\mathbb{Z}}$ be as in Sec. 1.6, $k \geqslant 0$, and P satisfy (H1). Then*

$$\|U(t)\triangle_k u_0\|_\infty \lesssim |t|^{-\theta}2^{k(n-m_1\theta)}\|u_0\|_1, \quad 0 \leqslant \theta \leqslant \frac{n-1}{2}. \tag{3.34}$$

Moreover, if P satisfies (H3), then

$$\|U(t)\triangle_k u_0\|_\infty \lesssim |t|^{-n/2}2^{k(n-\frac{m_1 n}{2}-\frac{\alpha_1-m_1}{2})}\|u_0\|_1. \tag{3.35}$$

(b) *Let $\{\triangle_k\}_{k\in\mathbb{Z}}$ be as in Sec. 1.6, $k < 0$, and P satisfy (H2). Then we have*

$$\|U(t)\triangle_k u_0\|_\infty \lesssim |t|^{-\theta}2^{k(n-m_2\theta)}\|u_0\|_1, \quad 0 \leqslant \theta \leqslant \frac{n-1}{2}. \tag{3.36}$$

Moreover, if P satisfies (H4), then

$$\|U(t)\triangle_k u_0\|_\infty \lesssim |t|^{-n/2}2^{k(n-\frac{m_2 n}{2}-\frac{\alpha_2-m_2}{2})}\|u_0\|_1. \tag{3.37}$$

(c) *Let Δ_0 be as in Sec. 1.3, and P satisfies (H2). Then we have*

$$\|U(t)\Delta_0 u_0\|_\infty \lesssim (1+|t|)^{-\theta}\|u_0\|_1, \quad 0 \leqslant \theta \leqslant \min\left(\frac{n}{m_2}, \frac{n-1}{2}\right). \tag{3.38}$$

Moreover, if P satisfies (H4) with $\alpha_2 = m_2$, then

$$\|U(t)\Delta_0 u_0\|_\infty \lesssim (1+|t|)^{-\theta}\|u_0\|_1, \quad 0 \leqslant \theta \leqslant \min\left(\frac{n}{m_2}, \frac{n}{2}\right). \tag{3.39}$$

Proof. The idea is to use the Bessel function. After making the polar coordinate transform, the estimate is reduced to an oscillating integration in one spatial dimension. Next, using the decay and cycle properties of the Bessel function, analogous to the 1D case, we can get the conclusion. Let $J_m(r)$ be the Bessel function

$$J_m(r) = \frac{(r/2)^m}{\Gamma(m+1/2)\pi^{1/2}} \int_{-1}^{1} e^{irt}(1-t^2)^{m-1/2}dt, \quad m > -1/2.$$

Let us state some properties on the Bessel functions, see [82] and [203].

Lemma 3.3. *For any $0 < r < \infty$, we have*

(i) $J_m(r) \leqslant Cr^m$,
(ii) $\frac{d}{dr}(r^{-m}J_m(r)) = -r^{-m}J_{m+1}(r)$.

(i) is obvious. Integrating by part to J_m, we can get (ii). It is known that, the Fourier transform for a radial f is also radial (see [202]):

$$\hat{f}(\xi) = 2\pi \int_0^\infty f(r)r^{n-1}(r|\xi|)^{-(n-2)/2}J_{\frac{n-2}{2}}(r|\xi|)dr. \tag{3.40}$$

By Lemma 3.3, for any $0 \leqslant s \leqslant 2$ and any $k \geqslant 0$,

$$\left|\frac{\partial^k}{\partial r^k}(\varphi(r)r^{n-1}(rs)^{-(n-2)/2}J_{\frac{n-2}{2}}(rs))\right| \leqslant C_k. \tag{3.41}$$

If $m = -\frac{n-2}{2}$, $J_m(r)$ has the following property (see [117, Chapter 1, (1.5)])

$$r^{-\frac{n-2}{2}}J_{\frac{n-2}{2}}(r) = c_n\mathscr{R}(e^{ir}h(r)), \tag{3.42}$$

where h satisfies

$$|\partial_r^k h(r)| \leqslant c_k(1+r)^{-\frac{n-1}{2}-k}. \tag{3.43}$$

So, for any $s \geqslant 2$ and $k \geqslant 0$,

$$|\partial_r^k(\varphi(r)r^{n-1}h(rs))| \leqslant c_k s^{-\frac{n-1}{2}}. \tag{3.44}$$

Now we prove (a). By Young's inequality,

$$\|U(t)\triangle_k u_0\|_\infty \lesssim \|\mathscr{F}^{-1}e^{itP(|\xi|)}\varphi(2^{-k}|\xi|)\|_\infty \|u_0\|_1.$$

Using (3.40), we have

$$\begin{aligned}
&\mathscr{F}^{-1}(e^{itP(|\xi|)}\varphi(2^{-k}|\xi|))(x)\\
&= 2^{kn}\mathscr{F}^{-1}(e^{itP(|2^k\xi|)}\varphi(|\xi|))(2^k|x|)\\
&= 2^{kn}\int_0^\infty e^{itP(2^k r)}\varphi(r)r^{n-1}(r2^k s)^{-(n-2)/2}J_{\frac{n-2}{2}}(r2^k s)dr\\
&=: I_k(2^k s),
\end{aligned}$$

where $s = |x|$ and $J_{\frac{n-2}{2}}(r)$ denotes the Bessel function. It suffices to show

$$\|I_k(s)\|_\infty \leqslant |t|^{-\theta}2^{k(n-m_1\theta)}.$$

By (i) of Lemma 3.3,

$$\|I_k(s)\|_\infty \lesssim 2^{kn}. \tag{3.45}$$

We consider the following two separated cases.

Case 1. $s \leqslant 2$. Denote $D^r = (\frac{1}{itP'(2^k r)2^k})\frac{d}{dr}$. We have $D^r(e^{itP(2^k r)}) = e^{itP(2^k r)}$. By the condition (H1), for any $m \geqslant 0$ and $r \sim 1$,

$$\frac{d^m}{dr^m}\left(\frac{1}{P'(2^k r)}\right) \leqslant C_m 2^{-k(m_1-1)}. \tag{3.46}$$

We write $\tilde{\varphi}(r) = \varphi(r)r^{n-1}$. Integrating by part to I_k, for any $q \in \mathbb{Z}^+$, we have

$$\begin{aligned}
I_k(s) &= 2^{kn}\int_0^\infty e^{itP(2^k r)}\tilde{\varphi}(r)(rs)^{-\frac{n-2}{2}}J_{\frac{n-2}{2}}(rs)dr\\
&= 2^{kn}\int_0^\infty D^r(e^{itP(2^k r)})\tilde{\varphi}(r)(rs)^{-\frac{n-2}{2}}J_{\frac{n-2}{2}}(rs)dr\\
&= -\frac{2^{kn}}{it2^k}\int_0^\infty e^{itP(2^k r)}\frac{d}{dr}\left(\frac{1}{P'(2^k r)}\tilde{\varphi}(r)(rs)^{-\frac{n-2}{2}}J_{\frac{n-2}{2}}(rs)\right)dr\\
&= \frac{2^{kn}}{(it2^k)^q}\sum_{m=0}^{q}\sum_{l_1,\ldots l_q\in\Lambda_m^q} C_{q,m}\\
&\quad\times\int_0^\infty e^{itP(2^k r)}\prod_{j=1}^{q}\partial_r^{l_j}\left(\frac{1}{P'(2^k r)}\right)\partial_r^{q-m}\left(\tilde{\varphi}(r)(rs)^{-\frac{n-2}{2}}J_{\frac{n-2}{2}}(rs)\right)dr,
\end{aligned} \tag{3.47}$$

where $\Lambda_m^q = \{l_1, \ldots, l_q \in \mathbb{Z}^+ : 0 \leqslant l_1 < \ldots < l_q \leqslant q, l_1 + \ldots l_q = m\}$. By (3.41), (3.46) and (3.47), we get that for any $q \in \mathbb{Z}^+$,

$$|I_k(s)| \lesssim |t|^{-q}2^{k(n-m_1 q)}. \tag{3.48}$$

Interpolating (3.48) with (3.45), we get for any $\theta \geqslant 0$, $|I_k(s)| \lesssim |t|^{-\theta} 2^{k(n-m_1\theta)}$.

Case 2. $s \geqslant 2$. By (3.42),

$$\begin{aligned} I_k(s) &= c_n 2^{kn} \int_0^\infty e^{itP(2^k r)} \tilde{\varphi}(r)(e^{irs} h(rs) + e^{-irs}\bar{h}(rs))dr \\ &= c_n 2^{kn} \int_0^\infty e^{i(tP(2^k r)+rs)} \tilde{\varphi}(r) h(rs) dr \\ &\quad + c_n 2^{kn} \int_0^\infty e^{i(tP(2^k r)-rs)} \tilde{\varphi}(r) \bar{h}(rs) dr \\ &=: B_1 + B_2. \end{aligned}$$

We can assume, without loss of generality that $t > 0$ and $P'(r) > 0$. We consider the estimate of B_1. Put $P_1(r) = tP(2^k r) + rs$. Noticing that $P_1'(r) = t2^k P'(2^k r) + s \geqslant ct2^{km_1}$, we see that (3.46) also holds if one replaces P by P_1. In view of (3.44), analogous to Case 1, we can get that for any $\theta \geqslant 0$,

$$|B_1| \lesssim |t|^{-\theta} 2^{k(n-m_1\theta)}.$$

We now consider the estimate of B_2. Put $P_2(r) = tP(2^k r) - rs$. Notice that if $s = t2^k P'(2^k r)$, then $P_2'(r) = 0$. We divide Case 2 into two subcases.

Case 2a. $s > 2\sup_{r\in[1/2,2]} t2^k P'(2^k r)$, or $s < \frac{1}{2}\inf_{r\in[1/2,2]} t2^k P'(2^k r)$. It is easy to see that $|P_2'(r)| \geqslant ct2^{km_1}$ if $r \sim 1$ and (3.46) still holds if one substitutes P by P_2. Using (3.44), we have for any $\theta \geqslant 0$,

$$|B_2| \lesssim |t|^{-\theta} 2^{k(n-m_1\theta)}.$$

Case 2b. $\frac{1}{2}\inf_{r\in[1/2,2]} t2^k P'(2^k r) \leqslant s \leqslant 2\sup_{r\in[1/2,2]} t2^k P'(2^k r)$. Using (3.44),

$$|B_2| \lesssim 2^{kn} s^{-\frac{n-1}{2}} \lesssim t^{-\frac{n-1}{2}} 2^{k(n-\frac{(n-1)m_1}{2})}. \tag{3.49}$$

Making an interpolation between (3.49) and (3.45), we get that for any $0 \leqslant \theta \leqslant \frac{n-1}{2}$,

$$|B_2| \leqslant t^{-\theta} 2^{k(n-m_1\theta)}. \tag{3.50}$$

If (H3) holds, then $|P_2''(r)| \geqslant t2^{k\alpha_1}$. Using Van der Corput lemma,

$$|B_2| \lesssim (t2^{k\alpha_1})^{-1/2} \int_0^\infty |\frac{d}{dr}(\tilde{\varphi}(r)h(rs))| dr \lesssim t^{-n/2} 2^{k(n-\frac{n}{2}(m_1+\frac{\alpha_1-m_1}{n}))}. \tag{3.51}$$

Thus, we finish the proof of (a).

The proof of (b) is similar to that of (a) and we omit the details. Finally, we prove (c). Let $0 \leqslant \theta \leqslant \min(\frac{n}{m_2}, \frac{n-1}{2})$. If $\theta < \frac{n}{m_2}$, then $n - m_2\theta > 0$. In view of (b), we immediately have

$$\|U(t)\triangle_0 u_0\|_\infty \lesssim \sum_{k=-\infty}^{2} |t|^{-\theta} 2^{k(n-m_2\theta)} \|\triangle_0 u_0\|_1$$
$$\lesssim |t|^{-\theta} \|\triangle_0 u_0\|_1.$$

Assume that $\frac{n-1}{2} \geqslant \frac{n}{m_2}$ and $\theta = \frac{n}{m_2}$. From the proof of (b), we see that for $k_0 < 0$ and $s \sim t2^{k_0 m_2} \geqslant 2$, there holds

$$|I_{k_0}(s)| \lesssim t^{-\frac{n-1}{2}} 2^{k_0(n - \frac{(n-1)m_2}{2})} \lesssim t^{-\frac{n}{m_2}}.$$

If $|k - k_0| > C \gg 1$, then

$$|I_k(s)| \lesssim t^{-\alpha} 2^{k(n-m_2\alpha)}, \ \forall\, \alpha \geqslant 0.$$

So, choosing α large enough, we have

$$|I_{\leqslant 0}(s)| \lesssim \sum_{|k-k_0| \leqslant C} |I_k(s)| + \sum_{|k-k_0|>C} |I_k(s)|$$
$$\lesssim t^{-\frac{n}{m_2}} + \sum_{2^k < t^{-\frac{1}{m_2}}} 2^{kn} + \sum_{2^k > t^{-\frac{1}{m_2}}} t^{-\alpha} 2^{k(n-m_2\alpha)}$$
$$\lesssim t^{-\frac{n}{m_2}},$$

which implies the result, as desired. If (H4) holds and $m_2 = \alpha_2$, the proof is analogous and the details are omitted. □

Using the above estimates to the Klein-Gordon equation

$$u_{tt} + u - \Delta u = f(t,x), \quad u(0,x) = u_0(x), \ u_t(0,x) = u_1(x),$$

which corresponds to the semi-group $G(t) = e^{\mathrm{i}t(I-\Delta)^{1/2}}$, we have the following

Proposition 3.8. *Let $G(t) = e^{\mathrm{i}t(I-\Delta)^{1/2}}$, $\theta \in [0,1]$, $2\sigma(\theta,\cdot) = (n+1+\theta)(1/2 - 1/\cdot)$, and $2/\beta(\theta,\cdot) = (n-1+\theta)(1/2-1/\cdot)$. Then we have*

$$\|G(t)f\|_{B^{s-2\sigma(\theta,p)}_{p,q}} \lesssim |t|^{-2/\beta(\theta,p)} \|f\|_{B^s_{p',q}}.$$

Remark 3.2. If the wave equation contains a damping term, say

$$u_{tt} - \Delta u + \alpha u_t = 0, \ \ u(0) = u_0, \ u_t(0) = u_1,$$

where $\alpha > 0$, the generating semi-group is quite different from the case $\alpha = 0$ and it is similar to $\mathscr{F}^{-1} e^{-\alpha t/2 - it|\xi| - i\alpha^2 t/|\xi|} \mathscr{F}$. In fact, there are many works have been devoted to study the decaying estimates for its solutions; cf. W.K. Wang and Yang [248], W.K. Wang and W.J. Wang [249], Ikehata, Nishihara and Zhao [112], Hayashi, Kaikina and Naumkin [103; 104], for details.

3.2 Strichartz inequalities: dual estimate techniques

Slip of bamboo for writing is eventually replaced by paper.

In this section we study the Strichartz estimates for a class of dispersive semi-groups. Such kinds of estimates go back to the pioneer work of R. S. Strichartz [206] in 1977 for the wave semi-group, afterwards there are a series of generalizations, see Kato [120], Cazenave and Weissler [26] for the Schrödinger equation; Pecher [191], Ginibre and Velo [80] for the wave equation; Brenner [22; 21] for the Klein-Gordon equation; Kenig-Ponce-Vega [126] for KdV and more general dispersive equations. Below, we give a unified approach to the Strichartz inequalities based on the dual estimate method, see [235; 234], which simplifies various Strichartz estimates for the above mentioned semi-groups. As corollaries, the Strichartz estimates for a class of higher order semi-groups are also obtained. Denote

$$U(t) = \mathscr{F}^{-1} e^{\mathrm{i}tP(\xi)} \mathscr{F}, \quad \mathcal{A} = \int_0^t U(t-\tau) f(\tau, \cdot) d\tau, \tag{3.52}$$

where $P(\cdot): \mathbb{R}^n \to \mathbb{R}$ is a smooth function. In what follows we always assume that

$$X = L^p, \quad \text{or } X = B^0_{p,2}, \quad 2 \leqslant p < \infty. \tag{3.53}$$

Assume that $U(t)$ satisfies the following estimate

$$\|U(t)f\|_{X^\alpha} \lesssim t^{-\theta} \|f\|_{X^*}, \tag{3.54}$$

where $\alpha \in \mathbb{R}$, $\theta \in (0,1)$, $X^\alpha := (I - \triangle)^{-\alpha/2} X$, and X^* is the dual space of X. Supposing that (3.54) holds, we can get some interesting estimates for $U(\cdot)$ and $\mathcal{A}$. Using (3.52) and (3.54), we have

$$\|\mathcal{A}f\|_{X^\alpha} \lesssim \int_0^t |t-\tau|^{-\theta} \|f(\tau)\|_{X^*} d\tau. \tag{3.55}$$

Using the Hardy-Littlewood-Sobolev inequality, we immediately obtain that

Lemma 3.4. *Assume that* (3.53) *and* (3.54) *are satisfied. For any* $T > 0$ *and* $s \in \mathbb{R}$, *we have*

$$\|\mathcal{A}f\|_{L^{2/\theta}(-T,T;X^{s+\alpha})} \lesssim \|f\|_{L^{(2/\theta)'}(-T,T;(X^*)^s)}, \tag{3.56}$$

where $(2/\theta)'$ *denotes the conjugate number of* $2/\theta$, *i.e.*, $\theta/2 + 1/(2/\theta)' = 1$.

For convenience, we denote by $\mathcal{D}_T$ the set of all of the functions which are defined in $(-T,T)$, valued in $\mathscr{S}$ and taken finitely many values, $I = (-T,T)$. It is easy to see that for $2 \leqslant p < \infty$, $\mathcal{D}_T$ is dense in $L^q(I,(X^*)^s)$ $(s \in \mathbb{R},\ 1 \leqslant q < \infty)$.

Lemma 3.5. *Assume that* (3.53) *and* (3.54) *are satisfied. Then*

$$\|U(t)f\|_{L^{2/\theta}(\mathbb{R},X^{s+\alpha/2})} \lesssim \|f\|_{H^s}. \tag{3.57}$$

Proof. First, we show that for any $T>0$, $I=(-T,T)$, $\varphi \in \mathscr{S}$, $\psi \in \mathcal{D}_T$,

$$\left|\int_{-T}^{T}(U(t)\varphi,\psi(t))dt\right| \lesssim \|\varphi\|_2\|\psi\|_{L^{(2/\theta)'}(I,(X^*)^{-\alpha/2})}. \tag{3.58}$$

In fact,

$$\left|\int_{-T}^{T}(U(t)\varphi,\psi(t))dt\right| \lesssim \|\varphi\|_2\left\|\int_{-T}^{T}U(-t)\psi(t)dt\right\|_2. \tag{3.59}$$

By Lemma 3.4,

$$\begin{aligned}
&\left\|\int_{-T}^{T}U(-t)\psi(t)dt\right\|_2^2\\
&=\left|\int_{-T}^{T}\left(\psi(t),\int_{-T}^{T}U(t-\tau)\psi(\tau)d\tau\right)dt\right|\\
&\lesssim \|\psi\|_{L^{(2/\theta)'}(I,(X^*)^{-\alpha/2})}\left\|\int_{-T}^{T}U(t-\tau)\psi(\tau)d\tau\right\|_{L^{2/\theta}(I,X^{\alpha/2})}\\
&\lesssim \|\psi\|^2_{L^{(2/\theta)'}(I,(X^*)^{-\alpha/2})}.
\end{aligned} \tag{3.60}$$

Combining (3.59) with (3.60), we see that (3.58) holds. Since $\mathscr{S}$ is dense in L^2, $\mathcal{D}_T$ is dense in $L^{(2/\theta)'}(I,(X^*)^{-\alpha/2})$, by (3.58) we immediately have

$$\|U(t)\varphi\|_{L^{2/\theta}(I,X^{\alpha/2})} \lesssim \|\varphi\|_2. \tag{3.61}$$

Noticing that the above estimates are independent of T, and letting $T\to\infty$, we get that those estimates also hold if $(-T,T)$ is substituted by $\mathbb{R}$. Taking $\varphi = (I-\Delta)^{s/2}f$, we can get the result, as desired. □

Lemma 3.6. *Assume that* (3.53) *and* (3.54) *are satisfied. Then for any* $T>0$, $I=(-T,T)$,

$$\|\mathcal{A}f\|_{L^\infty(I,H^{s+\alpha/2})} \lesssim \|f\|_{L^{(2/\theta)'}(I,(X^*)^s)}. \tag{3.62}$$

Proof. Analogous to Lemma 3.5, it suffices to show that for any $f \in \mathcal{D}_T$,

$$\|\mathcal{A}f\|_{L^\infty(I,H^{\alpha/2})} \lesssim \|f\|_{L^{(2/\theta)'}(I,X^*)}. \tag{3.63}$$

Let us write $J_s = (I-\Delta)^{s/2}$. We have

$$\begin{aligned}\|\mathcal{A}f\|^2_{H^{\alpha/2}} &= (\mathcal{A}J_{\alpha/2}f, \mathcal{A}J_{\alpha/2}f)\\ &\lesssim \|f\|_{L^{(2/\theta)'}(I,X^*)} \left\| \int_0^t U(\cdot - \tau)f(\tau)d\tau \right\|_{L^{2/\theta}(I,X^\alpha)}.\end{aligned} \tag{3.64}$$

Applying the same techniques as in Lemma 3.4, one has that

$$\left\| \int_0^t U(\cdot - \tau)f(\tau)d\tau \right\|_{L^{2/\theta}(I,X^\alpha)} \lesssim \|f\|_{L^{(2/\theta)'}(I,X^*)}. \tag{3.65}$$

Combining (3.64) and (3.65), we have (3.63). □

Lemma 3.7. *Assume that* (3.53) *and* (3.54)*are satisfied. Then for any* $T>0$, $I=(-T,T)$,

$$\|\mathcal{A}f\|_{L^{2/\theta}(I,X^{s+\alpha/2})} \lesssim \|f\|_{L^1(I,H^s)}. \tag{3.66}$$

Proof. Analogous to Lemma 3.5, it suffices to prove that for any $f \in \mathcal{D}_T$,

$$\|\mathcal{A}f\|_{L^{2/\theta}(I,X^{\alpha/2})} \lesssim \|f\|_{L^1(I,L^2)}. \tag{3.67}$$

Let ψ, $f \in \mathcal{D}_T$, we have

$$\left| \int_0^T (\mathcal{A}f(t), \psi(t))dt \right| \lesssim \|f\|_{L^1(0,T;L^2)} \left\| \int_\cdot^T U(\cdot - t)\psi(t)dt \right\|_{L^\infty(I,L^2)}. \tag{3.68}$$

Similar to Lemma 3.6, we have

$$\left\| \int_\cdot^T U(\cdot - t)\psi(t)dt \right\|_{L^\infty(I,L^2)} \lesssim \|\psi\|_{L^{(2/\theta)'}(I,(X^*)^{-\alpha/2})}. \tag{3.69}$$

One can similarly estimate $\int_{-T}^0 (\mathcal{A}f(t), \psi(t))dt$. So, (3.68) and (3.69) imply that

$$\left| \int_0^T (\mathcal{A}f(t), \psi(t))dt \right| \lesssim \|f\|_{L^1(0,T;L^2)} \|\psi\|_{L^{(2/\theta)'}(I,(X^*)^{-\alpha/2})}. \tag{3.70}$$

Using (3.70) and the duality, we can directly obtain (3.67). □

Remark 3.3. If (3.54) is replaced by

$$\|U(t)f\|_{\dot{X}^\alpha} \lesssim t^{-\theta}\|f\|_{\dot{X}^*}, \tag{3.71}$$

where $\dot{X} = L^p$ or $\dot{X} = \dot{B}^0_{p,2}$, then the results in Lemmas 3.4–3.7 also hold if one replaces the nonhomogeneous spaces by corresponding homogeneous spaces, for instance, the substitution of (3.56) in Lemma 3.5 is that

$$\|U(t)f\|_{L^{2/\theta}(I,\dot{X}^{s+\alpha/2})} \lesssim \|f\|_{\dot{H}^s}. \tag{3.72}$$

We omit the details of those results.

Now we apply the above results to a class of linear dispersive equations. We can use (3.14) and (3.15) to get some general results, however, every specific semi-group seems to be important in applications and so, we separately discuss every concrete semi-group, say $S(t)$, $W(t)$ and $U_m(t)$. First, we consider the Schrödinger semi-group $S(t) = e^{\mathrm{i}t\Delta}$ and the higher order semi-group $U_m(t) = e^{\mathrm{i}t(-\Delta)^{m/2}}$:

Theorem 3.1. *Let* $m \geqslant 2$,

$$m^* = \begin{cases} \infty, & n \leqslant m, \\ 2n/(n-m), & n > m, \end{cases} \tag{3.73}$$

$$\frac{1}{\gamma(\cdot)} = \frac{n}{m}\left(\frac{1}{2} - \frac{1}{\cdot}\right). \tag{3.74}$$

Assume that $2 \leqslant r, p < m^*$, $U_m(t) = e^{\mathrm{i}t(-\Delta)^{m/2}}$. *Then we have*

$$\|U_m(t)\phi\|_{L^{\gamma(p)}(I,\dot{B}^s_{p,2})} \lesssim \|\phi\|_{\dot{H}^s}, \tag{3.75}$$

$$\|\mathcal{A}_{U_m} f\|_{L^{\gamma(p)}(I,\dot{B}^s_{p,2})} \lesssim \|f\|_{L^{\gamma(r)'}(I,\dot{B}^s_{r',2})}, \tag{3.76}$$

where $I \subset \mathbb{R}$ *is an interval,* $\mathcal{A}_{U_m} := \int_0^t U_m(t-\tau)\cdot d\tau$. *In (3.75) and (3.76), replacing homogeneous Besov spaces by corresponding Besov spaces, Bessel potential spaces and Riesz potential spaces, respectively, the conclusions still hold.*

It is known that the solution of the wave equation

$$u_{tt} - \Delta u = f(t,x), \quad u(0,x) = u_0(x),\ u_t(0,x) = u_1(x) \tag{3.77}$$

is relevant to the semi-group $W(t)$, we have

Theorem 3.2. *Let* $W(t) = e^{\mathrm{i}t(-\Delta)^{1/2}}$, $n \geqslant 2$,

$$2^{**} = \begin{cases} \infty, & n = 2,3, \\ 2(n-1)/(n-3), & n > 3, \end{cases} \tag{3.78}$$

$$\frac{2\sigma(\cdot)}{n+1} = \frac{2}{(n-1)\beta(\cdot)} = \frac{1}{2} - \frac{1}{\cdot}. \tag{3.79}$$

Let $2 \leqslant r, p < 2^{**}$. *Then we have*

$$\|W(t)\phi\|_{L^{\beta(p)}(I,\dot{B}^{s-\sigma(p)}_{p,2})} \lesssim \|\phi\|_{\dot{H}^s}, \tag{3.80}$$

$$\|\mathcal{A}_W f\|_{L^{\beta(p)}(I,\dot{B}^{s-\sigma(p)}_{p,2})} \lesssim \|f\|_{L^{\beta(r)'}(I,\dot{B}^{s+\sigma(r)}_{r',2})}, \tag{3.81}$$

where $I \subset \mathbb{R}$ *is an arbitrary interval,* $\mathcal{A}_W := \int_0^t W(t-\tau)\cdot d\tau$. *In (3.80) and (3.81), substituting homogeneous Besov spaces by corresponding Riesz potential spaces, the results also hold.*

The solution of the Klein-Gordon equation

$$u_{tt} + u - \Delta u = f(t,x), \quad u(0,x) = u_0(x), \ u_t(0,x) = u_1(x) \tag{3.82}$$

is relevant to the semi-group $G(t) = e^{\mathrm{i}t(I-\Delta)^{1/2}}$, we have the following

Theorem 3.3. *Let $G(t) = e^{\mathrm{i}t(I-\Delta)^{1/2}}$ and $\theta \in [0,1]$. For $n > 3-\theta$, we write $2^{**} = 2(n-1+\theta)/(n-3+\theta)$ and for $n \leqslant 3-\theta$, we denote $2^{**} = \infty$. Let $2 \leqslant r,p < 2^{**}$, $2\sigma(\theta,\cdot) = (n+1+\theta)(1/2-1/\cdot)$ and $2/\beta(\theta,\cdot) = (n-1+\theta)(1/2-1/\cdot)$. Then we have the following estimates:*

$$\|G(t)\phi\|_{L^{\beta(\theta,p)}(I,B^{s-\sigma(\theta,p)}_{p,2})} \lesssim \|\phi\|_{H^s}, \tag{3.83}$$

$$\|\mathcal{A}_G f\|_{L^{\beta(\theta,p)}(I,B^{s-\sigma(\theta,p)}_{p,2})} \lesssim \|f\|_{L^{\beta(\theta,r)'}(I,B^{s+\sigma(\theta,r)}_{r',2})}, \tag{3.84}$$

where $I = (-T,T) \subset \mathbb{R}$ is arbitrary, $\mathcal{A}_G := \int_0^t G(t-\tau)\cdot d\tau$. In (3.83) and (3.84), replacing Besov spaces by corresponding Bessel potential spaces, the results also hold.

For the higher order Schrödinger semi-group, if we consider the smooth effect, we have

Theorem 3.4. *Let $U_m(t) = e^{\mathrm{i}t(-\Delta)^{m/2}}$, $m \geqslant 2$ and 2^* be as in (3.73). We write*

$$\frac{2}{\gamma(\cdot)} = n\left(\frac{1}{2} - \frac{1}{\cdot}\right), \tag{3.85}$$

$$2\sigma(m,\cdot) = n(2-m)\left(\frac{1}{2} - \frac{1}{\cdot}\right). \tag{3.86}$$

Let $2 \leqslant r,p < 2^*$. Then we have

$$\|U_m(t)\phi\|_{L^{\gamma(p)}(I,\dot{B}^{s-\sigma(m,p)}_{p,2})} \lesssim \|\phi\|_{\dot{H}^s}, \tag{3.87}$$

$$\|\mathcal{A}_{U_m} f\|_{L^{\gamma(p)}(I,\dot{B}^{s-\sigma(m,p)}_{p,2})} \lesssim \|f\|_{L^{\gamma(r)'}(I,\dot{B}^{s+\sigma(m,r)}_{r',2})}, \tag{3.88}$$

where $I \subset \mathbb{R}$ is arbitrary, $\mathcal{A}_{U_m} := \int_0^t U_m(t-\tau)\cdot d\tau$. In (3.87) and (3.88), replacing homogeneous Besov spaces by corresponding Riesz potential spaces, the conclusions also hold.

Proof. The proofs of Theorems 3.1–3.4 are analogous and we only prove Theorem 3.4. By Proposition 3.3,

$$\|U_m(t)f\|_{\dot{B}^{-2\sigma(m,p)}_{p,2}} \lesssim t^{-2/\gamma(p)}\|f\|_{\dot{B}^0_{p',2}}. \tag{3.89}$$

Putting $\dot{X} = \dot{B}^0_{p,2}$ and $\alpha = -2\sigma(m,p)$, in view of Lemma 3.5 we can get (3.87). Now we show that (3.88) holds. By Lemmas 3.4, 3.6, 3.7 and Remark 3.3,

$$\|\mathcal{A}_{U_m} f\|_{L^{\gamma(p)}(I,\dot{B}^{s-\sigma(m,p)}_{p,2})} \lesssim \|f\|_{L^{\gamma(p)'}(I,\dot{B}^{s+\sigma(m,p)}_{p',2})}, \tag{3.90}$$

$$\|\mathcal{A}_{U_m} f\|_{L^\infty(I,\dot{H}^s)} \lesssim \|f\|_{L^{\gamma(p)'}(I,\dot{B}^{s+\sigma(m,p)}_{p',2})}, \tag{3.91}$$

$$\|\mathcal{A}_{U_m} f\|_{L^{\gamma(p)}(I,\dot{B}^{s-\sigma(m,p)}_{p,2})} \lesssim \|f\|_{L^1(I,\dot{H}^s)}. \tag{3.92}$$

Case I. $p \in [2,r]$. Taking $\theta \in [0,1]$ such that $1/p = (1-\theta)/2 + \theta/r$, we deduce that

$$\frac{1}{\gamma(p)} = \frac{\theta}{\gamma(r)} + \frac{1-\theta}{\infty}, \quad \sigma(m,p) = \theta\sigma(m,r) + (1-\theta)\sigma(m,2). \tag{3.93}$$

In view of the convexity Hölder inequality, (3.90) and (3.91) imply that

$$\begin{aligned}\|\mathcal{A}_{U_m} f\|_{L^{\gamma(p)}(I,\dot{B}^{s-\sigma(m,p)}_{p,2})} &\leqslant \|\mathcal{A}_{U_m} f\|^{1-\theta}_{L^\infty(I,\dot{B}^{s-\sigma(m,2)}_{2,2})} \|\mathcal{A}_{U_m} f\|^{\theta}_{L^{\gamma(r)}(I,\dot{B}^{s-\sigma(m,r)}_{r,2})} \\ &\lesssim \|f\|_{L^{\gamma(r)'}(I,\dot{B}^{s+\sigma(m,r)}_{r',2})}. \end{aligned}\tag{3.94}$$

Case II. $p > r \geqslant 2$. Take $\theta \in (0,1)$ such that $1/r' = (1-\theta)/2 + \theta/p'$. It follows that

$$\frac{1}{\gamma(r)'} = \frac{\theta}{\gamma(p)'} + \frac{1-\theta}{1}, \quad \sigma(m,r) = \theta\sigma(m,p) + (1-\theta)\sigma(m,2). \tag{3.95}$$

Noticing the complex interpolation,

$$(L^{\gamma(p)'}(I,\dot{B}^{s+\sigma(m,p)}_{p',2}), \quad L^1(I,\dot{B}^s_{2,2}))_{[\theta]} = L^{\gamma(r)'}(I,\dot{B}^{s+\sigma(m,r)}_{r',2}), \tag{3.96}$$

(3.90) and (3.92) imply that

$$\mathcal{A}_{U_m} : L^{\gamma(r)'}(I,\dot{B}^{s+\sigma(m,r)}_{r',2}) \to L^{\gamma(p)}(I,\dot{B}^{s-\sigma(m,p)}_{p,2}) \tag{3.97}$$

is a bounded operator. Theorem 3.4 follows. □

3.3 Strichartz estimates at endpoints

Let us recall that the starting point in previous section is the estimate (3.54), where we always assume $\theta \in (0,1)$. We consider in this section the case $\theta = 1$. Assume that $U(t)$ is defined in (3.52), and for any $2 < p < \infty$, there exist $\alpha(p) \in \mathbb{R}$ and $\theta(p) > 0$ such that

$$\|U(t)f\|_{H^{\alpha(p)}_p} \lesssim t^{-\theta(p)}\|f\|_{p'}, \quad \forall\, 2 \leqslant p < \infty. \tag{3.98}$$

Suppose that there exists a $p_1 > 2$ satisfying $\theta(p_1) > 1$, then there exists a $r \in (2, p_1)$ such that $\theta(r) = 1$ (see (3.118)), i.e.,

$$\|U(t)f\|_{H_r^{\alpha(r)}} \lesssim t^{-1}\|f\|_{r'}. \tag{3.99}$$

(3.99) is said to be an endpoint estimate. Recall that the condition $\theta < 1$ in previous section aries from the Hardy-Littlewood-Sobolev (HLS) inequality, which is essential for (3.55) and (3.56). This is why we say that $\theta(r) = 1$ is the endpoint case. A natural question is what happens in the endpoint case $\theta(r) = 1$. If fact, if (3.99) occurs, then the Strichartz estimates still hold, i.e.,

$$\|U(t)\phi\|_{L^2(I,B^s_{r,2})} \lesssim \|\phi\|_{H^{s-\alpha(r)/2}}. \tag{3.100}$$

This estimate was essentially obtained by Keel and Tao [121],[3] where they used the techniques of the interpolation on bilinear operators.

According to the proof of (3.57), it suffices to show

$$\left|\int_{-T}^{T}(U(t)\varphi, \psi(t))dt\right| \lesssim \|\varphi\|_2\|\psi\|_{L^2(I,B^{-\alpha(r)/2}_{r',2})}. \tag{3.101}$$

In order to prove (3.101), analogous to (3.59) and (3.60), one needs to show that

$$\left\|\int_{-T}^{T}U(-t)\psi(t)dt\right\|_2^2 \lesssim \|\psi\|^2_{L^2(I,B^{-\alpha(r)/2}_{r',2})}. \tag{3.102}$$

The left hand side of (3.102) can be rewritten as

$$\int_{-T}^{T}\int_{-T}^{T}(U(-s)\psi(s),\ U(-t)\psi(t))dsdt. \tag{3.103}$$

By (3.103), it is natural to introduce the following bilinear operator

$$\mathcal{L}(F,G) := \iint_D (U(-s)F(s),\ U(-t)G(t))dsdt, \tag{3.104}$$

where

$$D := \{(s,t):\ s,t \in [-T,T],\ s \leqslant t\}. \tag{3.105}$$

Our goal is to prove that

$$|\mathcal{L}(F,G)| \lesssim \|F\|_{L^2(I,B^{-\alpha(r)/2}_{r',2})}\|G\|_{L^2(I,B^{-\alpha(r)/2}_{r',2})}. \tag{3.106}$$

[3]Keel and Tao did not consider the case $\alpha(r) \neq 0$, here our proof is a modification of Keel and Tao's proof.

Noticing that s and t in (3.102) have equal positions, and taking $F = G = \psi$, we have from (3.106) that (3.100) holds.

Now the question reduces to show the bilinear estimate (3.106). Keel and Tao's idea is to use the dyadic decomposition on $t-s$ to cancel the singularity at $t = s$. Put

$$D_j := \{(s,t) \in D : \ T2^j < t - s \leqslant T2^{j+1}\}. \tag{3.107}$$

It is easy to see that $D = \cup_{j\leqslant 0} D_j$. Denote

$$\mathcal{L}_j(F,G) := \iint_{D_j} (U(t-s)F(s), \ G(t))dsdt. \tag{3.108}$$

So, it suffices to show that

$$\sum_j |\mathcal{L}_j(F,G)| \lesssim \|F\|_{L^2(I,B^{-\alpha(r)/2}_{r',2})} \|G\|_{L^2(I,B^{-\alpha(r)/2}_{r',2})}. \tag{3.109}$$

In the following we estimate $\mathcal{L}_j(F,G)$.

Lemma 3.8. *Assume that* (3.98) *and* (3.99) *are satisfied, and* $P = (1/r, 1/r)$. *Let* $(1/a, 1/b) \in B(P,\varepsilon)$ *and* $\varepsilon > 0$ *is sufficiently small. Then we have*

$$|\mathcal{L}_j(F,G)| \lesssim (2^jT)^{-\beta(a,b)} \|F\|_{L^2(I,H^{-\alpha(a)/2}_{a'})} \|G\|_{L^2(I,H^{-\alpha(b)/2}_{b'})}, \tag{3.110}$$

where

$$\beta(a,b) = \frac{1}{2}(\theta(a) + \theta(b)) - 1. \tag{3.111}$$

Proof. We can assume that F and G are Schwartz functions which have compact support contained in $[-T,T]$ on the time variable. We will consider the following three cases:

(1) $a = b = p \in (2,\infty)$;

(2) $2 \leqslant a < r$, $b = 2$;

(3) $a = 2$, $2 \leqslant b < r$.

First, we consider case (1). Using Young's and Hölder's inequalities,

$$\begin{aligned} |\mathcal{L}_j(F,G)| &\lesssim \int_I \int_{t-s\sim 2^jT} (t-s)^{-\theta(p)} \|F(s)\|_{H^{-\alpha(p)/2}_{p'}} \|G(t)\|_{H^{-\alpha(p)/2}_{p'}} dsdt \\ &\lesssim (2^jT)^{1-\theta(p)} \|F\|_{L^2(I,\,H^{-\alpha(p)/2}_{p'})} \|G\|_{L^2(I,\,H^{-\alpha(p)/2}_{p'})}. \end{aligned} \tag{3.112}$$

It follows that (3.110) holds.

Next, we consider the case (2). By Hölder's inequality,

$$|\mathcal{L}_j(F,G)| \lesssim \left\| \int_{\cdot-2^{j+1}T}^{\cdot-2^jT} U(\cdot - s)F(s)ds \right\|_{L^2(I,L^2)} \|G\|_{L^2(I,\,L^2)}. \tag{3.113}$$

Applying (3.98) and Hölder's inequality, we have

$$\begin{aligned}
&\left\| \int_{\cdot-2^{j+1}T}^{\cdot-2^jT} U(\cdot - s)F(s)ds \right\|^2_{L^2(I,L^2)} \\
&\lesssim \int_I \int_{t-2^{j+1}T}^{t-2^jT} \int_{t-2^{j+1}T}^{t-2^jT} |s-\sigma|^{-\theta(a)} \|F(\sigma)\|_{H_{a'}^{-\alpha(a)/2}} \|F(s)\|_{H_{a'}^{-\alpha(a)/2}} d\sigma ds dt \\
&\lesssim \int_I \int_0^{2^jT} \int_0^{2^jT} |s-\sigma|^{-\theta(a)} \times \\
&\qquad \|F(\sigma + t - 2^{j+1}T)\|_{H_{a'}^{-\alpha(a)/2}} \|F(s + t - 2^{j+1}T)\|_{H_{a'}^{-\alpha(a)/2}} d\sigma ds dt \\
&\lesssim (2^jT)^{2-\theta(a)} \|F\|^2_{L^2(I,\, H_{a'}^{-\alpha(a)/2})}.
\end{aligned} \tag{3.114}$$

By (3.113) and (3.114),

$$|\mathcal{L}_j(F,G)| \lesssim (2^jT)^{1-\theta(a)/2} \|F\|_{L^2(I,H_{a'}^{-\alpha(a)/2})} \|G\|_{L^2(I,L^2)}. \tag{3.115}$$

Case (3) is similar to case (2).

We denote by Σ the open quadrilateral domain with the vertices $(1/p, 1/p)$ $(p \gg 1)$, $(1/r, 1/2)$, $(1/2, 1/2)$ and $(1/2, 1/r)$, which contains $B(P, \varepsilon)$. Now we show that (3.110) holds for any $(1/a, 1/b) \in \Sigma$. In fact, it suffices to consider the case that $1/b \geqslant 1/a$. We can choose $\eta \in (0,1)$, $(1/p_0, 1/p_0)$ and $(1/a_0, 1/2)$ satisfying (1) and (2), such that

$$\left(\frac{1}{a}, \frac{1}{b}\right) = \eta \left(\frac{1}{p_0}, \frac{1}{p_0}\right) + (1-\eta)\left(\frac{1}{a_0}, \frac{1}{2}\right). \tag{3.116}$$

Noticing that $U(t) : L^2 \to L^2$ and

$$\|U(t)\varphi\|_2 = \|\varphi\|_2, \tag{3.117}$$

i.e., $\theta(0) = \alpha(0) = 0$. From (3.117) and (3.98) we see that if $\alpha(p_0) \neq 0$, then $\alpha(a) \neq 0$ for any $2 < a < p_0$; if $\alpha(p_0) = 0$, then $\alpha(a) = 0$ for any $2 < a < p_0$. We consider the case $\alpha(p_0) \neq 0$. It is easy to see that

$$\frac{\theta(p)}{\theta(q)} = \frac{\alpha(p)}{\alpha(q)} = \frac{1/2 - 1/p}{1/2 - 1/q}. \tag{3.118}$$

(3.116) and (3.118) imply that

$$(\theta(a), \theta(b)) = \eta(\theta(p_0), \theta(p_0)) + (1-\eta)(\theta(a_0), \theta(0)), \tag{3.119}$$

$$(\alpha(a), \alpha(b)) = \eta(\alpha(p_0), \alpha(p_0)) + (1-\eta)(\alpha(a_0), \alpha(0)). \tag{3.120}$$

So, by the complex interpolation, we have the result, as desired. □

We need the following (see [13]):

Lemma 3.9. *Let*

$$T : \begin{cases} A_0 \times B_0 \to C_0, \\ A_0 \times B_1 \to C_1, \\ A_1 \times B_0 \to C_1 \end{cases} \tag{3.121}$$

be bounded bilinear operators, where (A_0, A_1), (B_0, B_1), (C_0, C_1) *are compatible Banach pairs*[4]. *Let* $0 < \eta, \eta_i < 1$ *and* $1 \leqslant p, q \leqslant \infty$ *satisfy* $\eta = \eta_1 + \eta_2$ *and* $1/p + 1/q \geqslant 1$. *Then*

$$T : (A_0, A_1)_{\eta_0,p} \times (B_0, B_1)_{\eta_1,q} \to (C_0, C_1)_{\eta,1} \tag{3.122}$$

is a bounded operator.

Theorem 3.5. *Let* $U(t)$ *and* $\mathcal{A}$ *be as in* (3.52) *and satisfy* (3.98) *and* (3.99). *Assume that* $2 \leqslant p, q \leqslant r$. *Then*

$$\|U(t)\phi\|_{L^{2/\theta(p)}(I, B^s_{p,2})} \lesssim \|\phi\|_{H^{s-\alpha(p)/2}}, \tag{3.123}$$

$$\|\mathcal{A}f\|_{L^{2/\theta(p)}(I, B^{s-\alpha(p)/2}_{p,2})} \lesssim \|f\|_{L^{(2/\theta(q))'}(I, \dot{B}^{s+\alpha(q)}_{q',2})}. \tag{3.124}$$

Proof. Now we can use Lemmas 3.8 and 3.9 to show (3.100). Take $p = q = 2$, $\eta = 2/3$ and $\eta_0 = \eta_1 = 1/3$ in Lemma 3.9. Choose $a_0 = b_0$ and $a_1 = b_1$ satisfying

$$\theta(a_0) = 1 + \epsilon, \quad \theta(a_1) = 1 - 2\epsilon, \tag{3.125}$$

where $\epsilon > 0$ is sufficiently small. Put

$$A_i = B_i = L^2(I, H^{-\alpha(a_i)/2}_{a_i'}), \quad C_i = \ell^{\beta(a_0,b_i)}_\infty, \quad i = 0, 1. \tag{3.126}$$

By $\theta(r) = 1$, it follows from (3.118) that

$$\begin{cases} 1/r = (1-\eta_0)/a_0 + \eta_0/a_1, \\ \alpha(r) = (1-\eta_0)\alpha(a_0) + \eta_0\alpha(a_1), \\ \beta(a_0, b_1) = \beta(a_1, b_0), \\ (1-\eta)\beta(a_0, b_0) + \eta\beta(a_0, b_1) = 0. \end{cases} \tag{3.127}$$

In view of $\alpha(p_0) \neq 0$, we have $\alpha(a_0) \neq \alpha(a_1)$. (3.127) implies that

$$(A_0, A_1)_{\eta_0,2} = (B_0, B_1)_{\eta_1,2} = L^2(I, B^{-\alpha(r)/2}_{r',2}), \quad (C_0, C_1)_{\eta,1} = \ell^0_1. \tag{3.128}$$

So, by Lemma 3.9,

$$\sum_j |\mathcal{L}_j(F, G)| \lesssim \|F\|_{L^2(I, B^{-\alpha(r)/2}_{r',2})} \|G\|_{L^2(I, B^{-\alpha(r)/2}_{r',2})}. \tag{3.129}$$

[4] (A_0, A_1) is said to be a compatible Banach pair, if there exists a linear topological space A such that $A_0, A_1 \subset A$.

We emphasize that in the right hand side of (3.129), the omitted constant is independent of T.

As $\alpha(p_0) = 0$, the proof is slight different from the above discussions and we omit the details. So, (3.100) is proved and the other cases of (3.123) have been discussed in the previous section.

Below we prove (3.124). In fact, (3.129) implies that

$$\|\mathcal{A}f\|_{L^2(I,B^{\alpha(r)/2}_{r,2})} \lesssim \|f\|_{L^2(I,B^{-\alpha(r)/2}_{r',2})}. \tag{3.130}$$

Using the same way as in the previous section, we have (see Lemmas 3.6, 3.7)

$$\|\mathcal{A}f\|_{L^\infty(I,L^2)} \lesssim \|G\|_{L^2(I,B^{-\alpha(r)/2}_{r',2})}, \tag{3.131}$$

$$\|\mathcal{A}f\|_{L^2(I,B^{\alpha(r)/2}_{r,2})} \lesssim \|f\|_{L^1(I,L^2)}. \tag{3.132}$$

Analogous to the proof of Theorem 3.4, we can get (3.124). □

Remark 3.4. We emphasize that Theorem 3.5 is independent of $T > 0$. Thus, by a standard limit argument, we can get that Theorem 3.5 also holds for $I = \mathbb{R}$.

Remark 3.5. One needs the condition $r < \infty$ in Theorem 3.5 and we may further ask what happens if $r = \infty$, which corresponds to the cases $n = 2$ and $n = 3$ for the Schrödinger equation and the wave equation, respectively. Generally speaking, Theorem 3.5 does not hold if $r = \infty$, however, for the radial functions, the conclusion of Theorem 3.5 is still true in the case $r = \infty$, see Tao [215].

Corollary 3.1. *Let $n \geqslant 3$, $S(t) = e^{\mathrm{i}t\Delta}$, $2^* = 2n/(n-2)$, $2 \leqslant r, p \leqslant 2^*$, and $2/\gamma(\cdot) = n\,(1/2 - 1/\cdot)$. Then we have*

$$\|S(t)\phi\|_{L^{\gamma(p)}(I,\dot{B}^s_{p,2})} \lesssim \|\phi\|_{\dot{H}^s}, \tag{3.133}$$

$$\left\| \int_0^t S(t-\tau)f(\tau)d\tau \right\|_{L^{\gamma(p)}(I,\dot{B}^s_{p,2})} \lesssim \|f\|_{L^{\gamma(r)'}(I,\dot{B}^s_{r',2})}, \tag{3.134}$$

where $I = (-T,T) \subset \mathbb{R}$ is an arbitrary interval. In (3.133) and (3.92), replacing homogeneous Besov spaces by corresponding Riesz potential spaces $\dot{H}^s_\rho$ $(\rho = r,p)$, the conclusion still holds.

Corollary 3.2. *Let $W(t) = e^{\mathrm{i}t(-\Delta)^{1/2}}$, $n > 3$, $2^{**} = 2(n-1)/(n-3)$, $2 \leqslant r, p \leqslant 2^{**}$, $2\sigma(\cdot) = (n+1)(1/2 - 1/\cdot)$, and $2/\beta(\cdot) = (n-1)(1/2 - 1/\cdot)$. Then we have*

$$\|W(t)\phi\|_{L^{\beta(p)}(I,\dot{B}^{s-\sigma(p)}_{p,2})} \lesssim \|\phi\|_{\dot{H}^s}, \tag{3.135}$$

$$\|\mathcal{A}_W f\|_{L^{\beta(p)}(I,\dot{B}^{s-\sigma(p)}_{p,2})} \lesssim \|f\|_{L^{\beta(r)'}(I,\dot{B}^{s+\sigma(r)}_{r',2})}, \tag{3.136}$$

where $I = (-T, T) \subset \mathbb{R}$ *is arbitrary,* $\mathcal{A}_W := \int_0^t W(t-\tau) \cdot d\tau$. *In* (3.135) *and* (3.136), *substituting homogeneous Besov spaces by relevant Riesz potential spaces* $\dot{H}^s_\rho$, *the result also holds.*

Corollary 3.3. *Let* $G(t) = e^{it(I-\Delta)^{1/2}}$, $\theta \in [0,1]$, $n > 3 - \theta$, $2^{**} = 2(n - 1 + \theta)/(n - 3 + \theta)$, $2 \leqslant r, p \leqslant 2^{**}$, $2\sigma(\theta, \cdot) = (n + 1 + \theta)(1/2 - 1/\cdot)$, *and* $2/\beta(\theta, \cdot) = (n - 1 + \theta)(1/2 - 1/\cdot)$. *Then*

$$\|G(t)\phi\|_{L^{\beta(\theta,p)}(I,B^{s-\sigma(\theta,p)}_{p,2})} \lesssim \|\phi\|_{H^s}, \tag{3.137}$$

$$\|\mathcal{A}_G f\|_{L^{\beta(\theta,p)}(I,B^{s-\sigma(\theta,p)}_{p,2})} \lesssim \|f\|_{L^{\beta(\theta,r)'}(I,B^{s+\sigma(\theta,r)}_{r',2})}, \tag{3.138}$$

where $I = (-T, T) \subset \mathbb{R}$ *is arbitrary,* $\mathcal{A}_G := \int_0^t G(t-\tau) \cdot d\tau$. *In* (3.137) *and* (3.138), *replacing Besov spaces by corresponding Bessel potential spaces, the conclusion also hold.*

Chapter 4

Local and global wellposedness for nonlinear dispersive equations

Stone from another mountain can be harder than jade. —— From the Book of Songs

In this chapter we will study the local and global wellposedness in H^s for a class of nonlinear dispersive equations. In order to solve a nonlinear dispersive equation, one needs to make a delicate balance between the solution and its survival space. The weak solution has too large existing space, which is hard to be unique. The smooth solution exists in a relatively small space and its existence is somehow a problem. So, one of our main task is to look for the most appropriate space to carry out the solution, which strongly associates with the energy and dispersive structures of the equation.

To some extent, Strichartz' inequalities realize the balance between the solution and its survival space, which enable us to obtain the global well-posed results in the energy space for a class of nonlinear dispersive equations, such as nonlinear Schrödinger equations, nonlinear Klein-Gordon equations and nonlinear wave equations and so on.

4.1 Why is the Strichartz estimate useful

Let us consider the Cauchy problem for the nonlinear Schrödinger equation (NLS)

$$\mathrm{i}u_t + \Delta u = f(u), \quad u(0, x) = u_0(x), \tag{4.1}$$

where $\mathrm{i} = \sqrt{-1}$, $f(u) = |u|^\alpha u$, $\alpha > 0$, $\Delta = \sum_{i=1}^n \partial^2_{x_i}$, $u(t, x)$ is a complex valued function of $(t, x) \in \mathbb{R} \times \mathbb{R}^n$, u_0 is the initial value of u at $t = 0$. We easily deduce that the solutions of (4.1) formally satisfy the conservations of mass and energy[1]:

$$\|u(t)\|_2^2 = \|u_0\|_2^2, \quad E(u(t)) = E(u_0), \tag{4.2}$$

where

$$E(u(t)) = \frac{1}{2}\|\nabla u(t)\|_2^2 + \frac{2}{2+\alpha}\|u(t)\|_{2+\alpha}^{2+\alpha}. \tag{4.3}$$

According to the conservation laws, it is natural to ask the wellposedness of solutions in $L^2(\mathbb{R}^n)$ and $H^1(\mathbb{R}^n)$, respectively.

For any $\alpha > 0$, using the compactness or the parabolic regularity method, we can easily prove the global existence of weak solutions of (4.1), however, the uniqueness, persistence and the continuity of the solution map $u_0 \to u$ are hard to obtain, see [159].

Noticed that $S(t) = e^{\mathrm{i}t\Delta} : L^2 \to L^2$, it is natural to solve NLS in H^k $(k = 0, 1, 2, ...)$ by using the boundedness of $S(t)$ in L^2. According to the standard semi-group theory, if one can solve NLS in H^k, then the uniqueness, persistence and the continuity of the solution map $u_0 \to u$ can also be obtained by a standard way. However, this method has much more assumptions on nonlinearity and initial data. Let us consider the equivalent integral form of (4.1):

$$u(t) = S(t)u_0 - \mathrm{i}\int_0^t S(t-\tau)f(u(\tau))d\tau, \tag{4.4}$$

here we only assume that $f(u)$ is a nonlinear function. Applying $\|S(t)f\|_2 = \|f\|_2$, we have

$$\begin{aligned} \|u(t)\|_2 &\leqslant \|S(t)u_0\|_2 + \int_0^t \|S(t-\tau)f(u(\tau))\|_2 d\tau \\ &= \|u_0\|_2 + \int_0^t \|f(u(\tau))\|_2 d\tau. \end{aligned} \tag{4.5}$$

If we want to get the solution in $C(0,T;L^2)$, then one needs

$$|f(u)| \leqslant C|u|,$$

from which we have

$$\|u\|_{C(0,T;L^2)} \lesssim \|u_0\| + T\|u\|_{C(0,T;L^2)}. \tag{4.6}$$

We can easily construct a contraction mapping by choosing $T > 0$ sufficiently small and show that NLS is well-posed in $C(0,T;L^2)$.

[1]Taking the inner product of (4.1) with u and u_t, and considering its imaginary and real part, respectively, we can get (4.2).

$|f(u)| \leqslant C|u|$ is a rather strong condition on the nonlinearity, which contains $f(u) = \sin u$ and $f(u) = (e^{-|u|^2} - 1)u$ as examples. The power nonlinearity is not included in the above discussions.

Similarly, say for $n = 1$, one can solve NLS in $C(0, T; H^1)$. Assume that

$$f(0) = 0, \quad |f'(u)| \leqslant C|u|^\alpha, \quad 0 < \alpha < \infty.$$

Using the embedding $H^1 \subset L^\infty$, we have for $k = 0, 1$,

$$\begin{aligned} \|\nabla^k u(t)\|_2 &\leqslant \|S(t)\nabla u_0\|_2 + \int_0^t \|S(t-\tau)\nabla^k f(u(\tau))\|_2 d\tau \\ &\lesssim \|\nabla^k u_0\|_2 + \int_0^t \||u|^\alpha \nabla^k u\|_2 d\tau \\ &\lesssim \|\nabla^k u_0\|_2 + \int_0^t \|u\|_\infty^\alpha \|\nabla^k u\|_2 d\tau. \end{aligned} \tag{4.7}$$

So,

$$\|\nabla^k u\|_{C(0,T;L^2)} \lesssim \|\nabla^k u_0\|_2 + T\|\nabla^k u\|_{C(0,T;L^2)}^{\alpha+1}. \tag{4.8}$$

Hence we obtain the local wellposedness in $C(0, T; H^1)$ for NLS. By the conservation of energy, one can extend it to a global one.

The above idea can be generalized to any spatial dimensions and we can solve NLS in $C(0, T; H^s)$ $(s > n/2)$ by assuming that

$$|f^{(k)}(u)| \leqslant C|u|^{\alpha+1-k}, \quad [n/2] < \alpha < \infty, \quad k = 0, 1, ..., [n/2] + 1,$$

where $s > n/2$ is essential to guarantee the inclusion $H^s \subset L^\infty$.

According to the above discussions, for $n \geqslant 2$, we can not solve NLS in H^1 if we only use the L^2 estimates.

Now, our question is how to show the wellposedness of NLS in the energy spaces in higher spatial dimensions. Using the Strichartz estimates as tools, we can obtain the global wellposedness results of NLS in H^1 for the power nonlinearity $|u|^\alpha u$ with $\alpha < 2^* - 2$ (2^* is as in (3.73)), see Kato [120]. If $\alpha = 2^* - 2$, the local wellposedness for NLS can be established by resorting to the Strichartz inequalities. On the basis of the local weposedness result, Bourgain [19] in 1999 developed the localized energy separation method and he obtained the global wellposedness of the radial solutions in 3D and 4D. Colliander, Keel, Staffilani, Takaoka and Tao [44] further developed Bourgain's technique and they removed the radial assumption. In higher spatial dimensions $n \geqslant 4$ and $\alpha = 2^* - 2$, the global wellposedness of NLS is solved by Ryckman and Visan [196] and Visan [232].

If $\alpha > 2^* - 2$, the global wellposedness of NLS is still open.

In order to indicate the general idea, now we briefly give an application of the Strichartz inequality to NLS in 2D; cf. Tsutsumi [226], Cazenave and Weissler [25]. We consider the integral equation:

$$u(t) = S(t)u_0 - \mathrm{i}\int_0^t S(t-\tau)|u(\tau)|^2 u(\tau)d\tau. \tag{4.9}$$

Recall the Strichartz inequalities:

$$\|S(t)u_0\|_{L^4_{x,t}(\mathbb{R}^{2+1})} \lesssim \|u_0\|_2, \tag{4.10}$$

$$\left\|\int_0^t S(t-\tau)f(\tau)d\tau\right\|_{L^4_{x,t}(\mathbb{R}^{2+1})} \lesssim \|f\|_{L^{4/3}_{x,t}(\mathbb{R}^{2+1})}. \tag{4.11}$$

So, we have

$$\begin{aligned}\|u\|_{L^4_{x,t}(\mathbb{R}^{2+1})} &\lesssim \|u_0\|_2 + \||u|^2 u\|_{L^{4/3}_{x,t}(\mathbb{R}^{2+1})} \\ &\lesssim \|u_0\|_2 + \|u\|^3_{L^4_{x,t}(\mathbb{R}^{2+1})}.\end{aligned} \tag{4.12}$$

So, we can solve (4.9) in the space $L^4_{x,t}(\mathbb{R}^{2+1})$ at least for small Cauchy data $u_0 \in L^2$. To realize (4.12), one needs to make a delicate balance between the nonlinearity $|u|^2u$ and the Strichartz space $L^4_{x,t}$. The above performance contains a very general idea to show the well posedness of NLS. In the next few sections we will mainly use this technique to study NLS and nonlinear Klein-Gordon equations.

4.2 Nonlinear mapping estimates in Besov spaces

In order to solve a nonlinear dispersive equation, after establishing the Strichartz inequalities for the relevant linear dispersive equation, we need to estimate the nonlinear terms in the spaces $L^{\gamma(r)'}(0,T;B^s_{r',2})$. The energy structure is not necessary for the local wellposedness and for the global wellposedness with small data. In this section, we consider a nonlinear mapping estimate in Besov spaces. The relevant estimates go back to the works of Pecher [190], Brenner [21], Ginibre and Velo [79]. For NLS, a general nonlinear estimate can be found in Cazenave and Weissler [26]. Wang [236; 234] obtained the nonlinear estimates for the nonlinear Klein-Gordon equations (NLKG). Now we prove a general result, which covers NLS, NLKG and their higher order versions as special cases, which was obtained in [241].

The nonlinear mapping estimates in Besov spaces rely upon two modified versions of the Hölder inequality, one is the convexity Hölder's inequality as in Proposition 1.21, another is the modified Hölder inequality in Sec. 1.7. We know that

$$D^\alpha f(u) = \sum_{q=1}^{|\alpha|} f^{(q)}(u) \sum_{\Lambda_\alpha^q} \prod_{i=1}^{q} D^{\alpha_i} u, \tag{4.13}$$

where $\Lambda_\alpha^q = \Big(\alpha_1 + \cdots + \alpha_q = \alpha,\ |\alpha_q| \geqslant \cdots \geqslant |\alpha_1| \geqslant 1\Big)$. Assume that $f(u)$ satisfies

$$|f^{(k)}(u)| \leqslant C|u|^{p+1-k}, \quad k = 0, 1, ..., |\alpha|. \tag{4.14}$$

Noticing that[2]

$$\prod_{i=1}^{N} a_i - \prod_{i=1}^{N} b_i = \sum_{i=1}^{N} \prod_{j=1}^{i-1} a_j \prod_{j=i+1}^{N} b_j (a_i - b_j),$$

we have for any $1 \leqslant r' < \infty$ and $|\alpha| \geqslant 1$,

$$\begin{aligned} &\| \triangle_h D^\alpha f(u)\|_{L^{r'}} \\ &\lesssim \sum_{q=1}^{|\alpha|} \sum_{\Lambda_\alpha^q} \Big\{ \Big\| (|u_h|^{p-q} + |u|^{p-q})|u_h - u| \prod_{i=1}^{q} D^{\alpha_i} u \Big\|_{L^{r'}} \\ &\quad + \sum_{i=1}^{q} \Big\| |u_h|^{p+1-q} \prod_{j=1}^{i-1} D^{\alpha_j} u_h \prod_{j=i+1}^{q} D^{\alpha_j} u D^{\alpha_i}(u_h - u) \Big\|_{L^{r'}} \Big\} \\ &=: \sum_{q=1}^{|\alpha|} \sum_{\Lambda_\alpha^q} \Big(\|I_q\|_{L^{r'}} + \sum_{i=1}^{q} \|II_q^i\|_{L^{r'}} \Big). \end{aligned} \tag{4.15}$$

Lemma 4.1. *Let $2 \leqslant r < \infty$ and $0 \leqslant \delta \leqslant s_0 \wedge s < \infty$. Assume that $f \in C^{\{s-\delta\}}$ satisfies the following condition:*

$$|f^{(k)}(u)| \lesssim |u|^{p+1-k}, \quad k = 0, 1, ..., \{s-\delta\}, \quad \{s-\delta\} \leqslant p+1, \tag{4.16}$$

where we assume that $\{a\} = 1 + [a]$ if a is not an integer; $\{a\} = a$ if a is an integer. Suppose that

$$p\Big(\frac{1}{r} - \frac{s_0}{n}\Big) + \frac{1}{r} - \frac{\delta}{n} = \frac{1}{r'}, \quad \frac{1}{r} - \frac{s_0}{n} > 0. \tag{4.17}$$

If $s - \delta \notin \mathbb{N}$, then we have

$$\|f(u)\|_{\dot B^{s-\delta}_{r',2}} \lesssim \|u\|^p_{\dot B^{s_0}_{r,2}} \|u\|_{\dot B^{s}_{r,2}}. \tag{4.18}$$

If $s - \delta \in \mathbb{N}$, substituting homogeneous Besov spaces by relevant Riesz potential spaces, (4.18) also holds.

[2]We assume $\prod_{j=1}^{0} a_j = \prod_{j=N+1}^{N} b_j = 0$.

Proof. If $s-\delta<1$, the proof is very easy and we omit the details of the proof. It suffices to consider the case $s-\delta\geqslant 1$. We assume that $s-\delta\notin\mathbb{N}$. By the equivalent norm in homogeneous Besov spaces,

$$\|f(u)\|_{\dot B^{s-\delta}_{r',2}}=\left(\int_0^\infty t^{-2v}\sum_{|\alpha|=[s-\delta]}\sup_{|h|\leqslant t}\|\ \triangle_h\ D^\alpha f(u)\|^2_{L^{r'}}\frac{dt}{t}\right)^{1/2}, \tag{4.19}$$

where $v=s-\delta-[s-\delta]$. In view of (4.16) we see that $\|\ \triangle_h\ D^\alpha f(u)\|_{L^{r'}}$ satisfies (4.15).

Step 1. we estimate

$$A_q:=\left(\int_0^\infty t^{-2v}\sup_{|h|\leqslant t}\|I_q\|^2_{L^{r'}}\frac{dt}{t}\right)^{1/2}. \tag{4.20}$$

In (4.16), we consider the estimate of $\|I_q\|_{L^{r'}}$. Put

$$a_0=(p-q)\Big(\frac1r-\frac{s_0}{n}\Big),\quad a_0'=\frac1r-\frac{s_0+\beta_0-v}{n},$$
$$a_i=\frac1r-\frac{s_0+\beta_i-|\alpha_i|}{n},\qquad i=1,\cdots,q,$$

where β_i will be chosen below.

If $s\leqslant s_0+1$, then we can take $\beta_0=\beta_1=\cdots=\beta_{q-1}=0$ and $\beta_q=s-s_0$. Since $s-\delta\geqslant 1$, we see that $v=s-\delta-[s-\delta]\leqslant s-\delta-1\leqslant s_0$, which implies $a_0'>0$. In view of $s-|\alpha_q|\leqslant s_0$, we obtain that $a_q\geqslant 1/r-s_0/n>0$. If $q\geqslant 2$, then $|\alpha_i|\leqslant s_0$ for all $i=1,\cdots,q-1$. Indeed, in the opposite case one has that $|\alpha_q|+|\alpha_{q-1}|>s_0+1\geqslant s$, which is impossible. So, we have $a_i>0$, $i=0,1,...,q$. Notice that $a_0'+\sum_{i=0}^q a_i=1/r'$. It follows from the modified Hölder inequality that

$$\|I_q\|_{L^{r'}}\leqslant C\|u\|^{p-1}_{\dot B^{s_0}_{r,2}}\|u_h-u\|_{L^{1/a_0'}}\|u\|_{\dot B^s_{r,2}}. \tag{4.21}$$

Hence,

$$A_q\leqslant C\|u\|^{p-1}_{\dot B^{s_0}_{r,2}}\|u\|_{\dot B^v_{1/a_0',2}}\|u\|_{\dot B^s_{r,2}}\leqslant C\|u\|^p_{\dot B^{s_0}_{r,2}}\|u\|_{\dot B^s_{r,2}}. \tag{4.22}$$

Below, we consider the case $s>s_0+1$. We will choose suitable β_i $(i=0,1,...,q)$ satisfying the following four conditions:

(a) $0\leqslant\beta_0\leqslant v,\quad 0\leqslant\beta_i\leqslant|\alpha_i|,\quad i=1,\cdots,q,$

(b) $s_0+\beta_0\geqslant v,\quad s_0+\beta_i\geqslant|\alpha_i|,\quad i=1,\cdots,q,$

(c) $\sum_{i=0}^q\beta_i=s-s_0,$

(d) $s_0+\beta_i\leqslant s,\quad i=0,\cdots,q.$

Conditions (a) and (c) imply condition (d), So, it suffices to show that there exist $\beta_0,\cdots,\beta_q$ satisfying (a)–(c).

If $q = 1$, then we can take $\beta_0 = v, \beta_1 = s - s_0 - v$. Due to $\delta \leqslant s_0$, we have $\beta_1 \leqslant [s - \delta]$ and $s_0 + \beta_1 \geqslant [s - \delta]$. So, conditions (a)–(c) are satisfied.

We consider the case $q \geqslant 2$ which is divided into the following three subcases.

First, we assume that $|\alpha_q| \geqslant s - s_0 - v$. Let $\beta_0 = v$, $\beta_1 = \cdots = \beta_{q-1} = 0$ and $\beta_q = s - s_0 - v$. Obviously, conditions (a) and (c) hold. On the other hand, it is easy to see that $s_0 + \beta_q \geqslant |\alpha_q|$ and $s_0 + \beta_0 \geqslant v$. If $|\alpha_{q-1}| > s_0$, then $|\alpha_q| + |\alpha_{q-1}| > s - v \geqslant [s - \delta]$, a contraction. So we have $s_0 + \beta_{q-1} \geqslant |\alpha_{q-1}|$, which implies the condition (b) holds.

Secondly, we consider the case $s_0 \leqslant |\alpha_q| < s - s_0 - v$. Put $\beta_0 = v$, $\beta_i = |\alpha_i|$ for $i = 1, \cdots, q-1$ and $\beta_q = |\alpha_q| + \delta - s_0$. A straightforward calculation yields $\sum_{i=0}^{q} \beta_i = s - s_0$. In view of $\delta \leqslant s_0$ and $|\alpha_q| \geqslant s_0$, we see that $0 \leqslant \beta_q \leqslant |\alpha_q|$, which leads the condition (a) holds. Noticing that $s_0 + \beta_q = |\alpha_q| + \delta$, we easily see that condition (b) holds.

Thirdly, we consider the case $|\alpha_q| \leqslant (s - s_0 - v) \wedge s_0$. In the current case, we easily see that $\sum_{i=1}^{q} |\alpha_i| = [s - \delta] = s - \delta - v \geqslant s - s_0 - v$. Hence, we can choose $\beta_i \in [0, |\alpha_i|]$ $(i = 1, \cdots, q)$ satisfying $\sum_{i=1}^{q} \beta_i = s - s_0 - v$. Put $\beta_0 = v$. Obviously, conditions (a) and (c) hold. From $|\alpha_i| \leqslant |\alpha_q| \leqslant s_0$ it follows that the condition (b) holds.

Therefore, we have chosen $\beta_0, \cdots, \beta_q$ satisfying conditions (a)–(d). We have,

$$\begin{aligned} a_0' + \sum_{i=0}^{q} a_i &= p\Big(\frac{1}{r} - \frac{s_0}{n}\Big) + \frac{1}{r} - \frac{s_0 + \beta_0 - v + \sum_{i=1}^{q}(\beta_i - |\alpha_i|)}{n} \\ &= p\Big(\frac{1}{r} - \frac{s_0}{n}\Big) + \frac{1}{r} - \frac{\delta}{n} = \frac{1}{r'}. \end{aligned}$$

Applying the modified Hölder inequality,

$$\|I_q\|_{L^{r'}} \leqslant C\|u\|^{p-q}_{\dot{B}^{s_0}_{r,2}} \|u_h - u\|_{L^{1/a_0'}} \prod_{i=1}^{q} \|u\|_{\dot{B}^{s_0+\beta_i}_{r,2}}. \tag{4.23}$$

By (4.23), we have

$$\begin{aligned} A_q &\leqslant C\|u\|^{p-q}_{\dot{B}^{s_0}_{r,2}} \|u\|_{\dot{B}^{v}_{1/a_0',2}} \prod_{i=1}^{q} \|u\|_{\dot{B}^{s_0+\beta_i}_{r,2}} \\ &\leqslant C\|u\|^{p-q}_{\dot{B}^{s_0}_{r,2}} \prod_{i=0}^{q} \|u\|_{\dot{B}^{s_0+\beta_i}_{r,2}}. \end{aligned} \tag{4.24}$$

One can choose θ_i $(i = 0, 1, \ldots, q)$ satisfying $s_0 + \beta_i = \theta_i s_0 + (1 - \theta_i)s$. It is easy to see that $0 \leqslant \theta_i \leqslant 1$, $\sum_{i=0}^{q} \theta_i = q$ and $\sum_{i=0}^{q}(1 - \theta_i) = 1$. Using the convexity Hölder inequality, we have

$$\prod_{i=1}^{q} \|u\|_{\dot{B}^{s_0+\beta_i}_{r,2}} \leqslant C\|u\|^q_{\dot{B}^{s_0}_{r,2}} \|u\|_{\dot{B}^{s}_{r,2}}. \tag{4.25}$$

So, we obtain the estimate of A_q.

Step 2. We estimate

$$B_q := \sum_{i=1}^{q} B_q^i := \sum_{i=1}^{q} \left(\int_0^\infty t^{-2v} \sup_{|h|\leqslant t} \|II_q^i\|^2_{L^{r'}} \frac{dt}{t} \right)^{1/2}. \tag{4.26}$$

The estimates of B_q are similar to those of A_q and we only sketch the proof.

(I) We consider the case $s \leqslant s_0 + 1$. If $q = 1$, the estimate of B_q is very easy and we omit the details. We now consider the estimates of B_q^i, $i \neq q$, $q \geqslant 2$. Put

$$\begin{aligned} a_0 &= (p+1-q)\Big(\frac{1}{r} - \frac{s_0}{n}\Big), \\ a_j &= \frac{1}{r} - \frac{s_0 - |\alpha_j|}{n}, \quad j \neq i, \quad j = 1, \cdots, q-1, \\ a_i &= \frac{1}{r} - \frac{s_0 - v - |\alpha_i|}{n}, \quad a_q = \frac{1}{r} - \frac{s - |\alpha_q|}{n}. \end{aligned}$$

It is easy to see that $s_0 \geqslant v + |\alpha_j|$. If not, then $s - \delta \geqslant |\alpha_q| + |\alpha_j| + v > s_0 + 1 \geqslant s$, which is impossible. So, $a_j > 0$, $j = 1, \cdots, q-1$. By $s \leqslant s_0 + 1$ we see that $a_q > 0$. Using the modified Hölder inequality, we can get the estimate of $\|II_q^i\|_{L^{r'}}$. Thus, we obtain the estimates of B_q^i for $i \neq q$. Put

$$\begin{aligned} a_0 &= (p+1-q)\Big(\frac{1}{r} - \frac{s_0}{n}\Big), \quad a_j = \frac{1}{r} - \frac{s_0 - |\alpha_j|}{n}, \quad j = 1, \cdots, q-1, \\ a_q &= \frac{1}{r} - \frac{s_0 - v - |\alpha_q|}{n}, \end{aligned}$$

then the estimate of B_q^q follows.

(II) We consider the case $s > s_0 + 1$. Put

$$a_0 = (p+1-q)\Big(\frac{1}{r} - \frac{s_0}{n}\Big), \quad a_j = \frac{1}{r} - \frac{s_0 + \beta_j - |\alpha_j|}{n}, \quad j = 1, \cdots, q,$$

where $\beta_1, \cdots, \beta_q$ can be chosen as in Step 1. So,

$$(a)\ 0 \leqslant \beta_i \leqslant |\alpha_i|; \quad (b)\ s_0 + \beta_i \geqslant |\alpha_i|; \quad (c)\ \sum_{j=1}^{q} \beta_j = s - s_0 - v.$$

Applying the modified Hölder inequality,

$$B_q^i \leqslant C\|u\|^{p+1-q}_{\dot{B}^{s_0}_{r,2}} \Big(\prod_{j\neq i} \|u\|_{\dot{B}^{s_0+\beta_j}_{r,2}} \Big) \|u\|_{\dot{B}^{s_0+\beta_i+v}_{r,2}}. \tag{4.27}$$

Since $s_0+\beta_j+v \leqslant s$, $1 \leqslant j \leqslant q$, we claim that there exist θ_i $(i=1,\ldots,q)$ satisfying

$$s_0+\beta_j=\theta_j s_0+(1-\theta_j)s, \quad j\neq i, j=1,...,q,$$
$$s_0+\beta_i+v=\theta_i s_0+(1-\theta_i)s.$$

Since $\sum_{i=1}^q \theta_i = q-1$ and $\sum_{i=1}^q (1-\theta_i)=1$, in view of the convexity Hölder inequality we obtain the estimates of B_q^i for $i=1,\cdots,q$. □

Remark 4.1.

(i) If $\delta = s_0 = 0$ and $s>0$, then the result of (4.18) can be slightly improved by

$$\|f(u)\|_{\dot{B}^s_{r',2}} \leqslant C\|u\|^p_{L^r}\|u\|_{\dot{B}^s_{r,2}}. \tag{4.28}$$

(ii) Lemma 4.1 covers the nonlinear estimates for NLS and NLKG as in [19; 26; 41; 173; 191; 236; 237]), where Cazenave and Weissler [26] considered the case $\delta=0, s_0=s$; Wang [236; 237] discussed the nonlinear estimates for NLKG.

(iii) If the nonlinearity has an exponential growth, say $\sinh u$, $(e^{|u|^2}-1)u$, the nonlinear mapping estimates can not be covered by Lemma 4.1, one can refer to Nakamura and Ozawa [173], and Wang [239].

4.3 Critical and subcritical NLS in H^s

4.3.1 *Critical NLS in H^s*

We consider the initial value problem for NLS,

$$iu_t+\Delta u=f(u), \quad u(0,x)=u_0(x), \tag{4.29}$$

where $f(u)=c|u|^\sigma u$. If u is a solution of (4.29), then $u_\lambda(t,x)=\lambda^{2/\sigma}u(\lambda^2 t,\lambda x)$ also solves (4.29) with initial data $\lambda^{2/\sigma}u_0(\lambda\cdot)$ at $t=0$. Let us observe that

$$\|u_\lambda\|_{\dot{H}^s(\mathbb{R}^n)}=\lambda^{2/\sigma+s-n/2}\|u\|_{\dot{H}^s(\mathbb{R}^n)}.$$

This implies that $\sigma=4/(n-2s)$ is the unique index such that the norm of u_λ in $\dot{H}^s$ is invariant for all $\lambda>0$. From this point of view, $\sigma=4/(n-2s)>0$ is said to be the critical power in $\dot{H}^s$ (or in H^s) for NLS. For $s=n/2-2/\sigma$, $\dot{H}^s$ (H^s) is said to be the critical space for NLS.

When $s<n/2$, corresponding to the critical case, $\sigma<4/(n-2s)$ is said to be a subcritical power in H^s. If $s\geqslant n/2$, $\sigma<\infty$ is said to be a subcritical power in H^s. $\sigma>4/(n-2s)$ is said to be a supercritical power in H^s for NLS.

4.3.2 Wellposedness in H^s

We consider the following equivalent integral equation with respect to (4.29),

$$u(t) = S(t)u_0 - \mathrm{i}\int_0^t S(t-\tau)f(u(\tau))d\tau, \tag{4.30}$$

where $S(t) = e^{\mathrm{i}t\Delta}$. According to the Strichartz inequalities, if $u_0 \in H^s$, then (3.75) and (3.76) indicate that the solution should belong to $L^{\gamma(p)}(I, B^s_{p,2})$. Assume that

$$|f^{(k)}(u)| \leqslant C|u|^{\sigma+1-k}, \quad k = 0, 1, ..., [s]+1, \tag{4.31}$$

where $0 \leqslant s < n/2$, $[s]$ denotes the integer part of s, $0 < \sigma \leqslant 4/(n-2s)$. We have (see [26])

Theorem 4.1. *Let $0 \leqslant s < n/2$, $0 < \sigma < 4/(n-2s)$. Assume that $f \in C^{[s]+1}$ satisfies (4.31), $[s] \leqslant \sigma$. If $u_0 \in H^s$, then there exists a $T^* := T^*(\|u_0\|_{H^s}) > 0$ such that (4.30) has a unique solution*

$$u \in L^{\gamma(r)}_{\mathrm{loc}}([0, T^*); B^s_{r,2}), \tag{4.32}$$

where

$$r = \frac{n(2+\sigma)}{n+s\sigma}. \tag{4.33}$$

Moreover, for any $2 \leqslant p \leqslant 2^*$ $(p \neq \infty)$, we have

$$u \in L^{\gamma(p)}_{\mathrm{loc}}([0, T^*); B^s_{p,2}), \tag{4.34}$$

and if $T^* < \infty$, then

$$\|u\|_{L^{\gamma(r)}([0,T^*);B^s_{r,2})} = \infty, \tag{4.35}$$

$$\|u(t)\|_{H^s} \gtrsim (T^*-t)^{1-\sigma(n-2s)/4}, \quad 0 < t < T^*. \tag{4.36}$$

In Theorem 4.1, we obtain the local wellposedness in H^s, where σ is a subcritical power in H^s. If σ is a critical power in H^s, we can get the global wellposedness for NLS with small data.

Theorem 4.2. *Let $0 \leqslant s < n/2$ and $\sigma = 4/(n-2s)$. Assume that $f \in C^{[s]+1}$ satisfies (4.31), $[s] \leqslant \sigma$. If $u_0 \in H^s$, then there exists a $T^* := T^*(u_0) > 0$ such that (4.30) has a unique solution u verifying (4.32), where $r = n(2+\sigma)/(n+s\sigma)$ and $\gamma(r) = 2+\sigma$. Moreover, for any $2 \leqslant p \leqslant 2^*$ $(p \neq \infty)$, the solution satisfies (4.34). If $T^* < \infty$, then (4.35) holds.*

If $\|u_0\|_{\dot H^s}$ is sufficiently small, then the above solution is a global one, i.e., $T^* = \infty$ and

$$\|u\|_{\bigcap_{2\leqslant p\leqslant 2^*,\, p\neq\infty} L^{\gamma(p)}(0,\infty;\dot B^s_{p,2})} \lesssim C\|u_0\|_{\dot H^s}, \tag{4.37}$$

$$\|u\|_{\bigcap_{2\leqslant p\leqslant 2^*,\, p\neq\infty} L^{\gamma(p)}(0,\infty;B^s_{p,2})} \lesssim C\|u_0\|_{H^s}. \tag{4.38}$$

Remark 4.2. If $T^* < \infty$ in Theorem 4.1, then T^* has a lower bound which depends on $\|u_0\|_{H^s}$, i.e., $T^* \geqslant \delta(\|u_0\|_{H^s}) > 0$. However, if $T^* < \infty$ in Theorem 4.2, then we can not get the lower bound of T^* which only depends on $\|u_0\|_{H^s}$, namely, T^* may depend on not only $\|u_0\|_{H^s}$, but also the choice of u_0 in H^s, which is easily seen from the scaling $u(t,x) \to u_\lambda(t,x) := \lambda^{2/\sigma} u(\lambda^2 t, \lambda x)$.

Proof. Now we prove Theorems 4.1 and 4.2. Let $T > 0$, $M > 0$ and $\delta > 0$ which will be chosen later. Put

$$\mathcal{D} = \Big\{ u \in L^{\gamma(r)}(0,T; B^s_{r,2}) : \|u\|_{L^{\gamma(r)}(0,T;\dot{B}^s_{r,2})} \leqslant \delta,\ \|u\|_{L^{\gamma(r)}(0,T;B^s_{r,2})} \leqslant M \Big\}, \tag{4.39}$$

which is equipped with the metric

$$d(u,v) = \|u - v\|_{L^{\gamma(r)}(0,T;L^r)}. \tag{4.40}$$

Considering the mapping

$$\mathscr{T} : u(t) \to S(t)u_0 - \mathrm{i} \int_0^t S(t-\tau) f(u(\tau))d\tau, \tag{4.41}$$

we show that $\mathscr{T} : (\mathcal{D}, d) \to (\mathcal{D}, d)$ is a contraction mapping for some $T, \delta, M > 0$. Since

$$\sigma\Big(\frac{1}{r} - \frac{s}{n}\Big) + \frac{1}{r} = \frac{1}{r'}, \tag{4.42}$$

by Lemma 3.1 we have

$$\|f(u)\|_{\dot{B}^s_{r',2}} \lesssim \|u\|^{\sigma+1}_{\dot{B}^s_{r,2}}, \tag{4.43}$$

$$\|f(u)\|_{B^s_{r',2}} \lesssim \|u\|^{\sigma}_{\dot{B}^s_{r,2}} \|u\|_{B^s_{r,2}}. \tag{4.44}$$

So,

$$\|f(u)\|_{L^{\gamma(r)'}(0,T;\dot{B}^s_{r',2})} \lesssim T^{1-\sigma(n-2s)/4} \|u\|^{\sigma+1}_{L^{\gamma(r)}(0,T;\dot{B}^s_{r,2})}. \tag{4.45}$$

In view of Theorem 3.1, one has that

$$\|\mathscr{T}u\|_{L^{\gamma(r)}(0,T;\dot{B}^s_{r,2})} \lesssim \|u_0\|_{\dot{H}^s} + T^{1-\sigma(n-2s)/4} \|u\|^{\sigma+1}_{L^{\gamma(r)}(0,T;\dot{B}^s_{r,2})}. \tag{4.46}$$

Similarly, we have

$$\|\mathscr{T}u\|_{L^{\gamma(r)}(0,T;B^s_{r,2})}$$

$$\lesssim \|u_0\|_{H^s} + T^{1-\sigma(n-2s)/4} \|u\|^{\sigma}_{L^{\gamma(r)}(0,T;\dot{B}^s_{r,2})} \|u\|_{L^{\gamma(r)}(0,T;B^s_{r,2})} \tag{4.47}$$

$$\|\mathscr{T}u - \mathscr{T}v\|_{L^{\gamma(r)}(0,T;L^r)},$$

$$\lesssim T^{1-\sigma(n-2s)/4}(\|u\|^\sigma + \|v\|^\sigma)_{L^{\gamma(r)}(0,T;\dot{B}^s_{r,2})}\|u-v\|_{L^{\gamma(r)}(0,T;L^r)}. \tag{4.48}$$

Case 1. $0 < \sigma < 4/(n-2s)$. Put

$$\delta = 2C\|u_0\|_{\dot{H}^s},\ M = 2C\|u_0\|_{H^s}. \tag{4.49}$$

We take $T > 0$ satisfying

$$2CT^{1-\sigma(n-2s)/4}M^\sigma \leqslant 1/2. \tag{4.50}$$

(4.46)–(4.48) imply that

$$\|\mathscr{T}u\|_{L^{\gamma(r)}(0,T;\dot{B}^s_{r,2})} \lesssim \delta, \tag{4.51}$$

$$\|\mathscr{T}u\|_{L^{\gamma(r)}(0,T;\dot{B}^s_{r,2})} \lesssim M, \tag{4.52}$$

$$\|\mathscr{T}u - \mathscr{T}v\|_{L^{\gamma(r)}(0,T;L^r)}\frac{1}{2}\|u-v\|_{L^{\gamma(r)}(0,T;L^r)}. \tag{4.53}$$

So, $\mathscr{T} : (\mathcal{D}, d) \to (\mathcal{D}, d)$ is a contraction mapping. So, there exists a $u \in \mathcal{D}$ satisfying (4.30). Again, in view of the Strichartz inequalities (3.75) and (3.76), we have

$$\|u\|_{L^{\gamma(p)}(0,T;\dot{B}^s_{p,2})} \lesssim \|u_0\|_{\dot{H}^s} + T^{1-\sigma(n-2s)/4}\|u\|^{\sigma+1}_{L^{\gamma(r)}(0,T;\dot{B}^s_{r,2})}, \tag{4.54}$$

from which we obtain that $u \in L^{\gamma(p)}(0,T;B^s_{p,2})$. By a standard argument, we can extend the solution above. Considering the mapping

$$\mathscr{T} : u(t) \to S(t-T)u(T) - \mathrm{i}\int_T^t S(t-\tau)f(u(\tau))d\tau, \tag{4.55}$$

and noticing that $u(T) \in H^s$, we can use the same way as in the above to solve (4.55). Repeating this argument step by step, we find a maximal $T^* > 0$ satisfying (4.33)–(4.36). The uniqueness of the solution can be shown by following the same way as that of (4.48) and we omit the details.

Case 2. $\sigma = 4/(n-2s)$. By Strichartz estimate (3.75), we see that

$$\|S(t)u_0\|_{L^{\gamma(r)}(0,T;B^s_{r,2})} \to 0, \text{ as } T \to 0. \tag{4.56}$$

Put

$$\mathcal{D}_0 = \left\{u \in L^{\gamma(r)}(0,T;B^s_{r,2}) : \|u\|_{L^{\gamma(r)}(0,T;B^s_{r,2})} \leqslant M\right\}, \tag{4.57}$$

which is equipped with the metric as in (4.40). Take $M > 0$ satisfying

$$2CM^\sigma \leqslant 1/2 \tag{4.58}$$

and $T > 0$ such that

$$\|S(t)u_0\|_{L^{\gamma(r)}(0,T;B^s_{r,2})} \leqslant M/2. \tag{4.59}$$

So, analogous to the above, one can show the local wellposedness of Theorem 4.2. For the global wellposedness with small data, it suffices to take $T = \infty$ in $(\mathcal{D}, d)$ as in Case 1. In fact, analogous to (4.46)–(4.48), we have

$$\|\mathscr{T}u\|_{L^{\gamma(r)}(0,\infty;\dot{B}^s_{r,2})} \lesssim \|u_0\|_{\dot{H}^s} + \|u\|^{\sigma+1}_{L^{\gamma(r)}(0,\infty;\dot{B}^s_{r,2})}, \tag{4.60}$$

$$\begin{aligned} &\|\mathscr{T}u\|_{L^{\gamma(r)}(0,\infty;B^s_{r,2})} \\ &\quad\lesssim \|u_0\|_{H^s} + \|u\|^{\sigma}_{L^{\gamma(r)}(0,\infty;\dot{B}^s_{r,2})}\|u\|_{L^{\gamma(r)}(0,\infty;B^s_{r,2})}, \end{aligned} \tag{4.61}$$

$$\begin{aligned} &\|\mathscr{T}u - \mathscr{T}v\|_{L^{\gamma(r)}(0,\infty;L^r)}, \\ &\quad\lesssim (\|u\|^\sigma + \|v\|^\sigma)_{L^{\gamma(r)}(0,\infty;\dot{B}^s_{r,2})}\|u - v\|_{L^{\gamma(r)}(0,\infty;L^r)}. \end{aligned} \tag{4.62}$$

Put

$$\delta = 2C\|u_0\|_{\dot{H}^s} \ll 1, \ M = 2C\|u_0\|_{H^s}. \tag{4.63}$$

Using the contraction mapping argument, we can get the result. □

4.4 Global wellposedness of NLS in L^2 and H^1

We consider the initial value problem for NLS:

$$iu_t + \Delta u = \lambda |u|^\sigma u, \quad u(0,x) = u_0(x), \tag{4.64}$$

where $\lambda \in \mathbb{R}$. $\lambda > 0$ is the defocusing case and $\lambda < 0$ is the focusing case. Recall that the solution of NLS formally satisfies

$$\|u(t)\|_2^2 = \|u_0\|_2^2, \quad E(u(t)) = E(u_0), \tag{4.65}$$

$$E(u(t)) = \frac{1}{2}\|\nabla u(t)\|_2^2 + \frac{2\lambda}{2+\alpha}\|u(t)\|^{2+\alpha}_{2+\alpha}. \tag{4.66}$$

From the conservation of energy we see that for the focusing NLS, the kinetic energy $\frac{1}{2}\|\nabla u\|_2^2$ and the potential energy $\frac{2\lambda}{2+\lambda}\|u\|^{2+\sigma}_{2+\sigma}$ have opposite signs and for the defocusing NLS, the kinetic energy and the potential energy are both positive. So, the defocusing case is better than the focusing case. Roughly speaking, the defocusing NLS has global solutions and the focusing NLS has blowup phenomena.

Theorem 4.3. *Let $0 < \sigma < 4/n$. If $u_0 \in L^2$, then (4.64) has a unique solution*

$$u \in C([0,\infty); L^2) \cap \left(\cap_{2<r\leqslant 2^*,\, r\neq\infty} L^{\gamma(r)}_{\mathrm{loc}}(0,\infty; L^r)\right) \tag{4.67}$$

and $\|u(t)\|_2 = \|u_0\|_2$.

Theorem 4.4. *Let 2^* be as in (3.73), $\lambda > 0$ and $0 < \sigma < 2^* - 2$. If $u_0 \in H^1$, then (4.64) has a unique solution*

$$u \in C([0, \infty); H^1) \cap \left(\cap_{2<r\leqslant 2^*,\, r\neq\infty} L^{\gamma(r)}_{\rm loc}(0, \infty; H^1_r) \right), \tag{4.68}$$

and (4.65) holds.

Theorems 4.3 and 4.4 are corollaries of Theorem 4.1 and the conservation laws in (4.65). Since we have gotten the local wellposedness in Theorem 4.1, using the standard argument, one can get that the local solution of NLS (4.64) satisfies (4.65) at any local time. By (4.36), we have $T^* = \infty$.

For the energy critical case $\sigma = 4/(n-2)$, $n \geqslant 3$, Theorem 4.4 also holds, see Chapter 7.

4.5 Critical and subcritical NLKG in H^s

We study the Cauchy problem for the nonlinear Klein-Gordon equation (NLKG):

$$u_{tt} + (m^2 - \Delta)u = f(u), \quad u(0,x) = u_0(x),\ u_t(0,x) = u_1(x), \tag{4.69}$$

where $m^2 > 0$. The main results in this section are the following

Theorem 4.5. *Let $n \geqslant 2$, $1/2 \leqslant s < n/2$ and $0 < \sigma < 4/(n-2s)$. Assume that $f \in C^{[s+1/2]}$ satisfies*

$$|f^{(k)}(u)| \leqslant C|u|^{\sigma+1-k}, \quad k = 0, 1, ..., [s+1/2], \tag{4.70}$$

$[s+1/2] \leqslant \sigma + 1$. If $(u_0, u_1) \in H^s \times H^{s-1}$, then there exists a $T^* := T^*(\|u_0\|_{H^s}, \|u_1\|_{H^{s-1}}) > 0$ such that (4.69) has a unique solution

$$u \in C([0, T^*); H^s) \cap \left(\cap_{r\in(2,2^*)} L^{\gamma(r)}_{\rm loc}([0, T^*); B^{s-\beta(r)}_{r,2}) \right), \tag{4.71}$$

where $2\beta(r) = (n+1)(1/2 - 1/r)$, $2/\gamma(r) = (n-1)(1/2 - 1/r)$, and

$$2^* = \begin{cases} 2(n-1)/(n-3), & n > 3, \\ \infty, & n = 2, 3. \end{cases}$$

Moreover, if $T^* < \infty$, then

$$\|u(t)\|_{H^s} \gtrsim (T^* - t)^{-\delta/p}, \quad 0 < t < T^*, \tag{4.72}$$

where $\delta = 1$ for $0 < \sigma < 2/(n-2s)$, $\delta = (4 - \sigma(n-2s))/2$ for $2/(n-2s) \leqslant \sigma < 4/(n-2s)$.

In Theorem 4.5 we get the local wellposedness in H^s, where σ is a subcritical power in H^s. If σ is a critical power in H^s, we have the following global wellposedness result with small data:

Theorem 4.6. *Let $n \geqslant 2$, $1/2 \leqslant s < n/2$ and $\sigma = 4/(n-2s)$. Assume that $f \in C^{[s+1/2]}$ satisfies (4.70), $[s+1/2] \leqslant \sigma + 1$. If $(u_0, u_1) \in H^s \times H^{s-1}$, then there exists a $T^* := T^*(u_0, u_1) > 0$ such that (4.69) has a unique solution u verifying (4.71). Moreover, if $\|(u_0, u_1)\|_{H^s \times H^{s-1}}$ is sufficiently small, then $T^* = \infty$.*

If the initial data belong to energy spaces, we have

Theorem 4.7. *Let $n \geqslant 2$, $0 < \sigma < 4/(n-2)$ and $f(u) = |u|^\sigma u$. If $(u_0, u_1) \in H^1 \times L^2$, then (4.69) has a unique solution*

$$u \in C([0,\infty); H^1) \cap \left(\cap_{r \in (2,2^*)} L^{\gamma(r)}_{\rm loc}([0,\infty); B^{s-\beta(r)}_{r,2}) \right), \tag{4.73}$$

and

$$E(u, u_t) := \frac{1}{2}\|\nabla u\|_2^2 + \frac{m^2}{2}\|u\|_2^2 + \frac{1}{2}\|u_t\|_2^2 + \frac{2}{2+\sigma}\|u\|_{2+\sigma}^{2+\sigma} = E(u_0, u_1). \tag{4.74}$$

Theorems 4.5 and 4.6 can be developed to the nonlinear wave equation (NLW), i.e., $m^2 = 0$ in (4.69). Indeed, substituting H^s, $B^a_{p,2}$ by relevant $\dot H^s$, $\dot B^a_{p,2}$, we see that Theorems 4.5 and 4.6 also hold for the following NLW

$$u_{tt} - \Delta u = f(u), \quad u(0,x) = u_0(x), \; u_t(0,x) = u_1(x). \tag{4.75}$$

We have similar results to Theorems 4.5–4.7 for NLKG in one spatial dimension, whose proofs are slightly different from the case $n \geqslant 2$. For NLW, one can use the L^2 estimates to get the local wellposedness in one spatial dimension. However, the Strichartz estimates fail in 1D and Theorem 4.6 can not be generalized to one spatial dimension. If $\sigma = 4/(n-2)$, the global wellposedness result will be discussed in Chapter 7.

Proof. [Sketch Proof of Theorem 4.5] We only consider the case $2/(n-2s) \leqslant \sigma \leqslant 4/(n-2s)$. Put

$$\rho = \frac{2(n-1)(2+\sigma)}{(n-1)(2+\sigma) + 4 - 2\sigma(n-2s)}.$$

ρ satisfies the following identity,

$$\sigma\left(\frac{1}{\rho} - \frac{s-\beta(\rho)}{n}\right) + \frac{1}{\rho} - \frac{1-2\beta(\rho)}{n} = \frac{1}{\rho'}.$$

By Lemma 4.1, we have

$$\|f(u)\|_{\dot{B}^{s-1+\beta(\rho)}_{\rho',2}} \lesssim \|u\|^{\sigma+1}_{\dot{B}^{s-\beta(\rho)}_{\rho,2}}. \tag{4.76}$$

Consider the equivalent integral equation of (4.69),

$$u(t) = K'(t)u_0 + K(t)u_1 + \int_0^t K(t-\tau)f(u(\tau))d\tau, \tag{4.77}$$

where $K(t) = (m^2 - \Delta)^{-1/2}\sin(m^2 - \Delta)^{1/2}$ and $K'(t) = \cos(m^2 - \Delta)^{1/2}$. Put

$$\mathcal{D} = \{u \in L^{\gamma(\rho)}(0,T; B^{s-\beta(\rho)}_{\rho,2}) : \|u\|_{L^{\gamma(\rho)}(0,T;B^{s-\beta(\rho)}_{\rho,2})} \leqslant M\},$$

and

$$d(u,v) = \|u - v\|_{L^{\gamma(\rho)}(0,T;L^\rho)}.$$

Using the Strichartz inequalities together with (4.76), we can show that

$$\mathcal{T} : u(t) \to K'(t)u_0 + K(t)u_1 + \int_0^t K(t-\tau)f(u(\tau))d\tau \tag{4.78}$$

is a contraction mapping from $(\mathcal{D}, d)$ into itself. For the details, see Wang [236; 237]. □

Chapter 5

The low regularity theory for the nonlinear dispersive equations

From PDE point of view, one of the main reasons that we like Fourier transform is that it makes the differential operation into fundamental algebraic multiplication.

This chapter is devoted to introduce some methods in studying the low regularity theory for the Cauchy problems of the nonlinear dispersive equations: $X^{s,b}$ method and I-method. By the low regularity theory we mean to study the Cauchy problems assuming rough initial data. For instance, whether the Cauchy problems are locally well-posed in H^s and how low can s be? How about global well-posedness?

The low regularity theory is important for several reasons: First, for Sobolev H^s, if s is smaller, then the space H^s is larger; thus the scope of initial data is wide, for example, it can include Dirac measure–δ function which belongs to $H^{-1/2-\epsilon}$, $\epsilon > 0$. Secondly, for some equations, conservation laws of L^2 and H^1 can be easily obtained, so the task of extending a local existence result to a global existence result can be easier if one is working at low regularities than high regularities.

In this chapter, we take the Korteweg de-Vries equation and the Schrödinger equation with derivative as examples to study the low regularity solution for the nonlinear dispersive equations.

5.1 Bourgain space

In this section, we introduce Bourgain space and its properties. The Cauchy problem for a general dispersive equation usually has the following form:

$$\begin{cases} \partial_t u - i\phi(D)u = f(u, \partial^\alpha u), & (x,t) \in \mathbb{R}^n \times \mathbb{R} \\ u(x,0) = u_0(x) \in H^s(\mathbb{R}^n). \end{cases} \tag{5.1}$$

where $u(x,t) : \mathbb{R}^n \to \mathbb{C}$ is a unknown function, $u_0(x)$ is a given data, $\partial^\alpha = (\partial_{x_1}^{\alpha_1}, ..., \partial_{x_n}^{\alpha_n})$, and $\phi(D)$ is a Fourier multiplier:

$$\phi(D)u = (2\pi)^{-n/2} \int_{\mathbb{R}^n} e^{ix\xi}\phi(\xi)\widehat{u}(\xi)d\xi.$$

Here $\phi(\xi)$ is a real-valued continuous function[1], which we refer as dispersion relation of the equation or phase function. Nonlinear term f is a function of multi-variables, for example f is a polynomials or usually of the form $|u|^p u$.

Many dispersive equations can be reduced to the form (5.1), for instance, Schrödinger equation (with dispersion relation $\phi(\xi) = |\xi|^2$), Korteweg de-Vries equation (with dispersion relation $\phi(\xi) = \xi^3$), and so on. In the previous several chapters we have discussed some space-time structures, such as Strichartz-type space $L_t^q L_x^r$, smoothing effect-type $L_x^q L_t^r$. In this chapter we will apply a new class of space-time structure $X^{s,b}$. To briefly introduce the motivation, we view (5.1) as a perturbation of a linear equation as before, hence consider first the corresponding linear equation:

$$\partial_t u - i\phi(D)u = 0. \tag{5.2}$$

Denote $S(t)f = \mathscr{F}_x^{-1} e^{it\phi(\xi)} \mathscr{F}_x f$. Taking Fourier transform with respect to both space and time on both side of (5.2), then it becomes

$$(\tau - \phi(\xi))\widehat{u}(\xi, \tau) = 0. \tag{5.3}$$

Unless particularly specified, $\widehat{u}$ or $\mathscr{F}u$ will always denote the Fourier transform of the function $u(x,t)$ with respect to x and t, $\mathscr{F}_x u$ denotes the one only on x.[2] From (5.3) we see that $\widehat{u}$ is supported on the surface $\{(\xi, \tau) : \tau = \phi(\xi)\}$. One can imagine that this surface is intimately related to the equation (5.1). By viewing x and t equally, we then introduce the Sobolev-type space for (5.1) as Sobolev space for the Laplacian equation.

Definition 5.1. Assume $\phi : \mathbb{R}^d \to \mathbb{R}$ is a continuous function. Let $s, b \in \mathbb{R}$. The space $X^{s,b}_{\tau=\phi(\xi)}(\mathbb{R}^d \times \mathbb{R})$, simply denoted as $X^{s,b}(\mathbb{R}^d \times \mathbb{R})$ or $X^{s,b}$, is then defined to be the closure of the Schwartz functions $\mathcal{S}(\mathbb{R}^{d+1})$ under the norm

$$\|u\|_{X^{s,b}} = \|\langle\xi\rangle^s \langle\tau - \phi(\xi)\rangle^b \mathcal{F}u\|_{L_\xi^2 L_\tau^2}. \tag{5.4}$$

---

[1] We emphasize that ϕ must be real-valued.

[2] If there is no confusion, we also denote by $\widehat{u}$ the Fourier transform of u for function $u : \mathbb{R}^n \to \mathbb{C}$.

These modern forms of the spaces were first used to systematically study nonlinear dispersive wave problems by Bourgain [15; 14]. Klainerman and Machedon [146] used similar ideas in their study of the nonlinear wave equation. The spaces appeared earlier in a dierent setting in the works [194], [7] of Rauch, Reed, and M. Beals. Then the bilinear estimates in these space were deeply studied by Kenig, Ponce and Vega [132], and by Tao [216] in an abstract setting. Bourgain spaces $X^{s,b}$ can exploit deeply dispersive effect of the equation, and analyze the frequency interactions of waves evolving from the equation. Thus it is usually a powerful tool in the well-posedness study, see [132; 218]. We take the nonlinear term uu_x as an example, then Bourgain space methods reduce to the following two crucial estimates:

$$\left\| \psi(t) \int_0^t S(t-s) f(s) ds \right\|_{X^{s,b}} \leqslant C\|f\|_{X^{s,b-1}}, \tag{5.5}$$

$$\|\partial_x(uv)\|_{X^{s,b-1}} \leqslant C\|u\|_{X^{s,b}}\|v\|_{X^{s,b}}, \tag{5.6}$$

where $\psi(t)$ is a smooth cut-off function. Intuitively, we can view these two inequalities in this way: s is the degree of regularity for the space variable, and b is that for the operator $\partial_t - i\phi(D)$. From (5.5) we see the nonlinear evolution $\int_0^t S(t-s) \cdot ds$ gains one derivative with respect to $\partial_t - i\phi(D)$, and (5.6) shows this gain can compensate the loss of derivative on the space variable. How much regularity can be obtained through this process depends on the strength of dispersive effect of the equation. More precisely, we will show (5.5) always hold for any ϕ when $1/2 < b < 1$, hence the main task is to show the second inequality which is the main topics of the next section.

This section devotes to prove some basic properties of Bourgain $X^{s,b}$. Assume $\psi \in C_0^\infty(\mathbb{R})$, supp$\psi \subset [-2,2]$, and equals to 1 on $[-1,1]$. Let $\psi_\delta(\cdot) = \psi(\cdot/\delta)$ for any $\delta > 0$. We will use the following lemma, which is proved in Chapter 3.

Lemma 5.1. *If* $0 < \alpha < 1$, $1 < p < \infty$, *then*

$$\|(-\Delta)^{\alpha/2}(fg)\|_{L^p} \lesssim \|f\|_{L^\infty}\|(-\Delta)^{\alpha/2}g\|_{L^p} + \|g\|_{L^\infty}\|(-\Delta)^{\alpha/2}f\|_{L^p}. \tag{5.7}$$

From the definition of $X^{s,b}$, it is easy to see that

Lemma 5.2. *Let* $s, b \in \mathbb{R}$. *Then*

$$\|f\|_{X^{s,b}} = \left\| \langle\xi\rangle^s \|e^{-it\phi(\xi)}(\mathscr{F}_x f)(\xi,t)\|_{H_t^b} \right\|_{L_\xi^2}.$$

Proposition 5.1. *Assume $s \in \mathbb{R}$, $1/2 < b < b' < 1$, $0 < \delta < 1$. Then*

$$\|\psi_\delta(t)f\|_{X^{s,b}} \lesssim \delta^{1/2-b}\|f\|_{X^{s,b}}, \quad \|\psi_\delta(t)f\|_{X^{s,b-1}} \lesssim \delta^{b'-b}\|f\|_{X^{s,b'-1}}.$$

Proof. We may assume $s = 0$. It follows from Lemma 5.2 and Lemma 5.1 that

$$\begin{aligned}\|\psi_\delta(t)f\|_{X^{0,b}} =& \big\|\|e^{-it\phi(\xi)}\psi_\delta(t)\mathscr{F}_x(f)\|_{H_t^b}\big\|_{L_\xi^2} \\ \lesssim& \big\|\|\psi_\delta(t)\|_{H_t^b}\|e^{-it\phi(\xi)}\mathscr{F}_x(f)(\xi,t)\|_{L_t^\infty}\big\|_{L_\xi^2} \\ &+ \big\|\|\psi_\delta(t)\|_{L_t^\infty}\|e^{-it\phi(\xi)}\mathscr{F}_x(f)(\xi,t)\|_{H_t^b}\big\|_{L_\xi^2}.\end{aligned}$$

By simple calculations one easily get for any $\lambda > 0$

$$\|f(\lambda t)\|_{\dot{H}^b} = \lambda^{b-1/2}\|f(t)\|_{\dot{H}^b}, \ \|f(\lambda t)\|_{H^b} \lesssim (\lambda^{1/2} + \lambda^{b-1/2})\|f(t)\|_{H^b}. \quad (5.8)$$

The first inequality follows from (5.8) and Sobolev's inequality $H^b \hookrightarrow L^\infty$.

Next we show the second inequality. For simplicity, let $c = 1 - b$ and $d = 1 - b'$, and then $0 \leqslant d < c < 1/2$. From Lemma 5.2 it suffices to prove

$$\|\psi_\delta(t)h\|_{H_t^{-c}} \lesssim \delta^{c-d}\|h\|_{H_t^{-d}}.$$

By duality it suffices to prove

$$\|\psi_\delta(t)g\|_{H_t^d} \lesssim \delta^{c-d}\|g\|_{H_t^c}, \quad \forall\ g \in H_t^c. \quad (5.9)$$

It follows from Hölder's inequality and Sobolev's inequality that

$$\|\psi_\delta(t)g\|_{L_t^2} \lesssim \delta^c\|g\|_{H_t^c}. \quad (5.10)$$

Then it suffices to prove

$$\|\psi_\delta(t)g\|_{\dot{H}_t^d} \lesssim \delta^{c-d}\|g\|_{H_t^c}. \quad (5.11)$$

From (5.10) and Gagliardo-Nirenberg inequality we get

$$\|\psi_\delta(t)g\|_{\dot{H}_t^d} \lesssim \|\psi_\delta(t)g\|_{\dot{H}_t^c}^{1-\theta}\|\psi_\delta(t)g\|_{L_t^2}^{\theta} \lesssim \delta^{c-d}\|\psi_\delta(t)g\|_{\dot{H}_t^c}^{1-\theta}\|g\|_{H_t^c}^{\theta},$$

where $\theta = (c-d)/c$, thus it suffices to show

$$\|\psi_\delta(t)g\|_{\dot{H}_t^c} \lesssim \|g\|_{H_t^c}. \quad (5.12)$$

Indeed,

$$\begin{aligned}\|\psi_\delta(t)g\|_{\dot{H}_t^c} \lesssim& (\|\psi_\delta\|_{L_t^\infty}\|g\|_{H_t^c} + \|((-\Delta)^{c/2}\psi_\delta)g\|_{L_t^2}) \\ \lesssim& (\|g\|_{H_t^c} + \|((-\Delta)^{c/2}\psi_\delta)g\|_{L_t^2}).\end{aligned}$$

By Hölder's inequality and Sobolev's inequality one get

$$\|((-\Delta)^{c/2}\psi_\delta)g\|_{L_t^2} \lesssim \|(-\Delta)^{c/2}\psi_\delta\|_{L^{1/c}}\|g\|_{H_t^c} \lesssim \|g\|_{H_t^c}. \quad (5.13)$$

In conclusion, the proof of the proposition is completed. □

Proposition 5.2. *(a) Let $s \in \mathbb{R}$, $u_0 \in H^s(\mathbb{R}^n)$, $1/2 < b < 1$, $0 < \delta < 1$. Then*

$$\|\psi_\delta(t)S(t)u_0\|_{X^{s,b}} \lesssim \delta^{1/2-b}\|u_0\|_{H^s}. \tag{5.14}$$

(b) Let $s \in \mathbb{R}$, $f \in X^{s,b-1}$ and $1/2 < b < 1$. Then

$$\left\|\psi_\delta(t)\int_0^t S(t-t')f(t')\right\|_{X^{s,b}} \lesssim \delta^{1/2-b}\|f\|_{X^{s,b-1}}. \tag{5.15}$$

Proof. First we show (a). It follows from Lemma 5.2 that

$$\|\psi_\delta(t)S(t)u_0\|_{X^{s,b}} = \|\psi_\delta(t)\|_{H_t^b}\|u_0\|_{H_x^s}.$$

Using (5.8), we have

$$\|\psi_\delta(t)\|_{H_t^b} \leqslant (\delta^{1/2-b} + \delta^{1/2})\|\psi(t)\|_{H_t^b} \lesssim \delta^{1/2-b}, \tag{5.16}$$

then we get (5.14).

For (b), It follows from Lemma 5.2 that the left-hand side of (5.15) equals to

$$\left\|\langle\xi\rangle^s\psi_\delta(t)\int_0^t e^{-it'\phi(\xi)}\mathscr{F}_x(f)(\xi,t')dt'\right\|_{L_\xi^2 H_t^b}.$$

Thus to prove (b) it suffices to prove for any $g \in H_t^{b-1}$

$$\left\|\psi_\delta(t)\int_0^t g(t')dt'\right\|_{H_t^b} \leqslant C\delta^{1/2-b}\|g\|_{H_t^{b-1}}.$$

Let $h(t) = \psi_\delta(t)\displaystyle\int_0^t g(t')dt'$. Simple calculation implies that

$$h(t) = \psi_\delta(t)\int_0^t\int_{\mathbb{R}} e^{it'\tau}\widehat{g}(\tau)d\tau dt' = \psi_\delta(t)\int_{\mathbb{R}}\frac{e^{it\tau}-1}{i\tau}\widehat{g}(\tau)d\tau.$$

To control $\|h(t)\|_{H_t^b}$, we divide h into two parts $h(t) = h_1(t) + h_2(t)$, where

$$h_1(t) = \psi_\delta(t)\int_{|\tau|\leqslant 1}\frac{e^{it\tau}-1}{i\tau}\widehat{g}(\tau)d\tau,\ h_2(t) = \psi_\delta(t)\int_{|\tau|\geqslant 1}\frac{e^{it\tau}-1}{i\tau}\widehat{g}(\tau)d\tau.$$

Since $\|t^n\psi_\delta(t)\|_{H^b} = \delta^n\|(t/\delta)^n\psi(t/\delta)\|_{H^b} \leqslant \delta^n\delta^{1/2-b}4^n$, then from Taylor's expansion we get

$$\begin{aligned}\|h_1\|_{H_t^b} &= \left\|\psi_\delta(t)\int_{|\tau|\leqslant 1}\sum_{n\geqslant 1}\frac{(it\tau)^n}{n!(i\tau)}\widehat{g}(\tau)d\tau\right\|_{H_t^b}\\ &\lesssim \sum_{n\geqslant 1}\frac{\|t^n\psi_\delta(t)\|_{H^b}}{n!}\left|\int_{|\tau|\leqslant 1}\frac{(i\tau)^n\widehat{g}(\tau)}{(i\tau)}d\tau\right| \lesssim \|g\|_{H_t^{b-1}}.\end{aligned} \tag{5.17}$$

Next we need to control $\|h_2\|_{H_t^b}$. Divide it into two parts

$$\|h_2\|_{H_t^b}\lesssim\left\|\psi_\delta(t)\int_{|\tau|\geqslant 1}\frac{\widehat{g}(\tau)d\tau}{i\tau}\right\|_{H_t^b}+\left\|\psi_\delta(t)\int_{|\tau|\geqslant 1}\frac{e^{it\tau}\widehat{g}(\tau)}{i\tau}d\tau\right\|_{H_t^b}:=I+II.$$

For the term I we have

$$I\lesssim\|\psi_\delta\|_{H^b}\|g\|_{H_t^{b-1}}\Big(\int_{|\tau|\geqslant 1}\frac{\langle\tau\rangle^{2(1-b)}}{|\tau|^2}d\tau\Big)^{1/2}\lesssim\delta^{1/2-b}\|g\|_{H_t^{b-1}}. \tag{5.18}$$

For the term II we have

$$\begin{aligned}II=&\left\|\psi_\delta(t)\mathscr{F}_t^{-1}\Big(\frac{\chi_{|\tau|\geqslant 1}\hat{g}(\tau)}{i\tau}\Big)(t)\right\|_{H_t^b}\\ \lesssim&\|\psi_\delta\|_{H^b}\left\|\mathscr{F}_t^{-1}\Big(\frac{\chi_{|\tau|\geqslant 1}\hat{g}(\tau)}{i\tau}\Big)\right\|_{L^\infty}+\|\psi_\delta\|_{L^\infty}\left\|\mathscr{F}_t^{-1}\Big(\frac{\chi_{|\tau|\geqslant 1}\hat{g}(\tau)}{i\tau}\Big)\right\|_{H_t^b}\\ \lesssim&\delta^{1/2-b}\|g\|_{H_t^{b-1}}.\end{aligned} \tag{5.19}$$

Combining (5.17), (5.18)(5.19), then we obtain (5.15) if $1/2<b<1$. □

Lemma 5.3. *Assume Y is a space-time Banach space such that for all $u_0\in L^2(\mathbb{R}^n)$ and $\tau_0\in\mathbb{R}$*

$$\|e^{it\tau_0}S(t)u_0\|_Y\lesssim\|u_0\|_{L^2(\mathbb{R}^n)}.$$

Then when $1/2<b<1$, for any $u\in X^{0,b}$

$$\|u\|_Y\lesssim\|u\|_{X^{0,b}}.$$

Proof. From the inverse Fourier transform and change of variables we get

$$\begin{aligned}u(x,t)&=(2\pi)^{-(n+1)/2}\int_{\mathbb{R}^{n+1}}\widehat{u}(\xi,\tau)e^{ix\xi+it\tau}d\xi d\tau\\&=(2\pi)^{-(n+1)/2}\int_{\mathbb{R}}e^{it\tau}\int_{\mathbb{R}^n}\widehat{u}(\xi,\tau+\phi(\xi))e^{ix\xi}e^{it\phi(\xi)}d\xi d\tau.\end{aligned}$$

Thus from Minkowski's inequality and the hypothesis on Y we get

$$\|u\|_Y\lesssim\int_{\mathbb{R}}\|\widehat{u}(\xi,\tau+\phi(\xi))\|_{L^2_\xi}d\tau\lesssim\int_{\mathbb{R}}\langle\tau\rangle^{-b}\|\langle\tau\rangle^b\widehat{u}(\xi,\tau+\phi(\xi))\|_{L^2_\xi}d\tau.$$

The lemma follows from the condition $b>1/2$ and Cauchy-Schwartz inequality. □

If $b = 1/2$, there are counter-examples to show both Proposition 5.2 (b) and Lemma 5.3 fail. We leave this to the readers as an exercise. When $b = 1/2$, one has a good substitute for $X^{s,b}$: Besov-type Bourgain space F^s. We will use this space combined with a special low frequency structure to solve the $H^{-3/4}$ GWP problem for the KdV equation in Section 5.4. Now we introduce the space F^s. For convenience, for $k \in \mathbb{N} \cup \{0\}$, denote

$$I_0 = [-2, 2]; \quad I_k = [2^{k-1}, 2^{k+1}], \ k \geqslant 1. \tag{5.20}$$

Assume $\{\varphi_k\}_{k=0}^{\infty}$ is the sequence of functions for the non-homogeneous dyadic decomposition which are constructed in Section 1.3, and $\{\triangle_k\}_{k=0}^{\infty}$ are the corresponding Littlewood-Paley projector operators. For $k \in \mathbb{N} \cup \{0\}$ we define the dyadic $X^{s,b}$-type space $X_k(\mathbb{R}^2)$:

$$X_k = \left\{ \begin{array}{l} f \in L^2(\mathbb{R}^2): \quad f(\xi, \tau) \text{ is supported in } I_k \times \mathbb{R} \text{ and} \\ \|f\|_{X_k} = \sum_{j=0}^{\infty} 2^{j/2} \|\varphi_j(\tau - \phi(\xi)) \cdot f(\xi, \tau)\|_{L^2_{\xi,\tau}} < \infty \end{array} \right\}. \tag{5.21}$$

Then we resemble these space in a Littlewood-Paley manner

$$\|u\|_{F^s}^2 = \sum_{k \geqslant 0} 2^{2sk} \|\varphi_k(\xi) \mathscr{F} u\|_{X_k}^2, \tag{5.22}$$

$$\|u\|_{N^s}^2 = \sum_{k \geqslant 0} 2^{2sk} \|\langle \tau - \phi(\xi) \rangle^{-1} \varphi_k(\xi) \mathscr{F} u\|_{X_k}^2. \tag{5.23}$$

These Besov-type $X^{s,b}$ space was first introduced by Tataru [221], and then widely used by Ionescu and Kenig [113; 115], Tao [219] and the authors [96; 90; 91; 92]. F^s is a good substitute of $X^{s,1/2}$ for the following two reasons: Firstly, it is easy to see $X^{s,1/2+} \subset F^s \subset X^{s,1/2}$, and moreover, F^s space can control many space-time norm such as Strichartz-type space $L_t^p L_x^q$ (while $X^{s,1/2}$ fails logarithmicly), secondly, F^s has the same scale in time as $X^{s,1/2}$. This is very similar as the comparison between Besov space $B_{2,1}^{n/2}$ and Sobolev space $H^{n/2}$, $H^{n/2+}$.

First we prove the linear estimates in the space F^s.

Proposition 5.3. *(a) Let $s \in \mathbb{R}$ and $u_0 \in H^s$. Then*

$$\|\psi(t) S(t) u_0\|_{F^s} \lesssim \|u_0\|_{H^s}.$$

(b) Assume $s \in \mathbb{R}$, $k \in \mathbb{N} \cup \{0\}$ and $\langle \tau - \phi(\xi) \rangle^{-1} \mathscr{F} u \in X_k$. Then there exists $C > 0$ independent of k such that

$$\left\| \mathscr{F} \left[\psi(t) \int_0^t S(t-s) u(s) ds \right] \right\|_{X_k} \leqslant C \|\langle \tau - \phi(\xi) \rangle^{-1} \mathscr{F} u\|_{X_k}.$$

Proof. To prove (a), from the definition of F^s it suffices to show that for any $k \in \mathbb{N} \cup \{0\}$

$$\|\varphi_k(\xi)\mathscr{F}(\psi(t)S(t)u_0)\|_{X_k} \lesssim \|\varphi_k(\xi)\widehat{u}_0(\xi)\|_{L^2}.$$

By the definition of X_k we get

$$\begin{aligned}\|\varphi_k(\xi)\mathscr{F}(\psi(t)S(t)u_0)\|_{X_k} &= \sum_{j \geqslant 0} 2^{j/2}\|\varphi_k(\xi)\varphi_j(\tau)\widehat{\psi}(\tau)\widehat{u_0}(\xi)\|_{L^2}\\ &\lesssim \|\psi\|_{B^{1/2}_{2,1}}\|\varphi_k(\xi)\widehat{u}_0(\xi)\|_{L^2},\end{aligned}$$

which implies (a).

Next we show (b). It follows from simple calculation that

$$\begin{aligned}&\mathscr{F}\left[\psi(t)\int_0^t S(t-s)u(s)ds\right](\xi,\tau)\\ &= C\int_{\mathbb{R}} \mathscr{F}u(\xi,\tau')\frac{\widehat{\psi}(\tau-\tau')-\widehat{\psi}(\tau-\phi(\xi))}{\tau'-\phi(\xi)}d\tau'.\end{aligned}$$

Let $f(\xi,\tau') = \mathscr{F}u(\xi,\tau')\langle\tau'-\phi(\xi)\rangle^{-1}$. For $f_k \in X_k$ we define the operator T:

$$T(f_k)(\xi,\tau) = \int_{\mathbb{R}} f_k(\xi,\tau')\frac{\widehat{\psi}(\tau-\tau')-\widehat{\psi}(\tau-\phi(\xi))}{\tau'-\phi(\xi)}\langle\tau'-\phi(\xi)\rangle d\tau'.$$

Then it suffices to prove that

$$\|Tf_k\|_{X_k} \leqslant C\|f_k\|_{X_k}$$

hold uniformly for $k \geqslant 0$.

Denote $f_k^\sharp(\xi,\tau) = f_k(\xi,\tau+\phi(\xi))$, $(Tf_k)^\sharp(\xi,\tau) = (Tf_k)(\xi,\tau+\phi(\xi))$. By change of variables we get

$$(Tf_k)^\sharp(\xi,\tau) = \int_{\mathbb{R}} f_k^\sharp(\xi,\tau')\frac{\widehat{\psi}(\tau-\tau')-\widehat{\psi}(\tau)}{\tau'}\langle\tau'\rangle d\tau'.$$

It is easy to see that

$$\left|\frac{\widehat{\psi}(\tau-\tau')-\widehat{\psi}(\tau)}{\tau'}\langle\tau'\rangle\right| \leqslant C[(1+|\tau|)^{-4}+(1+|\tau-\tau'|)^{-4}].$$

For $j \in \mathbb{N}\cup\{0\}$, define $f_{k,j} = f_k(\xi,\tau)\varphi_j(\tau-\phi(\xi))$ and $f_{k,j}^\sharp = f_{k,j}(\xi,\tau+\phi(\xi))$. Then we have

$$(Tf_{k,j})^\sharp(\xi,\tau)|\lesssim(1+|\tau|)^{-4}2^{j/2}\left[\int_{I_j}|f_{k,j}^\sharp(\xi,\tau')|^2d\tau'\right]^{1/2}$$

$$+\sum_{l=-2}^{2}\varphi_{j+l}(\tau)\int_{I_j}|f^{\sharp}_{k,j}(\xi,\tau')|(1+|\tau-\tau'|)^{-4}d\tau'$$
$$:=I+II.$$

Hence

$$\|Tf_k\|_{X_k}\lesssim\sum_{j'\geqslant 0}2^{j'/2}\|\varphi_{j'}(\tau)(Tf_k)^{\sharp}(\xi,\tau)\|_{L^2}$$
$$\lesssim\sum_{j',j\geqslant 0}2^{j'/2}\|\varphi_{j'}(\tau)(Tf_{k,j})^{\sharp}(\xi,\tau)\|_{L^2}.$$

For the term I, it is obvious that

$$\sum_{j',j\geqslant 0}2^{j'/2}\|\varphi_{j'}(\tau)I\|_{L^2}\lesssim\|f_k\|_{X_k}.$$

For the term II, it follows from Young's inequality that

$$\sum_{|j'-j|\lesssim 1}2^{j'/2}\|\varphi_{j'}(\tau)II\|_{L^2}\lesssim\|f_k\|_{X_k}.$$

Therefore, the proposition is proved. □

We give an analogue of Lemma 5.3 in the case $b=1/2$ in the following lemma which can be proved similarly.

Lemma 5.4. *Assume Y is space-time Banach space such that for all $u_0\in L^2(\mathbb{R}^d)$ and $\tau_0\in\mathbb{R}$*

$$\|e^{it\tau_0}S(t)u_0\|_Y\leqslant C\|u_0\|_{L^2(\mathbb{R}^n)}.$$

Then for any $k\in\mathbb{N}\cup\{0\}$ and $u\in F^0$

$$\|\triangle_k(u)\|_Y\lesssim\|\widehat{\triangle_k(u)}\|_{X_k}.$$

5.2 Local smoothing effect and maximal function estimates

In the last Section, we already see that the elements of Bourgain space $X^{s,b}$ are very close to the linear solutions $u=S(t)u_0$ with initial data $u_0\in H^s$. Thus to use $X^{s,b}$ we need to get the estimates for the linear solutions. In studying the dispersive equations with derivatives in the nonlinear terms, there are two important kinds of estimates known as Local smoothing effect and maximal function estimates. In this section, we introduce these two estimates using $S^\alpha(t)=\mathscr{F}_x^{-1}e^{it|\xi|^\alpha\xi}\mathscr{F}_x$, $1\leqslant\alpha\leqslant 2$ as examples. The results were due to Kenig-Ponce-Vega, for example see [127; 126].

Theorem 5.1. *Assume* $1\leqslant\alpha\leqslant 2$. *Then*

$$\sup_x\left(\int_{\mathbb{R}}|S^\alpha(t)u_0|^2dt\right)^{1/2}\lesssim\left(\int_{\mathbb{R}}\frac{|\widehat{u_0}(\xi)|^2}{|\xi|^\alpha}d\xi\right)^{1/2}, \tag{5.24}$$

$$\left(\int_{\mathbb{R}}\sup_{t\in\mathbb{R}}|S^\alpha(t)u_0|^4dx\right)^{1/4}\lesssim\left(\int_{\mathbb{R}}|\widehat{u_0}(\xi)|^2|\xi|^{1/2}d\xi\right)^{1/2}. \tag{5.25}$$

Proof. Let $\omega(\xi)=|\xi|^\alpha\xi$. Then ω is invertible and the inverse is denoted by ω^{-1}. By change of variable $\eta=\omega(\xi)$, we have

$$S^\alpha(t)u_0=C\int_{\mathbb{R}}e^{ix\xi}e^{it\omega(\xi)}\hat{u}_0(\xi)d\xi=C\int_{\mathbb{R}}e^{ix\omega^{-1}(\eta)}e^{it\eta}\frac{\hat{u}_0(\omega^{-1}(\eta))}{\omega'(\omega^{-1}(\eta))}d\eta.$$

Using Plancherel's equality and making change of varible $\eta=\omega(\xi)$ (noting $\omega'(\xi)=(\alpha+1)|\xi|^\alpha$), we get

$$\|S^\alpha(t)u_0\|_{L_t^2}^2\lesssim\int_{\mathbb{R}}\frac{|\widehat{u_0}(\omega^{-1}(\eta))|^2}{|\omega'(\omega^{-1}(\eta))|^2}d\eta\lesssim\int_{\mathbb{R}}\frac{|\widehat{u_0}(\xi)|^2}{|\xi|^\alpha}d\xi.$$

Therefore, (5.24) is proved.

Next we prove (5.25). The main idea is to permute the position of x,t, namely, view t as space variable and x is time variable, then $L_x^4L_t^\infty$ becomes Strichartz-type estimate, and $(4,\infty)$ is usually admissible pair. From the proof of the first inequality we see $S^\alpha(t)u_0$ can be viewed as the linear solution to the dispersive equation with dispersive relation $\omega^{-1}(\eta)$, and with the initial data whose Fourier transform is $\frac{\widehat{u_0}(\omega^{-1}(\eta))}{\omega'(\omega^{-1}(\eta))}$. It is easy to check that $\omega^{-1}(\xi)$ satisfies (H1), (H2), (H3), (H4) in Section 2.1, and $m_1=m_2=\alpha_1=\alpha_2=1/(\alpha+1)$. Then we get the Strichartz estimate

$$\|\mathscr{F}_x^{-1}e^{it\omega^{-1}(\xi)}\mathscr{F}_xf\|_{L_t^4L_x^\infty}\lesssim\|f\|_{\dot{H}^{\frac{1}{2}-\frac{1}{4(\alpha+1)}}}.$$

Thus we have

$$\|S^\alpha(t)u_0\|_{L_x^4L_t^\infty}\lesssim\left(\int_{\mathbb{R}}\left||\eta|^{\frac{1}{2}-\frac{1}{4(\alpha+1)}}\frac{\hat{u}_0(\omega^{-1}(\eta))}{\omega'(\omega^{-1}(\eta))}\right|^2d\eta\right)^{1/2}.$$

By change of variable $\eta=\omega(\xi)$ we get

$$\|S^\alpha(t)u_0\|_{L_x^4L_t^\infty}\lesssim\|u_0\|_{\dot{H}^{1/4}},$$

which is (5.25) as desired. □

The proof we gave here is a simplified argument for the concrete example. In [126], Kenig, Ponce and Vega proved a more precise estimate for a general class of dispersive equation

$$\left(\int_{\mathbb{R}} \sup_{t\in\mathbb{R}} |S(t)u_0(x)|^4 dx\right)^{\frac{1}{4}} \lesssim \left(\int_{\mathbb{R}} |\widehat{u_0}(\xi)|^2 \left(\frac{\omega'(\xi)}{\omega''(\xi)}\right)^{1/2} d\xi\right)^{1/2}.$$

Next we give another maximal function estimate for $S^\alpha(t)$, which was proved in [127].

Theorem 5.2. *Assume $s > \frac{\alpha+1}{4}$, $\alpha \geqslant 1$ and $u_0 \in H^s$. Then*

$$\left(\int_{-\infty}^{\infty} \sup_{t\in[-T,T]} |S^\alpha(t)u_0|^2 dx\right)^{1/2} \leqslant C\|u_0\|_{H^s}, \tag{5.26}$$

where the constant $C > 0$ depends on T.

To prove Theorem 5.2, we need the following lemma.

Lemma 5.5. *Assume φ is a C^∞ function supported in $[2^{k-1}, 2^{k+1}]$ and $k \in \mathbb{N}$. For $\alpha \geqslant 1$, define the function $H_k^\alpha(\cdot)$*

$$H_k^\alpha(x) = \begin{cases} 2^k, & |x| \leqslant 1, \\ 2^{k/2}|x|^{-1/2}, & 1 \leqslant |x| \leqslant C2^{\alpha k}, \\ 1/(1+x^2), & |x| > C2^{\alpha k}. \end{cases}$$

Then, if $|t| \leqslant 2$ there exists $C > 0$ independent of x, t and k such that

$$\left|\int_{\mathbb{R}} e^{i(t|\xi|^\alpha\xi + x\xi)}\varphi(\xi)d\xi\right| \leqslant CH_k^\alpha(x). \tag{5.27}$$

Moreover, if $k = 0$ and φ is smooth function supported in $[-2, 2]$, then the same conclusion holds.

Proof. Let $u(x,t) = \int_{\mathbb{R}} e^{i(t|\xi|^\alpha\xi + x\xi)}\varphi(\xi)d\xi$. First we have the trivial estimate for any $x, t \in \mathbb{R}$

$$|u(x,t)| \leqslant C2^k.$$

Thus it remains to consider the case $|x| \geqslant 1$.

Assume $k = 0$. It follows from integration by parts and $|t| \leqslant 2$ that

$$|u(x,t)| \lesssim \left|\int_{\mathbb{R}} x^{-2} e^{ix\xi} \frac{d^2}{d\xi^2}\left(e^{i(t|\xi|^\alpha\xi)}\varphi(\xi)\right) d\xi\right| \lesssim |x|^{-2}.$$

Thus it remains to consider the case $k \geqslant 1$.

Define the set $\Omega = \{\xi \in \operatorname{supp}\varphi : |(\alpha+1)t|\xi|^\alpha + x| \leqslant |x|/2\}$. Choose a cut-off function $\zeta \in C^\infty$ supported in Ω such that $\zeta = 1$ when $|(\alpha + 1)t|\xi|^\alpha + x| \leqslant |x|/3$, for example we may set

$$\zeta = \eta\left(\frac{(\alpha+1)t|\xi|^\alpha + x}{|x|}\right),$$

where η is a fixed smooth cut-off function.

Since $|t| \leqslant 2$, if $\xi \in \Omega$, then $|x| \sim \alpha t|\xi|^\alpha \leqslant C2^{\alpha k}$, hence $h(\xi, x) = t\xi|\xi|^\alpha + x\xi$ satifies

$$|h''(\xi)| = Ct|\xi|^{\alpha-1} \geqslant C2^{-k}|x|.$$

It follows from Van der Corput's lemma that

$$\left|\int e^{ih(\xi,x)}\varphi(\xi)\zeta d\xi\right| \leqslant C2^{k/2}|x|^{-1/2}.$$

If $\xi \in \operatorname{supp}(1-\zeta)$, then $|h'(\xi)| = |(\alpha+1)t|\xi|^\alpha + x| \geqslant |x|/3$. Integrating by parts, we get for $|x| \geqslant 1$

$$\left|\int e^{ih(\xi,x)}(1-\zeta)\varphi(\xi)d\xi\right| \leqslant \frac{C}{1+x^2}.$$

In view of the estimates above, the lemma is proved. □

Using the lemma above, we can prove the following lemma which implies Theorem 5.2 immediately.

Lemma 5.6. *If $s > \frac{1+\alpha}{4}$, $\alpha \geqslant 1$, then*

$$\left(\sum_{j=-\infty}^{\infty} \sup_{|t|\leqslant 1} \sup_{j<x<j+1} |S^\alpha(t)u_0|^2\right)^{1/2} \leqslant C\|u_0\|_{H^s}. \tag{5.28}$$

Proof. Define $S_k^\alpha(t)u_0 = \mathscr{F}_x^{-1}e^{it\xi|\xi|^\alpha}\varphi_k(\xi)\mathscr{F}_x u_0$, where $\{\varphi_k\}_{k=0}^\infty$ is non-homogeneous dyadic functions. It suffices to prove

$$\left(\sum_{j=-\infty}^{\infty} \sup_{|t|\leqslant 1} \sup_{j<x<j+1} |S_k^\alpha(t)u_0|^2\right)^{1/2} \lesssim 2^{\frac{(1+\alpha)k}{4}}\|u_0\|_{L^2}.$$

By duality we need to prove

$$\left\|\int_{-1}^{1} S_k^\alpha(t)g(\cdot,t)dt\right\|_{L^2} \lesssim 2^{\frac{(1+\alpha)k}{4}}\left[\sum_{j=-\infty}^{\infty}\left(\int_{-1}^{1}\int_j^{j+1}|g(x,t)|dxdt\right)^2\right]^{1/2}. \tag{5.29}$$

By duality (see [223]), (5.29) follows from the following:

$$\Big(\sum_{j=-\infty}^{\infty}\sup_{|t|\leqslant 1}\sup_{j<x<j+1}\Big|\int_{-1}^{1}S_k^\alpha(t-t')g(\cdot,t')dt'\Big|^2\Big)^{1/2}$$
$$\lesssim 2^{\frac{(1+\alpha)k}{2}}\Big[\sum_{j=-\infty}^{\infty}\Big(\int_{-1}^{1}\int_{j}^{j+1}|g(x,t)|dxdt\Big)^2\Big]^{1/2}. \tag{5.30}$$

To prove (5.30), we get from Lemma 5.5 that

$$\Big|\int_{-1}^{1}S_k^\alpha(t-t')g(x,t')dt'\Big|\lesssim\int H_k^\alpha(y)\int_{-1}^{1}|g(x-y,t')|dt'dy$$
$$\lesssim\sum_{l=-\infty}^{\infty}H_k^\alpha(l)\int_{l}^{l+1}\int_{-1}^{1}|g(x-y,t')|dt'dy.$$

Thus the left-hand side of (5.30) is bounded by

$$\Big[\sum_{j=-\infty}^{\infty}\Big(\sup_{j\leqslant x<j+1}\sum_{l=-\infty}^{\infty}H_k^\alpha(l)\int_{l}^{l+1}\int_{-1}^{1}|g(x-y,t')|dt'dy\Big)^2\Big]^{1/2} \tag{5.31}$$

By change of variables we get

$$(5.31)\lesssim\Big[\sum_{j=-\infty}^{\infty}\Big(\sum_{l=-\infty}^{\infty}H_k^\alpha(l)\int_{l-j-1}^{l-j+2}\int_{-1}^{1}|g(z,t')|dt'dz\Big)^2\Big]^{1/2}.$$

It follows from Minkowski's inequality that

$$(5.31)\lesssim\sum_{l=-\infty}^{\infty}H_k^\alpha(l)\Big[\sum_{j=-\infty}^{\infty}\Big(\int_{l-j-1}^{l-j+2}\int_{-1}^{1}|g(z,t')|dt'dz\Big)^2\Big]^{1/2}$$
$$\lesssim\sum_{l=-\infty}^{\infty}H_k^\alpha(l)\Big[\sum_{j=-\infty}^{\infty}\Big(\int_{j}^{j+1}\int_{-1}^{1}|g(x,t')|dt'dx\Big)^2\Big]^{1/2}$$
$$\lesssim 2^{\frac{(\alpha+1)k}{2}}\Big[\sum_{j=-\infty}^{\infty}\Big(\int_{-1}^{1}\int_{j}^{j+1}|g(x,t')|dxdt'\Big)^2\Big]^{1/2},$$

where in the last inequality we use the following simple fact:

$$\sum_{l=1}^{[2^{\alpha k}]}l^{-1/2}\lesssim\sum_{l=1}^{[2^{\alpha k}]}\int_{l}^{l+1}x^{-1/2}dx\lesssim 2^{\alpha k/2}.$$

Thus we complete the proof of the lemma. □

The maximal function associated to dispersive equation was first proposed by Carleson [24]. We take Schrödinger equation as an example,

$$\begin{cases} iu_t + \Delta u = 0, \\ u(x,0) = u_0(x). \end{cases}$$

From Plancherel's equality we know as $t \to 0$ the solution $u = e^{it\Delta}u_0 \to u_0$ in H^s. Carleson asked the following question: under what condition one has $u(x,t) \to u_0(x)$ almost everywhere? To answer the question, he asked further: what is least regularity on $u_0 \in H^s$ so that $\sup_{|t|\leqslant 1} |e^{it\Delta}u_0(x)|$ is locally integrable?

This problem was completely solved in one dimension, but remains unclear in high dimension. A general conjecture states as following: in any dimension $u(x,t) \to u_0(x)$ almost everywhere if and only if $s \geqslant 1/4$. In one dimension we already show if $s > 1/2$

$$\|e^{it\partial_{xx}}u_0\|_{L_x^2 L_{|t|\leqslant 1}^\infty} \lesssim \|u_0\|_{H^s}.$$

A conjecture of Kenig and Vega states that it still hold when $s = 1/2$. Actually we show the inequality holds when $s = 1/2$ and $\|u\|_{H^s}$ is replaced by $B^s_{2,1}$. For these problems we refer the readers to [126; 127].

5.3 Bilinear estimates for KdV and local well-posedness

In this section we consider Korteweg-de Vries (KdV) equation, and prove the local well-posedness by showing the bilinear estimates in $X^{s,b}$. KdV equation is given by the following form:

$$\begin{cases} \partial_t u + \partial_x^3 u + 6u\partial_x u = 0, \ (x,t) \in \mathbb{R}^2, \\ u(x,0) = u_0(x) \in H^s(\mathbb{R}). \end{cases} \tag{5.32}$$

Then the dispersion relation for KdV equation is $\phi(\xi) = \xi^3$. KdV equation is the fundamental equation in shallow water theory, which was founded in 1895 by Korteweg and de Vries. From the discussion in Section 5.1 we know the main task is to prove the bilinear estimate. First we prove

Theorem 5.3. If $-3/4 < s < -1/2$, then there exists $b \in (1/2, 1)$ such that for any $b' \in (1/2, b]$ with $b - b' \leqslant \min\{-s - 1/2, 1/4 + s/3\}$ we have

$$\|\partial_x(u_1 u_2)\|_{X^{s,b-1}} \leqslant C\|u_1\|_{X^{s,b'}}\|u_2\|_{X^{s,b'}}. \tag{5.33}$$

We will prove Theorem 5.3 by adopting the $[k; Z]$-multiplier ideas. The method is quite standard now. From the definition of $X^{s,b}$, (5.33) is equal to

$$\|\langle\xi\rangle^s\langle\tau-\xi^3\rangle^{b-1}\xi(\widehat{u}*\widehat{v})(\xi,\tau)\|_{L^2_{\xi,\tau}}\lesssim\|\langle\xi\rangle^s\langle\tau-\xi^3\rangle^{b'}\widehat{u}\|_{L^2_{\xi,\tau}}\|\langle\xi\rangle^s\langle\tau-\xi^3\rangle^{b'}\widehat{v}\|_{L^2_{\xi,\tau}}.$$

For the convenience, we denote

$$\Lambda_{s,b}(\xi,\tau)=\langle\xi\rangle^s\langle\tau-\xi^3\rangle^b.$$

Then (5.33) is equal to

$$\left\|\xi\Lambda_{s,b-1}(\xi,\tau)[(\Lambda^{-1}_{s,b}u)*(\Lambda^{-1}_{s,b}v)](\xi,\tau)\right\|_{L^2_{\xi,\tau}}\lesssim\|u\|_{L^2(\mathbb{R}^2)}\|v\|_{L^2(\mathbb{R}^2)}. \quad (5.34)$$

Assume $\{\varphi_k\}_{k=0}^\infty(\{\chi_k\}_{k=-\infty}^\infty)$ is the nonhomogeneous (homogeneous) dyadic functions, the intervals $\{I_k\}_{k=0}^\infty$ are given by (5.20). Decompose ξ_i, $\tau_i-\xi_i^3$ into dyadic pieces, then by duality we get (5.34) is equivalent to

$$\sum_{k_i\in\mathbb{Z},j_i\in\mathbb{Z}_+,}\frac{\langle 2^{k_3}\rangle^s 2^{j_3(b-1)}2^{k_3}}{\langle 2^{k_1}\rangle^s 2^{j_1b'}\langle 2^{k_2}\rangle^s 2^{j_2b'}}$$

$$\times\int[1_{D_{k_3,j_3}}(u_{k_1,j_1}*v_{k_2,j_2})f](\xi,\tau)d\xi d\tau\lesssim\|u\|_{L^2}\|v\|_{L^2}\|f\|_{L^2}, \quad (5.35)$$

where $L^2=L^2(\mathbb{R}^2)$, for $k\in\mathbb{Z}, j\in\mathbb{Z}_+$ we denote

$$D_{k,j}=\{(\xi,\tau):\xi\in[2^{k-1},2^{k+1}],\tau-\xi^3\in I_j\},$$

and

$$u_{k_1,j_1}=\chi_{k_1}(\xi)\varphi_{j_1}(\tau-\xi^3)u.$$

Thus to prove Theorem 5.3, we need to prove first

$$\left|\int_{\mathbb{R}^2}1_{D_{k_3,j_3}}(\xi,\tau)(u_{k_1,j_1}*v_{k_2,j_2})\cdot fd\xi d\tau\right|.$$

So far, the issues reduce to some elementary calculus. We rewrite it in a symmetric form, then the bilinear estimates will follow from the trilinear estimates below

$$\int_{\Gamma_3}f_1(\xi_1,\tau_1)f_2(\xi_2,\tau_2)f_3(\xi_3,\tau_3)\lesssim C(k_i,j_i)\prod_{i=1}^3\|f_i\|_{L^2}, \quad (5.36)$$

where f_i are non-negative functions supported in D_{k_i,j_i}, and

$$\Gamma_3=\{\xi_1+\xi_2+\xi_3=0,\ \tau_1+\tau_2+\tau_3=0\},$$

endowed with induced measure [3]. The estimates of the form (5.36) were first studied by Bourgain [14; 15] using the relations between $X^{s,b}$ and

[3] $\int_{\Gamma_3}\Pi_{i=1,2,3}f_i(\xi_i,\tau_i)=\int_{\mathbb{R}^4}f_1(\xi_1,\tau_1)f_2(\xi_2,\tau_2)f_3(-\xi_1-\xi_2,-\tau_1-\tau_2)d\xi_1d\xi_2d\tau_1d\tau_2.$

other space-time norm, then by Kenig, Ponce, Vega [132] using elementary calculus. Later, Tao [216] systematically studied that in an abstract setting by using dyadic decomposition.

We will study the estimate of (5.36) in the next lemma, while for a general class of dispersion relation $\phi(\xi) = |\xi|^\alpha \xi$ we refer the readers to [92]. Our proofs are simplified version of the $[k; Z]$-multiplier, quite elementary but a little complicated. We suggest the readers to see the crucial ideas in the calculation. For convenience, if $a_1, a_2, a_3 \in \mathbb{R}$, we use $a_{min} \leqslant a_{med} \leqslant a_{max}$ to denote the maximum, medium, minimum among a_1, a_2, a_3. We also denote

$$(f_1 * f_2) \cdot f_3 = f_1 * f_2 \cdot f_3.$$

Lemma 5.7. *Assume $k_i \in \mathbb{Z}$, $j_i \in \mathbb{Z}_+$, and $f_{k_i,j_i} \in L^2(\mathbb{R}^2)$ are non-negative function supported in $\cup_{l=0}^{j_i} D_{k_i,l}$ and $\|f_{k_i,j_i}\|_2 \leqslant 1$, $i = 1,2,3$. Then*

(a) for any $k_1, k_2, k_3 \in \mathbb{Z}$ and $j_1, j_2, j_3 \in \mathbb{Z}_+$

$$\int_{\mathbb{R}^2} f_{k_1,j_1} * f_{k_2,j_2} \cdot f_{k_3,j_3} d\xi d\tau \leqslant C 2^{j_{min}/2} 2^{k_{min}/2}; \tag{5.37}$$

(b) if $k_{min} \leqslant k_{max} - 10$, then for $i = 1,2,3$

$$\int_{\mathbb{R}^2} f_{k_1,j_1} * f_{k_2,j_2} \cdot f_{k_3,j_3} d\xi d\tau \leqslant C 2^{(j_1+j_2+j_3)/2} 2^{-k_{max}/2} 2^{-(j_i+k_i)/2}; \tag{5.38}$$

(c) if $k_{min} \geqslant k_{max} - 10 \geqslant 10$, then

$$\int_{\mathbb{R}^2} f_{k_1,j_1} * f_{k_2,j_2} \cdot f_{k_3,j_3} d\xi d\tau \leqslant C 2^{j_{min}/2} 2^{j_{med}/4} 2^{-k_{max}/4}. \tag{5.39}$$

Proof. For $f, g, h \in L^2(\mathbb{R}^2)$, let $J(f,g,h) = \int_{\mathbb{R}^2} f * g \cdot h d\xi d\tau$. By change of variables we get

$$|J(f,g,h)| = |J(g,f,h)| = |J(\widetilde{f}, h, g)| = |J(f, \widetilde{g}, h)|,$$

where $\widetilde{f}(\xi,\mu) = f(-\xi,-\mu)$. Note that $\phi(\xi) = \xi^3$ is an odd function[4], Thus $\widetilde{f}_{k_i,j_i}$ and f_{k_i,j_i} have the same support. By Cauchy-Schwartz inequality and support properties of functions f_{k_i,j_i} we get

$$\begin{aligned} & J(f_{k_1,j_1}, f_{k_2,j_2}, f_{k_3,j_3}) \\ \lesssim & 2^{j_{min}/2} \int_{\mathbb{R}^2} \|f_{k_1,j_1}(\xi_1,\cdot)\|_{L^2_\tau} \|f_{k_2,j_2}(\xi_2,\cdot)\|_{L^2_\tau} \|f_{k_3,j_3}(\xi_1+\xi_2,\cdot)\|_{L^2_\tau} d\xi_1 d\xi_2 \end{aligned}$$

[4]If not odd, there are less symmetries, but the methods still work.

$$\lesssim 2^{k_{min}/2} 2^{j_{min}/2} \prod_{i=1}^{3} \|f_{k_i,j_i}\|_{L^2},$$

which is (a).

For (b), from the support properties we know $J(f_{k_1,j_1}, f_{k_2,j_2}, f_{k_3,j_3}) = 0$ unless

$$|k_{max} - k_{med}| \leqslant 5. \tag{5.40}$$

From symmetry we may assume $k_1 \leqslant k_2 \leqslant k_3$, hence $|k_2 - k_3| \leqslant 5$. Then we divide it into three cases: $j_1 = j_{max}, j_2 = j_{max}, j_3 = j_{max}$. If $j_3 = j_{max}$, $J(f_{k_1,j_1}, f_{k_2,j_2}, f_{k_3,j_3})$ can be rewritten into the following form:

$$J(f_{k_1,j_1}, f_{k_2,j_2}, f_{k_3,j_3}) = \int_{\mathbb{R}^4} f^\sharp_{k_1,j_1}(\xi_1,\tau_1) f^\sharp_{k_2,j_2}(\xi_2,\tau_2) \\ \times f^\sharp_{k_3,j_3}(\xi_1+\xi_2, \tau_1+\tau_2+\Omega(\xi_1,\xi_2)) d\xi_1 d\xi_2 d\tau_1 d\tau_2,$$

where $f^\sharp_{k_i,j_i}(\xi,\tau) = f_{k_i,j_i}(\xi,\tau+\xi^3)$, $i = 1,2,3$, and

$$\Omega(\xi_1,\xi_2) = \xi_1^3 + \xi_2^3 - (\xi_1+\xi_2)^3 = -3\xi_1\xi_2(\xi_1+\xi_2).$$

By change of variable $\xi_2' = \xi_1 + \xi_2$ and Hölder's inequality we get

$$\begin{aligned} &J(f_{k_1,j_1}, f_{k_2,j_2}, f_{k_3,j_3}) \\ &= \int_{\mathbb{R}^4} f^\sharp_{k_1,j_1}(\xi_1,\tau_1) f^\sharp_{k_2,j_2}(\xi_2-\xi_1,\tau_2) \\ &\quad \times f^\sharp_{k_3,j_3}(\xi_2, \tau_1+\tau_2+\Omega(\xi_1,\xi_2-\xi_1)) d\xi_1 d\xi_2 d\tau_1 d\tau_2 \\ &\lesssim \int_{\mathbb{R}^2} \|f^\sharp_{k_1,j_1}(\cdot,\tau_1)\|_{L^2(\mathbb{R})} \|f^\sharp_{k_2,j_2}(\cdot,\tau_2)\|_{L^2(\mathbb{R})} \\ &\quad \times \|f^\sharp_{k_3,j_3}(\xi_2, \tau_1+\tau_2+\Omega(\xi_1,\xi_2-\xi_1))\|_{L^2(|\xi_i|\sim N_i,\, i=1,2)} d\tau_1 d\tau_2. \end{aligned} \tag{5.41}$$

By change of variable $\mu_2 = \tau_1 + \tau_2 + \Omega(\xi_1, \xi_2 - \xi_1)$ and noting $\left|\partial_{\xi_1}[\Omega(\xi_1,\xi_2-\xi_1)]\right| \sim N_2^2$, we get

$$\|f^\sharp_{k_3,j_3}(\xi_2, \tau_1+\tau_2+\Omega(\xi_1,\xi_2-\xi_1))\|_{L^2(|\xi_i|\sim N_i,\, i=1,2)} \lesssim N_2^{-1} \|f^\sharp_{k_3,j_3}\|_{L^2(\mathbb{R}^2)}. \tag{5.42}$$

It follows from (5.41), (5.42) and Hölder's inequality that

$$J(f_{k_1,j_1}, f_{k_2,j_2}, f_{k_3,j_3}) \lesssim 2^{-k_3} \prod_{i=1}^{3} \|f_{k_i,j_i}\|_{L^2(\mathbb{R}^2)}.$$

If $j_2 = j_{max}$, from symmetry it is identical to the case $j_3 = j_{max}$. If $j_1 = j_{max}$, it follows from symmetry that

$$J(f_{k_1,j_1}, f_{k_2,j_2}, f_{k_3,j_3})$$
$$= \int_{\mathbb{R}^4} f^\sharp_{k_1,j_1}(\xi_1, \tau_2 + \tau_3 + \Omega(\xi_2, \xi_1 - \xi_2)) f^\sharp_{k_2,j_2}(\xi_2, \tau_2) f^\sharp_{k_3,j_3}(\xi_1 - \xi_2, \tau_3).$$

Note that $\left|\partial_{\xi_2}\left[\Omega(\xi_2, \xi_1 - \xi_2)\right]\right| \sim N_2 N_1$, then repeating the argument we get

$$J(f_{k_1,j_1}, f_{k_2,j_2}, f_{k_3,j_3}) \lesssim 2^{-(k_1+k_3)/2} \prod_{i=1}^{3} \|f_{k_i,j_i}\|_{L^2(\mathbb{R}^2)},$$

hence (b) is proved.

Now we prove (c). For the simplicity of notations, we denote $f_i = f^\sharp_{k_i,j_i}$, $i = 1, 2, 3$. We may assume $j_1 \leqslant j_2 \leqslant j_3$ and rewrite $J(f_1, f_2, f_3)$ into the following form

$$\int_{\mathbb{R}^4} f_1(\xi_1, \mu_1) f_2(\xi_2 - \xi_1, \mu_2 - \mu_1 - \xi_1^3 - (\xi_2 - \xi_1)^3)$$
$$\times f_3(\xi_2, \mu_2 - \xi_2^3) d\xi_1 d\mu_1 d\xi_2 d\mu_2.$$

It follows from Cauchy-Schwartz inequality that $J(f_1, f_2, f_3)$ is bounded by

$$\left\| \int_{\mathbb{R}^2} f_1(\xi_1, \mu_1) f_2(\xi_2 - \xi_1, \mu_2 - \mu_1 - \xi_1^3 - (\xi_2 - \xi_1)^3) d\xi_1 d\mu_1 \right\|_{L^2_{\xi_2,\mu_2}} \cdot \|f_3\|_{L^2}.$$

The integral area in $L^2_{\xi_2,\mu_2}$ is

$$E = \{(\xi_1, \mu_1) : |\mu_1| \lesssim 2^{j_1}, \xi_1^3 + (\xi_2 - \xi_1)^3 = \mu_2 - \mu_1 + O(2^{j_2})\},$$

where $|\xi_2| \sim 2^{k_2}$. Since $\xi_1^3 + (\xi_2 - \xi_1)^3 = 3\xi_2(\xi_1 - \xi_2/2)^2 + \xi_2^3/4$, then

$$3\xi_2(\xi_1 - \xi_2/2)^2 + \xi_2^3/4 - \mu_2 + \mu_1 = O(2^{j_2}),$$

moreover, if $\xi_2^3/4 - \mu_2 + \mu_1 \geqslant 0$ then

$$|\xi_1 - \xi_2/2| \lesssim 2^{(j_2-k_2)/2},$$

or, if $\xi_2^3/4 - \mu_2 + \mu_1 > 0$, for some $\theta > 0$,

$$(\xi_1 - \xi_2/2 + \theta)(\xi_1 - \xi_2/2 - \theta) = O(2^{j_2-k_2}),$$

and then $|\xi_1 - \xi_2/2 + \theta| \lesssim 2^{(j_2-k_2)/2}$ or $|\xi_1 - \xi_2/2 - \theta| \lesssim 2^{(j_2-k_2)/2}$. In any case we always have $|E| \lesssim 2^{j_1} 2^{(j_2-k_2)/2}$. Thus from Cauchy-Schwartz inequality we have

$$J(f_1, f_2, f_3)$$
$$\lesssim 2^{j_1/2} 2^{(j_2-k_2)/4} \|f_1(\xi_1, \mu_1) f_2(\xi_2 - \xi_1, \mu_2 - \mu_1 - \xi_1^3 - (\xi_2 - \xi_1)^3)\|_{L^2_{\xi_2,\mu_2,\xi_1,\mu_1}}$$
$$\lesssim 2^{j_1/2} 2^{(j_2-k_2)/4},$$

which is (c), and therefore, the lemma is proved. □

Now we prove Theorem 5.3 using the lemma. In order to better understand how the conditions are imposed, we divide the proof into many cases. For each case one has a condition, then intersection of all conditions would ensure the theorem.

Proof. [Proof of Theorem 5.3] From the previous discussion it suffices to prove (5.35). We may assume $\|u\|_2 = \|v\|_2 = 1$. Then we need to show

$$\sup_{k_{max}\geqslant 1} \sum_{k_{min}\leqslant k_{max}, j_i\in\mathbb{Z}_+,} \frac{\langle 2^{k_3}\rangle^s 2^{j_3(b-1)}2^{k_3}}{\langle 2^{k_1}\rangle^s 2^{j_1 b'}\langle 2^{k_2}\rangle^s 2^{j_2 b'}} \left\|1_{D_{k_3,j_3}} \cdot u_{k_1,j_1} * v_{k_2,j_2}\right\|_{L^2_{\xi,\tau}} \tag{5.43}$$

is bounded. Actually, from the support properties of u_{k_1,j_1}, v_{k_2,j_2} we see $1_{D_{k_3,j_3}} \cdot u_{k_1,j_1} * v_{k_2,j_2} = 0$ unless

$$|k_{max} - k_{med}| \leqslant 5, \ 2^{j_{max}} \sim \max(2^{j_{med}}, 2^{k_1+k_2+k_3}). \tag{5.44}$$

Thus one may assume (5.44) in (5.43). Now we prove (5.35) using (5.43). Divide the left-hand side of (5.35) into several parts, we may assume $(k_2, k_3) = (k_{med}, k_{max})$, and $k_2, k_3 \geqslant 1$, then from (5.43) we get

$$\sum_{|k_2-k_3|\leqslant 5, j_i\in\mathbb{Z}_+,} \frac{\langle 2^{k_3}\rangle^s 2^{j_3(b-1)}2^{k_3}}{\langle 2^{k_1}\rangle^s 2^{j_1 b'}\langle 2^{k_2}\rangle^s 2^{j_2 b'}} \int [1_{D_{k_3,j_3}}(u_{k_1,j_1} * v_{k_2,j_2})f](\xi,\tau)d\xi d\tau$$
$$\lesssim \sum_{|k_2-k_3|\leqslant 5} \|u\|_{L^2}\|\triangle_{k_2}v\|_{L^2}\|\triangle_{k_3}f\|_{L^2} \lesssim \|u\|_{L^2}\|v\|_{L^2}\|f\|_{L^2},$$

which is (5.35). Next we study (5.43) according to the frequency interactions. Fix k_{max}, then the summation below $\sum_{k_i\in\mathbb{Z}}$ means $\sum_{k_{min}\in\mathbb{Z}}$.

Case 1 (low-low interactions): $k_{max} \leqslant 100$. It follows from Lemma 5.7 (a) that

$$(5.43) \lesssim \sum_{k_i\in\mathbb{Z}, j_i\in\mathbb{Z}_+} \frac{\langle 2^{k_3}\rangle^s 2^{j_3(b-1)}2^{k_3}}{\langle 2^{k_1}\rangle^s 2^{j_1 b'}\langle 2^{k_2}\rangle^s 2^{j_2 b'}} 2^{j_{min}/2}2^{k_{min}/2} \lesssim 1.$$

Case 2 (high-low interaction): $k_2 \geqslant 100, |k_3 - k_2| \leqslant 5, k_1 \leqslant k_2 - 10$ (or k_1, k_2 exchange). If $j_3 = j_{max}$, then we get from Lemma 5.7 (b) that

$$(5.43) \lesssim \sum_{k_i\in\mathbb{Z}, j_i\in\mathbb{Z}_+} \frac{2^{j_3(b-1)}2^{k_3}}{\langle 2^{k_1}\rangle^s 2^{j_1 b'} 2^{j_2 b'}} \min(2^{(j_1+j_2)/2}2^{-k_2}, 2^{j_1/2}2^{k_1/2}).$$

Divide the summation into three parts, the first part is $k_1 \leqslant -2k_2$, denoted by I, the second part is $k_1 \geqslant -2k_2$ and $2^{k_1/2} \leqslant 2^{j_2/2-k_2}$, denoted by II, the third part is $k_1 \geqslant -2k_2$ and $2^{k_1/2} > 2^{j_2/2-k_2}$, denoted by III. It is easy to see that $I \lesssim 1$. For II, summing on j_1, j_2, j_3 we get

$$II \lesssim \sum_{-2k_2 \leqslant k_1 \leqslant k_2} \frac{2^{(k_1+2k_2)(b-1)}}{\langle 2^{k_1} \rangle^s}.$$

Then it is easy to get that $II \lesssim 1$ if $1+s < b \leqslant 1+s/3$ by considering $k_1 \geqslant 1$ and $k_1 \leqslant 1$. The term III can be similarly handled and we get

$$III \lesssim \sum_{-2k_2 \leqslant k_1 \leqslant k_2} \frac{2^{(k_1+2k_2)(b-1)}}{\langle 2^{k_1} \rangle^s}.$$

Thus if $1+s < b \leqslant 1+s/3$ then $III \lesssim 1$.

If $j_2 = j_{max}$, then from a similar argument and using Lemma 5.7 (b) we can get if $0 \leqslant b - b' \leqslant 1/2 + s/3$,

$$(5.43) \lesssim \sum_{k_i \in \mathbb{Z}, j_i \in \mathbb{Z}_+} \frac{2^{j_3(b-1)} 2^{k_3}}{\langle 2^{k_1} \rangle^s 2^{j_1 b'} 2^{j_2 b'}} \min(2^{(j_1+j_3)/2} 2^{-k_2}, 2^{j_1/2} 2^{k_1/2}) \lesssim 1.$$

If $j_1 = j_{max}$, it follows from Lemma 5.7 (b) that if $0 \leqslant b - b' \leqslant 1/4$,

$$(5.43) \lesssim \sum_{k_i \in \mathbb{Z}, j_i \in \mathbb{Z}_+} \frac{2^{j_3(b-1)} 2^{k_3}}{\langle 2^{k_1} \rangle^s 2^{j_1 b'} 2^{j_2 b'}} \min(2^{(j_2+j_3)/2} 2^{-(k_2+k_1)/2}, 2^{j_2/2} 2^{k_1/2}) \lesssim 1.$$

Case 3 (high-low interaction II): $k_2 \geqslant 100$, $|k_3 - k_2| \leqslant 10$, $k_1 \geqslant k_2 - 10$. Similarly, it follows from Lemma 5.7 (c) that if $s > -3/4$, $1/2 < b < 3/4 + s/3$,

$$(5.43) \lesssim \sum_{k_i \in \mathbb{Z}, j_i \in \mathbb{Z}_+,} \frac{2^{j_{max}(b-1)} 2^{k_3(1-s)} 2^{j_{min}/2} 2^{j_{med}/4}}{2^{j_{min} b'} 2^{j_{med} b'} 2^{k_{max}/4}} \lesssim \sum_{k \geqslant 1} 2^{k(3/4-s+3(b-1))} \lesssim 1.$$

Case 4 (high-high interactions): $k_2 \geqslant 100$, $|k_1 - k_2| \leqslant 5$, $k_3 \leqslant k_2 - 10$. If $j_3 = j_{max}$, from Lemma 5.7 (c) we get $1/2 < b < 5/4 + s$,

$$\begin{aligned}(5.43) &\lesssim \sum_{k_i \in \mathbb{Z}, j_i \in \mathbb{Z}_+} \frac{\langle 2^{k_3} \rangle^s 2^{j_3(b-1)} 2^{k_3}}{\langle 2^{k_1} \rangle^s 2^{j_1 b'} \langle 2^{k_2} \rangle^s 2^{j_2 b'}} 2^{(j_1+j_2)/2} 2^{-k_1/2} 2^{-k_3/2} \\ &\lesssim \sum_{k_i \in \mathbb{Z}} \frac{\langle 2^{k_3} \rangle^s \max(1, 2^{k_3+2k_1})^{b-1} 2^{k_3/2}}{2^{2sk_1} 2^{k_1/2}} \lesssim 1. \end{aligned} \tag{5.45}$$

If $j_1 = j_{max}$, from Lemma 5.7 (c) we get if $b - b' \leqslant -1/2 - s$,

$$(5.43) \lesssim \sum_{k_i \in \mathbb{Z}, j_i \in \mathbb{Z}_+} \frac{\langle 2^{k_3} \rangle^s 2^{j_3(b-1)} 2^{k_3}}{\langle 2^{k_1} \rangle^s 2^{j_1 b'} \langle 2^{k_2} \rangle^s 2^{j_2 b'}} 2^{(j_2+j_3)/2} 2^{-k_1}$$

$$\lesssim \sum_{k_i \in \mathbb{Z}} \frac{\langle 2^{k_3}\rangle^s \max(1, 2^{k_3+2k_1})^{b-b'-1/2} 2^{k_3-k_1}}{2^{2sk_1}} \lesssim 1. \tag{5.46}$$

The case $j_2 = j_{max}$ is identical to the case $j_1 = j_{max}$.

Taking all the conditions, we proved that if $s \in (-3/4, -1/2)$, $1/2 < b < 3/4 + s/3$, $1/2 < b' \leqslant b$ with $b - b' \leqslant -1/2 - s$, then (5.33) holds. Therefore, Theorem 5.3 is proved. □

If s is large, one has similar estimates. Actually, the same methods above also shows the following theorem. For $a \in \mathbb{R}$, $a+$ denotes $a + \epsilon$ for a fixed $0 < \epsilon \ll 1$.

Theorem 5.4. *If $-3/4 < s \leqslant 0$, then there exists $b \in (1/2, 1)$ such that*

$$\|\partial_x(u_1 u_2)\|_{X^{s,b-1}} \leqslant C\|u_1\|_{X^{s,b}} \|u_2\|_{X^{s,b}} \tag{5.47}$$

and

$$\|\partial_x(u_1 u_2)\|_{X^{s,b-1}} \lesssim \|u_1\|_{X^{-3/4+,b}} \|u_2\|_{X^{s,b}} + \|u_1\|_{X^{s,b}} \|u_2\|_{X^{-3/4+,b}}. \tag{5.48}$$

Remark 5.1. The condition $s > -3/4$ in Theorem 5.4 is necessary. When $s \leqslant -3/4$, (5.47) fails for any $b \in \mathbb{R}$. The case $s < -3/4$ was due to Kenig, Ponce and Vega [132], and the case $s = -3/4$ was due to Nakanishi, Takaoka and Tsutsumi [180].

Now we prove the local well-posedness for KdV equation using the bilinear estimates. First assume $u_0 \in H^s$, $-3/4 < s < -1/2$, and define the operator and set:

$$\Phi_{u_0}(u) = \psi_1(t) S(t) u_0 - \psi_1(t) \int_0^t S(t-t')\psi_T(t')\partial_x(u^2)(t')dt',$$

$$\mathcal{B} = \{u \in X^{s,b} : \|u\|_{X^{s,b}} \leqslant 2C\|u_0\|_{H^s}\}.$$

We will show if T is sufficiently small, the map Φ_{u_0} is a contraction mapping in $\mathcal{B}$.

From Proposition 5.2, Proposition 5.1 and Theorem 5.3, we have for $1/2 < b < b' < 1$

$$\begin{aligned}\|\Phi_{u_0}(u)\|_{X^{s,b}} &\leqslant C\|u_0\|_{H^s} + CT^{b'-b}\|u\|^2_{X^{s,b}} \\ &\leqslant C\|u_0\|_{H^s} + 4C^2 T^{b'-b}\|u_0\|^2_{H^s}.\end{aligned}$$

Thus choose T such that

$$4CT^{b'-b}\|u_0\|_{H^s} \leqslant 1/2,$$

then we have $\Phi_{u_0}(\mathcal{B}) \subset \mathcal{B}$. Assume $(u_1, u_2) \in \mathcal{B}$, and we have

$$\|\Phi_{u_0}(u_1) - \Phi_{u_0}(u_2)\|_{X^{s,b}} \leqslant 1/2\|u_1 - u_2\|_{X^{s,b}}.$$

Thus, Φ_{u_0} is a contraction mapping in $\mathcal{B}$ and there exists a unique $u \in \mathcal{B}$ such that $\Phi_{u_0}(u) = u$:

$$u(t) = \psi_1(t)S(t)u_0 - \psi_1(t)\int_0^t S(t-t')\psi_T(t')\partial_x(u^2)(t')dt'.$$

It is easy to see when $t \in [-T, T]$

$$u(t) = S(t)u_0 - \int_0^t S(t-t')\partial_x(u^2)(t')dt',$$

hence u is a solution to KdV equation (5.32) in $[-T, T]$ and $u \in X_T^{s,b}$, where $X_T^{s,b}$ is defined as following:

$$\|u\|_{X_T^{s,b}} = \inf\{\|\widetilde{u}\|_{X^{s,b}} : \widetilde{u}(t) = u(t) t \in [-T, T]\}.$$

Next we will show u is the unique solution in $X_T^{s,b}$ by using the ideas in [165]. Assume $u_1, u_2 \in X_T^{s,b}$ are two solutions to KdV equation with the same initial data u_0, we will prove $u_1(t) = u_2(t)$, $t \in [-T, T]$. From symmetry it suffices to prove $u_1(t) = u_2(t)$, $t \in [0, T]$. For $\delta > 0$ which will be determined later, $i = 1, 2$, define $\widetilde{u_i}$

$$\widetilde{u_i} = \begin{cases} u_i(t), \ t \in [0, \delta], \\ u_i(2\delta - t), \ t \in [\delta, 2\delta], \\ u_0, \ \text{otherwise}. \end{cases} \tag{5.49}$$

Thus $t \to \widetilde{u_i}(t)$ is continuous, and $\psi(t)\widetilde{u_i}(t) \in X^{s,b}$, $\widetilde{u}_1(t) - \widetilde{u}_2(t) = 0$ if $t \in \mathbb{R} \setminus [0, 2\delta]$.

Since u_1, u_2 are solutions to KdV equation, then when $t \in [0, \delta]$

$$u_1(t) - u_2(t) = -\psi_1(t)\int_0^t S(t-t')\psi_\delta(t')\partial_x[(\widetilde{u}_1 - \widetilde{u}_2)(u_1 + u_2)](t')dt'.$$

For $T > 0$ define

$$\|u\|_{X_{T+}^{s,b}} = \inf\{\|\widetilde{u}\|_{X^{s,b}} : \widetilde{u}(t) = u(t) \text{ if } t \in [0, T]\}.$$

Thus we get from Proposition 5.2, Proposition 5.1 and Theorem 5.3 that

$$\|u_1 - u_2\|_{X_{\delta+}^{s,b}} \leqslant \delta^{b'-b}(\|u\|_{X_T^{s,b}} + \|v\|_{X_T^{s,b}})\|\widetilde{u}_1 - \widetilde{u}_2\|_{X^{s,b}}.$$

From the constructing we know

$$\|\widetilde{u}_1 - \widetilde{u}_2\|_{X^{s,b}} \leqslant 2\|u_1 - u_2\|_{X_{\delta+}^{s,b}},$$

then we have

$$\|u_1 - u_2\|_{X^{s,b}_{\delta+}} \leqslant C\delta^{b'-b}(\|u\|_{X^{s,b}_T} + \|v\|_{X^{s,b}_T})\|u_1 - u_2\|_{X^{s,b}_{\delta+}}.$$

Choose δ such that $C\delta^{b'-b}(\|u\|_{X^{s,b}_T} + \|v\|_{X^{s,b}_T}) < 1/2$, then we get $u_1(t) = u_2(t)$ if $t \in [0, \delta]$. Repeating this procedure, we obtained the uniqueness in $[0, T]$.

If s is large, then we use the scaling invariance. It is easy to see that KdV equation has the following scaling invariance: for any $\lambda > 0$

$$u(x,t) \to \lambda^2 u(\lambda x, \lambda^3 t), \; u_0(x) \to \lambda^2 u_0(\lambda x). \tag{5.50}$$

Then $\dot{H}^{-3/2}$ is critical in the sense that: $\|\lambda^2 u_0(\lambda \cdot)\|_{\dot{H}^{-3/2}} = \|u_0\|_{\dot{H}^{-3/2}}$. From the fact

$$\|\lambda^2 u_0(\lambda x)\|_{H^{-3/4+}} \lesssim \lambda^{\frac{3}{2}+}\|u_0\|_{H^{-3/4+}} + \lambda^{3/4+}\|u_0\|_{H^{-3/4+}},$$

thus choosing λ sufficiently small, we may assume

$$\|\phi\|_{H^{-3/4+}} \leqslant \epsilon_0 \ll 1. \tag{5.51}$$

The rest argument is similar to the case $-3/4 < s < -1/2$, and we leave it to the readers. Therefore, we prove

Theorem 5.5. *Assume $s \in (-3/4, 0]$, $u_0 \in H^s$, Then there exists $T = T(\|u_0\|_{H^{-3/4+}}) > 0$ and $b > 1/2$ such that KdV equation* (5.32) *has a unique solution $u(x,t) \in X^{s,b}_T \subset C([-T,T]; H^s)$. Moreover, $\forall\, R > 0$, the map $u_0 \to u(t)$ is Lipschitz continuous from $\{\phi \in H^s, \|\phi\|_{H^s} \leqslant R\}$ to $C([-T,T]; H^s)$.*

5.4 Local well-posedness for KdV in $H^{-3/4}$

In Section 5.3, we consider the local well-posedness of Korteweg-de Vries equation in H^s with $s > -3/4$. It is known that the bilinear estimate (5.47) is invalid if $s \leqslant -3/4$, one can refer to [132]. Moreover, KdV equation is not locally well-posed in H^s for $s < -3/4$, the solution operator fails to be uniformly continuous with respect to the H^s norm [47; 134]. Thus there exists a natural question: whether is the Cauchy problem of KdV equation locally well-posed in $H^{-3/4}$? In this section, we give the answer, one can refer to [90].

From the results in Section 5.3, it follows that the space $X^{s,b}$ with $b > 1/2$ is unsuitable to be chosen as the work space for the local well-popsedness in $H^{-3/4}$. We try to choose a new space instead of $X^{s,b}$ with

$b > 1/2$. From the definition of Bourgain's spaces $X^{s,b}$, it is expected to obtain more elaborate estimates if H_t^b with $b > 1/2$ is replaced by $B_{2,1}^{1/2}$ in spaces $X^{s,b}$. In Section 5.1, we have introduced this space F^s, where $\phi(\xi) = \xi^3$. The key approach is prove whether the following bilinear in spaces F^s holds

$$\|\partial_x(uv)\|_{N^s} \leqslant C(\|u\|_{F^s}\|v\|_{F^s} + \|v\|_{F^s}\|u\|_{F^s}). \tag{5.52}$$

By the definitions of F^s and N^s, in order to obtain the bilinear estimate (5.52), it suffices to show that the following dyadic bilinear estimate

$$\|\langle\tau - \xi^3\rangle^{-1}\varphi_{k_3}(\xi)\xi(\widehat{\Delta_{k_1}u} * \widehat{\Delta_{k_2}v})\|_{X_{k_3}} \leqslant C(k_1, k_2, k_3)\|\widehat{\Delta_{k_1}u}\|_{X_{k_1}}\|\widehat{\Delta_{k_2}v}\|_{X_{k_2}}.$$

By examining the supports of the functions, $\varphi_{k_3}(\xi)(\widehat{\Delta_{k_1}u} * \widehat{\Delta_{k_2}v}) \equiv 0$ unless

$$|\max(k_1, k_2, k_3) - \mathrm{med}(k_1, k_2, k_3)| \leqslant 5.$$

Thus we have the following several cases depending on the relative sizes k_1, k_2, k_3: *high* $\times$ *low* $\to$ *high*, *low* $\times$ *high* $\to$ *high*, *high* $\times$ *high* $\to$ *low*, *high* $\times$ *high* $\to$ *high*, *low* $\times$ *low* $\to$ *low*. For each case, we have the corresponding estimtaes.

Using Lemma 5.4, Theorem 5.1 and the proof of Lemma 5.6, we have the following proposition.

Proposition 5.4. *If $k \in \mathbb{Z}_+$, $j \in \mathbb{N}$, $I \subset \mathbb{R}$ and $|I| \lesssim 1$, then*

$$\begin{aligned}
&\|\Delta_k(u)\|_{L_t^\infty L_x^2} \lesssim \|\mathscr{F}[\Delta_k(u)]\|_{X_k},\\
&\|\Delta_k(u)\|_{L_x^2 L_{t\in I}^\infty} \lesssim 2^{3k/4}\|\mathscr{F}[\Delta_k(u)]\|_{X_k},\\
&\|\Delta_k(u)\|_{L_x^4 L_t^\infty} \lesssim 2^{k/4}\|\mathscr{F}[\Delta_k(u)]\|_{X_k},\\
&\|\Delta_j(u)\|_{L_x^\infty L_t^2} \lesssim 2^{-j}\|\mathscr{F}[\Delta_j(u)]\|_{X_j}.
\end{aligned}$$

Now we prove dyadic bilinear estimates according to frequency interactions.

Proposition 5.5. *(a) If $k \geqslant 10$, $|k - k_2| \leqslant 5$, then*

$$\|\langle\tau - \xi^3\rangle^{-1}\varphi_k(\xi)\xi(\widehat{\Delta_0 u} * \widehat{\Delta_{k_2}v})\|_{X_k} \lesssim \|\widehat{\Delta_0 v}\|_{X_0}\|\widehat{\Delta_{k_2}v}\|_{X_{k_2}}. \tag{5.53}$$

(b) If $k \geqslant 10$, $|k - k_2| \leqslant 5$$1 \leqslant k_1 \leqslant k - 9$, then

$$\|\langle\tau - \xi^3\rangle^{-1}\varphi_k(\xi)\xi(\widehat{\Delta_{k_1}u} * \widehat{\Delta_{k_2}v})\|_{X_k} \lesssim k^3 2^{-k/2-k_1}\|\widehat{\Delta_{k_1}u}\|_{X_{k_1}}\|\widehat{\Delta_{k_2}v}\|_{X_{k_2}}. \tag{5.54}$$

Proof. For the simplicity of notations, we assume $k = k_2$. We always denote

$$u_{k,j} = \varphi_k(\xi)\varphi_j(\tau - \xi^3)\widehat{u}.$$

First we show (a). From the definition of X_k we get

$$\|\langle\tau - \xi^3\rangle^{-1}\varphi_k(\xi)\xi(\widehat{\Delta_0 u} * \widehat{\Delta_k v})\|_{X_k} \lesssim 2^k \sum_{j,j_1,j_2\geqslant 0} 2^{-j/2}\|1_{D_{k,j}}(u_{0,j_1} * v_{k,j_2})\|_2.$$

Similar to the proof of Lemma 5.7 (b), we get

$$\|1_{D_{k,j}}(u_{0,j_1} * v_{k,j_2})\|_2 \lesssim 2^{-k}2^{(j_1+j_2)/2}\|u_{0,j_1}\|_2\|v_{k,j_2}\|_2.$$

Thus

$$\begin{aligned}\|\langle\tau - \xi^3\rangle^{-1}\varphi_k(\xi)\xi(\widehat{\Delta_0 u} * \widehat{\Delta_k v})\|_{X_k} &\lesssim 2^k \sum_{j,j_1,j_2\geqslant 0} 2^{-j/2}\|1_{D_{k,j}}(u_{0,j_1} * v_{k,j_2})\|_2\\ &\lesssim \|\widehat{\Delta_0 v}\|_{X_0}\|\widehat{\Delta_{k_2} v}\|_{X_{k_2}}.\end{aligned}$$

Then (a) is proved.

For (b), from the definition of X_k we have

$$\|\langle\tau - \xi^3\rangle^{-1}\varphi_k(\xi)\xi(\widehat{\Delta_{k_1} u} * \widehat{\Delta_k v})\|_{X_k} \lesssim 2^k \sum_{j_i\geqslant 0} 2^{-j_3/2}\|1_{D_{k,j_3}}(u_{k_1,j_1} * v_{k,j_2})\|_2. \tag{5.55}$$

From the support properties we may assume $j_{max} \geqslant 2k + k_1 - 10$ in the right-hand side of (5.55). We may also assume $j_1, j_2, j_3 \leqslant 10k$, otherwise (b) follows from Lemma 5.7 (a) immediately. It follows from Lemma 5.7 (b) that

$$\begin{aligned}&2^k \sum_{j_3,j_1,j_2\geqslant 0} 2^{-j_3/2}\|1_{D_{k,j}}(u_{k_1,j_1} * v_{k,j_2})\|_2\\ &\lesssim 2^k \sum_{j_3,j_1,j_2\geqslant 0} 2^{-j/2}2^{j_{min}/2}2^{-k/2}2^{-k_1/2}2^{j_{med}/2}\|u_{k_1,j_1}\|_2\|v_{k,j_2}\|_2\\ &\lesssim 2^k \sum_{j_{max}\geqslant 2k+k_1-10} k^3 2^{-k/2}2^{-k_1/2}2^{-j_{max}/2}\|\widehat{\Delta_{k_1} u}\|_{X_{k_1}}\|\widehat{\Delta_k v}\|_{X_k}\\ &\lesssim k^3 2^{-k/2}2^{-k_1}\|\widehat{\Delta_{k_1} u}\|_{X_{k_1}}\|\widehat{\Delta_k v}\|_{X_k}.\end{aligned}$$

Therefore, the proposition is proved. □

Proposition 5.6. *If $k \geqslant 10$, $|k - k_2| \leqslant 5$ and $k - 9 \leqslant k_1 \leqslant k + 10$, then*

$$\|\langle\tau - \xi^3\rangle^{-1}\varphi_{k_1}(\xi)\xi(\widehat{\Delta_k u} * \widehat{\Delta_{k_2} v})\|_{X_{k_1}} \lesssim 2^{-3k/4}\|\widehat{\Delta_k u}\|_{X_k}\|\widehat{\Delta_{k_2} v}\|_{X_{k_2}}. \tag{5.56}$$

Proof. We assume $k = k_2$. From the definition of X_{k_1} we get

$$\begin{aligned}&\|\langle\tau-\xi^3\rangle^{-1}\varphi_{k_1}(\xi)\xi(\widehat{\triangle_k u}*\widehat{\triangle_k v})\|_{X_{k_1}}\\&\lesssim 2^{k_1}\sum_{j_i\geqslant 0}2^{-j_1/2}\|1_{D_{k_1,j_1}}(u_{k,j_2}*v_{k,j_3})\|_2.\end{aligned}\tag{5.57}$$

As before, we may assume $j_{max}\geqslant 3k-20$ and $j_1,j_2,j_3\leqslant 10k$. It follows from Lemma 5.7 (c) that

$$\begin{aligned}&2^{k_1}\sum_{j_1,j_2,j_3\geqslant 0}2^{-j_1/2}\|1_{D_{k_1,j_1}}(u_{k,j_2}*v_{k,j_3})\|_2\\&\lesssim\Big(\sum_{j_1=j_{max}}+\sum_{j_2=j_{max}}+\sum_{j_3=j_{max}}\Big)2^{-j_1/2}2^{3k/4}2^{j_{min}/2}2^{j_{med}/4}\|u_{k,j_2}\|_2\|v_{k,j_3}\|_2\\&:=I+II+III.\end{aligned}$$

The term I is easy to control, we omit the details. By symmetry we only need to control II, and dividing it into two parts we get

$$\begin{aligned}II&\lesssim\Big(\sum_{j_2=j_{max},j_1\leqslant j_3}+\sum_{j_2=j_{max},j_1\geqslant j_3}\Big)2^{-j_1/2}2^{3k/4}2^{j_{min}/2}2^{j_{med}/4}\|u_{k,j_2}\|_2\|v_{k,j_3}\|_2\\&:=II_1+II_2.\end{aligned}$$

For II_1, by summing on j_1 we get

$$\begin{aligned}II_1&\lesssim\sum_{j_2=j_{max},j_1\leqslant j_3}2^{-j_1/2}2^{3k/4}2^{j_1/2}2^{j_3/4}\|u_{k,j_2}\|_2\|v_{k,j_3}\|_2\\&\lesssim\sum_{j_2\geqslant 3k-20,j_3\geqslant 0}2^{3k/4}2^{j_3/2}\|u_{k,j_2}\|_2\|v_{k,j_3}\|_2\\&\lesssim 2^{-3k/4}\|\widehat{\triangle_k u}\|_{X_k}\|\widehat{\triangle_{k_2} v}\|_{X_{k_2}}.\end{aligned}$$

For II_2 we have

$$\begin{aligned}II_2&\lesssim\sum_{j_2=j_{max},j_1\geqslant j_3}2^{-j_1/2}2^{3k/4}2^{j_3/2}2^{j_1/4}\|u_{k,j_2}\|_2\|v_{k,j_3}\|_2\\&\lesssim 2^{-3k/4}\|\widehat{\triangle_k u}\|_{X_k}\|\widehat{\triangle_{k_2} v}\|_{X_{k_2}}.\end{aligned}$$

Therefore, the proposition is proved. □

Proposition 5.7. *If* $0\leqslant k_1,k_2,k_3\leqslant 100$, *then*

$$\|\langle\tau-\xi^3\rangle^{-1}\varphi_{k_1}(\xi)\xi(\widehat{\triangle_{k_2} u}*\widehat{\triangle_{k_3} v})\|_{X_{k_1}}\lesssim\|\widehat{\triangle_{k_2} u}\|_{X_{k_2}}\|\widehat{\triangle_{k_3} v}\|_{X_{k_3}}.\tag{5.58}$$

Proof. By the definition we get

$$\|\langle\tau-\xi^3\rangle^{-1}\varphi_{k_1}(\xi)\xi(\widehat{\Delta_{k_2}u}*\widehat{\Delta_{k_3}v})\|_{X_{k_1}}\lesssim 2^{k_1}\sum_{j_i\geqslant 0}2^{-j_1/2}\|1_{D_{k_1,j_1}}(u_{k_2,j_2}*v_{k_3,j_3})\|_2.$$

From the support properties we have $1_{D_{k_1,j}}(u_{k_2,j_1}*v_{k_3,j_2})=0$ unless $|j_{max}-j_{med}|\leqslant 10$ or $j_{max}\leqslant 1000$. It follows from Lemma 5.7 (a) that

$$\|\langle\tau-\xi^3\rangle^{-1}\varphi_{k_1}(\xi)\xi(\widehat{\Delta_{k_2}u}*\widehat{\Delta_{k_3}v})\|_{X_{k_1}}\lesssim\|\widehat{\Delta_{k_2}u}\|_{X_{k_2}}\|\widehat{\Delta_{k_3}v}\|_{X_{k_3}}.$$

Therefore, the proposition is proved. □

Actually, for the low-low interactions, we can prove a stronger result.

Proposition 5.8. *If $0\leqslant k_1,k_2,k_3\leqslant 100$, then*

$$\|\langle\tau-\xi^3\rangle^{-1}\varphi_{k_1}(\xi)\xi(\mathscr{F}[\psi(t)\Delta_{k_2}u]*\widehat{\Delta_{k_3}v})\|_{X_{k_1}}\lesssim\|\Delta_{k_2}u\|_{L_t^\infty L_x^2}\|\Delta_{k_3}v\|_{L_t^\infty L_x^2}.$$

Proof. It follows from the definition of X_{k_1}, Plancherel's equality and Bernstein's inequality that

$$\begin{aligned}&\|\langle\tau-\xi^3\rangle^{-1}\varphi_{k_1}(\xi)\xi(\mathscr{F}[\psi(t)\Delta_{k_2}u]*\widehat{\Delta_{k_3}v})\|_{X_{k_1}}\\ &\lesssim 2^{k_1}\sum_{j_3\geqslant 0}2^{-j_3/2}\|\psi(t)\Delta_{k_2}u\cdot\Delta_{k_3}v\|_{L_t^2L_x^2}\lesssim\|\Delta_{k_2}u\|_{L_t^\infty L_x^2}\|\Delta_{k_3}v\|_{L_t^\infty L_x^2},\end{aligned}$$

which implies the proposition. □

The last one is the high-high to low interactions.

Proposition 5.9. *(a) If $k\geqslant 10|k-k_2|\leqslant 5$, then*

$$\|\langle\tau-\xi^3\rangle^{-1}\varphi_0(\xi)\xi(\widehat{\Delta_k u}*\widehat{\Delta_{k_2}v})\|_{X_0}\lesssim k2^{-3k/2}\|\widehat{\Delta_k u}\|_{X_k}\|\widehat{\Delta_{k_2}v}\|_{X_{k_2}}.$$

(b) If $k\geqslant 10$, $|k-k_2|\leqslant 5$, $1\leqslant k_1\leqslant k-9$, then

$$\begin{aligned}&\|\langle\tau-\xi^3\rangle^{-1}\varphi_{k_1}(\xi)\xi(\widehat{\Delta_k u}*\widehat{\Delta_{k_2}v})\|_{X_{k_1}}\\ &\lesssim(2^{-3k/2}+k2^{-2k+k_1/2})\|\widehat{\Delta_k u}\|_{X_k}\|\widehat{\Delta_{k_2}v}\|_{X_{k_2}}.\end{aligned}$$

Proof. First we show (a), assuming $k=k_2$. The left-hand side of the inequality in (a) is bounded by

$$\sum_{k_3=-\infty}^{0}2^{k_3}\sum_{j_1,j_2,j_3\geqslant 0}2^{-j_3/2}\|1_{D_{k_3,j_3}}\cdot(u_{k,j_1}*v_{k,j_2})\|_2. \tag{5.59}$$

We may assume $k_3\geqslant -10k$ and $j_1,j_2,j_3\leqslant 10k$. We only consider the worst case $|j_3-2k-k_3|\leqslant 10$. It follows from Lemma 5.7 (b) that

$$\|\langle\tau-\xi^3\rangle^{-1}\varphi_0(\xi)\xi(\widehat{\Delta_k u}*\widehat{\Delta_k v})\|_{X_0}$$

$$\lesssim \sum_{k_3=-10k}^{0} \sum_{j_1,j_2\geqslant 0} 2^{-k}2^{-k_3/2}2^{k_3}2^{-k/2}2^{-k_3/2}2^{j_1/2}2^{j_2/2}\|u_{k,j_1}\|_2\|v_{k,j_2}\|_2$$

$$\lesssim k2^{-3k/2}\|\widehat{\triangle_k u}\|_{X_k}\|\widehat{\triangle_k v}\|_{X_k}. \tag{5.60}$$

Thus, (a) is proved.

Now we prove (b), assuming $k = k_2$. From definition we have

$$\|\langle\tau-\xi^3\rangle^{-1}\varphi_{k_1}(\xi)\xi(\widehat{\triangle_k u} * \widehat{\triangle_k v})\|_{X_{k_1}} \lesssim 2^{k_1}\sum_{j_i\geqslant 0} 2^{-j_1/2}\|1_{D_{k_1,j_1}}(u_{k,j_2} * v_{k,j_3})\|_2. \tag{5.61}$$

We may assume $j_{max} \geqslant 2k + k_1 - 10$ and $j_1, j_2, j_3 \leqslant 10k$. We will control the right-hand side of (5.61) case by case. If $j_1 = j_{max}$, it follows from Lemma 5.7 (b) that

$$2^{k_1}\sum_{j_1,j_2,j_3\geqslant 0} 2^{-j_1/2}\|1_{D_{k_1,j_1}}\cdot(u_{k,j_2} * v_{k,j_3})\|_2$$

$$\lesssim 2^{k_1}\sum_{j_1\geqslant 2k+k_1-10}\sum_{j_2,j_3\geqslant 0} 2^{-j_1/2}2^{-k/2}2^{-k_1/2}2^{(j_2+j_3)/2}\|u_{k,j_2}\|_2\|v_{k,j_3}\|_2$$

$$\lesssim 2^{-3k/2}\|\widehat{\triangle_k u}\|_{X_k}\|\widehat{\triangle_{k_2} v}\|_{X_{k_2}}.$$

If $j_2 = j_{max}$, we have better estimate for the characterization multiplier. It follows from Lemma 5.7 (b) that

$$2^{k_1}\sum_{j_1,j_2,j_3\geqslant 0} 2^{-j_1/2}\|1_{D_{k_1,j_1}}\cdot(u_{k,j_2} * v_{k,j_3})\|_2$$

$$\lesssim 2^{k_1}\sum_{j_2\geqslant 2k+k_1-10}\sum_{j_1,j_3\geqslant 0} 2^{-j_1/2}2^{-k}2^{(j_1+j_3)/2}\|u_{k,j_2}\|_2\|v_{k,j_3}\|_2$$

$$\lesssim k2^{-2k}2^{k_1/2}\|\widehat{\triangle_k u}\|_{X_k}\|\widehat{\triangle_{k_2} v}\|_{X_{k_2}},$$

where in the last inequality we use $j_1 \leqslant 10k$. By symmetry, the case $j_3 = j_{max}$ is identical to the case $j_2 = j_{max}$. Thus the proposition is proved. □

Next we prove the bilinear estimates in F^s.

Theorem 5.6. *Fix $s \in (-3/4, 0]$. Then $\forall\ s \leqslant \sigma \leqslant 0$, there exists $C > 0$ such that $\forall\ u, v \in F^\sigma$*

$$\|\partial_x(uv)\|_{N^\sigma} \leqslant C(\|u\|_{F^s}\|v\|_{F^\sigma} + \|v\|_{F^s}\|u\|_{F^\sigma}). \tag{5.62}$$

Proof. From definition we have

$$\|\partial_x(uv)\|_{N^\sigma}^2 = \sum_{k_3\in\mathbb{Z}_+} 2^{2\sigma k_3}\|\langle\tau-\xi^3\rangle^{-1}\varphi_{k_3}(\xi)\xi(\widehat{u}*\widehat{v})\|_{X_{k_3}}^2.$$

Applying Littlewood-Paley decomposition to u, v we get

$$\begin{aligned}&\|\langle\tau-\xi^3\rangle^{-1}\varphi_{k_3}(\xi)\xi(\widehat{u}*\widehat{v})\|_{X_{k_3}}\\ &\lesssim \sum_{k_1,k_2\in\mathbb{Z}_+}\|\langle\tau-\xi^3\rangle^{-1}\varphi_{k_3}(\xi)\xi(\widehat{\triangle_{k_1}u}*\widehat{\triangle_{k_2}v})\|_{X_{k_3}}.\end{aligned} \tag{5.63}$$

We may assume $|k_{max}-k_{med}|\leqslant 5$. By symmetry we may also assume $k_1\leqslant k_2$. Then the right-hand side of (5.63) is bounded

$$\Big(\sum_{j=1}^4\sum_{k_1,k_2\in A_j}\Big)\|\langle\tau-\xi^3\rangle^{-1}\varphi_{k_3}(\xi)\xi(\widehat{\triangle_{k_1}u}*\widehat{\triangle_{k_2}v})\|_{X_{k_3}}, \tag{5.64}$$

where A_j, $j=1,2,3,4$, is defined as following

$$\begin{aligned}A_1 &= \{k_2\geqslant 10, |k_2-k_3|\leqslant 5, k_1\leqslant k_2-10\};\\ A_2 &= \{k_2\geqslant 10, |k_2-k_3|\leqslant 5, k_2-9\leqslant k_1\leqslant k_2+10\};\\ A_3 &= \{k_2\geqslant 10, |k_2-k_1|\leqslant 5, k_3\leqslant k_1-10\};\\ A_4 &= \{k_1,k_2,k_3\leqslant 100\}.\end{aligned}$$

Thus, (5.62) follows from Propositions 5.5-5.9, the condition $-3/4<s\leqslant 0$ and discrete Young's inequality. □

With this bilinear estimates and the scaling (5.50), we can also obtain the LWP of KdV equation in H^s for $s>-3/4$. From the proof of Theorem 5.6 we see the condition $s>-3/4$ was only needed in Proposition 5.9 (a). Thus to study the endpoint case $s=-3/4$, one naturally asks whether Theorem 5.6 holds at $s=-3/4$, more precisely, whether the bound in Proposition 5.9 (a) can be improved to $2^{-3k/2}$? The answer is negative by the following counter-example which was given by N. Kishimoto [141] using the one in [180].

Proposition 5.10. *If $k\geqslant 200$ and $|k-k_2|\leqslant 5$, then there exists $u, v\in F^{-3/4}$ such that*

$$\|\langle\tau-\xi^3\rangle^{-1}\varphi_0(\xi)\xi(\widehat{\triangle_k u}*\widehat{\triangle_{k_2}v})\|_{X_0}\gtrsim\log(k)2^{-\frac{3k}{2}}\|\widehat{\triangle_k u}\|_{X_k}\|\widehat{\triangle_{k_2}v}\|_{X_{k_2}}. \tag{5.65}$$

Proof. Let $N = 2^k$, choose $m \in \mathbb{N}$ sufficiently large such that $2^m \ll N^{1/2}$. For $j = 0, 1, \cdots, m$, define $R_j^0 \subset \mathbb{R}^2$ to be the parallelogram with vortex

$(\tau, \xi) = (0,0),\ (1,0),\ (3\cdot 2^{-j}N^{3/2}+1, 2^{-j}N^{-1/2}),\ (3\cdot 2^{-j}N^{3/2}, 2^{-j}N^{-1/2}).$

Then let $R_j := ((N+2^jN^{-1/2})^3, N+2^jN^{-1/2}) + R_j^0$. Choose

$$\widehat{u} = \widehat{v} = N\sum_{j=0}^{m} 2^{j/2} a_j 1_{R_j \cup (-R_j)},$$

where $a_j > 0$. It is easy to see that $\cup_{j=0}^m R_j \subset \{(\tau,\xi) : |\tau - \xi^3| \leqslant 10\}$, thus we get

$$N^{-3/2}\|\widehat{\triangle_k u}\|_{X_k}\|\widehat{\triangle_{k_2} v}\|_{X_{k_2}} \sim N^2\sum_{j=0}^{m}[N^{-3/4}2^{j/2}a_j|R_j|^{1/2}]^2 \sim \sum_{j=0}^{m} a_j^2.$$

On the other hand, if $1 \leqslant j \leqslant m$

$$1_{R_j} * 1_{-R_0} \gtrsim |R_j| 1_{(\tau^{(j)},\xi^{(j)})-1/2R_0^0} \sim 2^{-j}N^{-1/2}1_{(\tau^{(j)},\xi^{(j)})-1/2R_0^0}, \tag{5.66}$$

where

$(\tau^{(j)}, \xi^{(j)}) = ((N+2^jN^{-1/2})^3, N+2^jN^{-1/2}) - ((N+N^{-1/2})^3, N+N^{-1/2}).$

Thus $\|\langle \tau - \xi^3\rangle^{-1}\varphi_0(\xi)\xi\widehat{\triangle_k u} * \widehat{\triangle_{k_2} v}\|_{X_0}$ is larger than

$$\sum_{j'=0}^{\infty} 2^{-j'/2}\|\varphi_{j'}(\tau - \xi^3)\varphi_0(\xi)\xi(N\sum_{j=1}^{m} 2^{j/2}a_j 1_{R_j}) * (Na_0 1_{-R_0})\|_{L^2}. \tag{5.67}$$

From (5.66) we get (5.67) is larger than

$$N^{3/2}a_0\sum_{j'=0}^{\infty} 2^{-j'/2}\|\varphi_{j'}(\tau - \xi^3)\varphi_0(\xi)\xi\sum_{j=1}^{m} 2^{-j/2}a_j 1_{(\tau^{(j)},\xi^{(j)})-1/2R_0^0}\|_{L^2}. \tag{5.68}$$

It is easy to see that if $(\tau,\xi) \in (\tau^{(j)},\xi^{(j)}) - 1/2R_0$ $(j \geqslant 1)$, then $|\xi| \sim 2^jN^{-1/2}|\tau| \sim 2^jN^{3/2}$, hence (5.68) is equivalent to

$$\begin{aligned} &N^{3/2}a_0 \sum_{j'\geqslant 0, 2^jN^{3/2}\sim 2^{j'}} 2^{-j'/2}\|\varphi_{j'}(\tau - \xi^3)\varphi_0(\xi)\xi 2^{-j/2}a_j 1_{(\tau^{(j)},\xi^{(j)})-1/2R_0^0}\|_{L^2} \\ &\sim N^{3/2}a_0\sum_{j=1}^{m}(2^jN^{3/2})^{-1/2}2^jN^{-1/2}2^{-j/2}a_j|R_0^0|^{1/2} \sim a_0\sum_{j=1}^{m} a_j. \end{aligned}$$

Taking $a_j = 1/(j+1)$, we complete the proof of the proposition. $\square$

The proposition above shows that if $s=-3/4$, one can not obtain local well-posedness by using F^s or $X^{s,b}$. From the discussion we see the only problem is the low frequency structure $P_{\leqslant 0}u$. Fortunately, by using a new low frequency structure, we can overcome the logarithmic divergence in Proposition 5.9 (a). We will use the following structure:

$$\|u\|_{\bar{X}_0}=\|u\|_{L_x^2L_t^\infty}.$$

From Proposition 5.4 we have

$$\|\varphi_0(t)\triangle_0 u\|_{\bar{X}_0}\lesssim\|\triangle_0 u\|_{X_0}. \tag{5.69}$$

Thus it is weaker than X_0, but on the other hand, we have for any $1\leqslant q\leqslant\infty$ and $2\leqslant r\leqslant\infty$

$$\|\triangle_0 u\|_{L^q_{|t|\leqslant T}L^r_x\cap L^r_xL^q_{|t|\leqslant T}}\lesssim_T\|\triangle_0 u\|_{L_x^2L^\infty_{|t|\leqslant T}}. \tag{5.70}$$

For $-3/4\leqslant s\leqslant 0$, define

$$\bar{F}^s=\{u\in\mathcal{S}'(\mathbb{R}^2):\|u\|^2_{\bar{F}^s}=\sum_{k\geqslant 1}2^{2sk}\|\varphi_k(\xi)\mathscr{F}u\|^2_{X_k}+\|\triangle_0 u\|^2_{\bar{X}_0}<\infty\}.$$

Assume $T\geqslant 0$, define the local space $\bar{F}^s(T)$:

$$\|u\|_{\bar{F}^s(T)}=\inf_{w\in\bar{F}^s}\{\|\triangle_0 u\|_{L_x^2L^\infty_{|t|\leqslant T}}+\|(I-\triangle_0)w\|_{\bar{F}^s},\ w(t)=u(t),t\in[-T,T]\}.$$

Now we will show local well-posedness for KdV equation at $s=-3/4$.

Theorem 5.7. *Assume $s\geqslant -3/4$ and $\phi\in H^s$. Then*

(a) Existence. There exists $T=T(\|u_0\|_{H^{-3/4}})>0$ and a solution u to the Cauchy problem (5.32) *such that*

$$u\in\bar{F}^s(T)\subset C([-T,T]:H^s).$$

(b) Uniqueness. The solution map $S_T:u_0\to u$ is the unique continuous extension of the smooth solution map $H^\infty\to C([-T,T]:H^\infty)$.

(c) Lipschitz continuity. For any $R>0$, the map $u_0\to u$ is Lipschitz continuous from $\{u_0\in H^s:\|u_0\|_{H^s}<R\}$ to $C([-T,T]:H^s)$.

(d) Persistence. If $u_0\in H^s$ for $\sigma>s$, then $u\in H^\sigma$.

Now we indicate our ideas in constructing $\bar{F}^s$. The starting point is the bilinear estimates in F^s:

$$\|\partial_x(uv)\|_{N^s}\leqslant C\|u\|_{F^s}\|v\|_{F^s}.$$

This estimate fails at $s=-3/4$. On the other hand, we expect contraction principle still work at $s=-3/4$. Thus we need a new space $\bar{F}^{-3/4}$, which of course satisfies the following

$$\bar{F}^{-3/4}\subset C(\mathbb{R}:H^{-3/4}).$$

From the equivalent integral equation of (5.32):

$$u = S(t)u_0 + C\int_0^t S(t-s)\partial_x(u^2)(s)ds.$$

Localizing in time, we get

$$u = T_{u_0}(u) = \psi(t)S(t)u_0 + C\psi(t)\int_0^t S(t-s)\partial_x(u^2)(s)ds, \qquad (5.71)$$

where $\psi(t) = \varphi_0(t)$. By an iteration,

$$u^{(0)} = 0; \cdots ; \ u^{(n+1)} = T_{u_0}(u^{(n)}); \cdots$$

we obtain a sequence $\{u^{(n)}\}$. If contraction principle works in $\bar{F}^{-3/4}$ and u_0 satisfies $\|u_0\|_{H^{-3/4}} \ll 1$, then $\{u^{(n)}\}$ is a Cauchy sequence in $C(\mathbb{R} : H^{-3/4})$. Then a basic requirement is

$$u^{(n)} \in C(\mathbb{R} : H^{-3/4}), \quad \forall\, n \in \mathbb{N}. \qquad (5.72)$$

Next we check (5.72) for $\{u^{(n)}\}$. For $n = 0, 1$, it is obvious that $u^{(0)}, u^{(1)} \in C(\mathbb{R}; H^{-3/4})$. The case $n = 2$ is nontrivial. From the definition we have

$$u^{(2)} = \psi(t)S(t)u_0 + C\psi(t)\int_0^t S(t-s)\partial_x(S(s)u_0 \cdot S(s)u_0)ds.$$

It suffices to prove $(I - \Delta_0)(u^{(2)}) \in C(\mathbb{R}; H^{-3/4})\Delta_0(u^{(2)}) \in C(\mathbb{R}; L^2)$. For the high frequency part

$$\begin{aligned}
&\psi(t)\int_0^t S(t-s)\partial_x(S(s)u_0 \cdot S(s)u_0)ds \\
&= \psi(t)\int_0^t S(t-s)\partial_x[\psi(s/2)S(s)u_0 \cdot \psi(s/2)S(s)u_0]ds.
\end{aligned}$$

From the linear estimate

$$\|\psi(t/2)S(t)u_0\|_{F^s} \lesssim \|u_0\|_{H^s},$$

then by dyadic bilinear estimates and Proposition 5.3, we get for $k \in \mathbb{N}$, $\Delta_k(u^{(2)}) \in X_k \subset C(\mathbb{R}; H^{-3/4})$.

For the low frequency part, we can not obtain $\Delta_0(u^{(2)}) \in X_0$, due to the logarithmic divergence. By calculation we get

$$\begin{aligned}
&\mathscr{F}_x\Big[\psi(t)\int_0^t S(t-s)\Delta_0\partial_x[\Delta_{k_1}u(s)\Delta_{k_2}v(s)]ds\Big](\xi) \\
&= \psi(t)\eta_0(\xi)i\xi\int_0^t e^{i(t-s)\xi^3}\int_{\xi=\xi_1+\xi_2} e^{is\xi_1^3}\widehat{\Delta_{k_1}u_0}(\xi_1)e^{is\xi_2^3}\widehat{\Delta_{k_2}v_0}(\xi_2)ds
\end{aligned}$$

$$= \psi(t)\eta_0(\xi)e^{it\xi^3}\xi \int_{\xi=\xi_1+\xi_2} \frac{1-e^{-it(\xi^3-\xi_1^3-\xi_2^3)}}{\xi^3-\xi_1^3-\xi_2^3}\widehat{\triangle_{k_1}u_0}(\xi_1)\widehat{\triangle_{k_2}v_0}(\xi_2)$$

$$:= \mathscr{F}_x(I) + \mathscr{F}_x(II).$$

Since in the hyperplane $\xi = \xi_1 + \xi_2$ one has $\xi^3 - \xi_1^3 - \xi_2^3 = 3\xi\xi_1\xi_2$, then we get

$$\mathscr{F}_x(I) = \psi(t)\varphi_0(\xi)e^{it\xi^3} \int_{\xi=\xi_1+\xi_2} \frac{\widehat{\triangle_{k_1}u_0}(\xi_1)\widehat{\triangle_{k_2}v_0}(\xi_2)}{3\xi_1\xi_2}.$$

Applying Theorem 5.2 we get

$$\|I\|_{L_x^2L_t^\infty} \leqslant C\left\|\int_{\xi=\xi_1+\xi_2} \frac{\widehat{\triangle_{k_1}u_0}(\xi_1)\widehat{\triangle_{k_2}v_0}(\xi_2)}{3\xi_1\xi_2}\right\|_{L_\xi^2} \leqslant C2^{-3k_1/2}\|u_0\|_{L^2}\|v_0\|_{L^2}.$$

For the term II we have

$$\mathscr{F}_x(II) = \psi(t)\varphi_0(\xi)\int_{\xi=\xi_1+\xi_2} \frac{-e^{it(\xi_1^3+\xi_2^3)}}{3\xi_1\xi_2}\widehat{\triangle_{k_1}u_0}(\xi_1)\widehat{\triangle_{k_2}v_0}(\xi_2).$$

Applying Theorem 5.2 we get

$$\begin{aligned}\|II\|_{L_x^2L_t^\infty} \leqslant& C\|e^{t\partial_x^3}\partial_x^{-1}\triangle_{k_1}u_0 \cdot e^{t\partial_x^3}\partial_x^{-1}\triangle_{k_2}v_0\|_{L_x^2L_t^\infty}\\ \leqslant& C\|e^{t\partial_x^3}\partial_x^{-1}\triangle_{k_1}u_0\|_{L_x^4L_t^\infty}\|e^{t\partial_x^3}\partial_x^{-1}\triangle_{k_2}v_0\|_{L_x^4L_t^\infty}\\ \leqslant& C2^{-3k_1/2}\|u_0\|_{L^2}\|v_0\|_{L^2}.\end{aligned}$$

Thus, we proved $\triangle_0(u^{(2)}) \in L_x^2L_t^\infty$.

Next we consider the case $n \geqslant 3$. If $n \geqslant 3$, do we have $(I - \triangle_0)(u^{(n)}) \in F^s$ and $\triangle_0(u^{(n)}) \in L_x^2L_t^\infty$? To answer this, we have the following proposition:

Proposition 5.11 ($\bar{X}_0$ estimate). *Assume $|k_1 - k_2| \leqslant 5$ and $k_1 \geqslant 10$. Then $u, v \in \bar{F}^0$*

$$\begin{aligned}&\left\|\psi(t)\int_0^t S(t-s)\triangle_0\partial_x[\triangle_{k_1}u(s)\triangle_{k_2}v(s)]ds\right\|_{L_x^2L_t^\infty}\\ &\lesssim 2^{-\frac{3k_1}{2}}\|\widehat{\triangle_{k_1}u}\|_{X_{k_1}}\|\widehat{\triangle_{k_2}u}\|_{X_{k_2}}.\end{aligned}$$

Proof. Denote $Q(u,v) = \psi(t)\int_0^t S(t-s)\triangle_0\partial_x[\triangle_{k_1}u(s)\triangle_{k_2}v(s)]ds$. Direct computations show that

$$\mathscr{F}\left[Q(u,v)\right](\xi,\tau) = c\int_{\mathbb{R}} \frac{\widehat{\psi}(\tau-\tau') - \widehat{\psi}(\tau-\xi^3)}{\tau'-\xi^3}\varphi_0(\xi)i\xi$$

$$\times\, d\tau' \int_{\xi=\xi_1+\xi_2,\tau'=\tau_1+\tau_2} \widehat{\triangle_{k_1}u}(\xi_1,\tau_1)\widehat{\triangle_{k_2}v}(\xi_2,\tau_2).$$

Fix $\xi \in \mathbb{R}$, divide $\Gamma := \{\xi = \xi_1+\xi_2, \tau' = \tau_1+\tau_2\}$ as following

$$\begin{aligned}
\Gamma_1 =& \{|\xi| \lesssim 2^{-2k_1}\} \cap \Gamma;\\
\Gamma_2 =& \{|\xi| \gg 2^{-2k_1}, |\tau_i - \xi_i^3| \ll 3\cdot 2^{2k_1}|\xi|, i=1,2\} \cap \Gamma;\\
\Gamma_3 =& \{|\xi| \gg 2^{-2k_1}, |\tau_1 - \xi_1^3| \gtrsim 3\cdot 2^{2k_1}|\xi|\} \cap \Gamma;\\
\Gamma_4 =& \{|\xi| \gg 2^{-2k_1}, |\tau_2 - \xi_2^3| \gtrsim 3\cdot 2^{2k_1}|\xi|\} \cap \Gamma.
\end{aligned}$$

Then we get

$$\mathscr{F}\left[\psi(t)\cdot \int_0^t S(t-s)\triangle_0\partial_x[\triangle_{k_1}u(s)\triangle_{k_2}v(s)]ds\right](\xi,\tau) = A_1 + \ldots + A_4,$$

where

$$A_i = C\int_{\mathbb{R}} \frac{\widehat{\psi}(\tau-\tau') - \widehat{\psi}(\tau-\xi^3)}{\tau'-\xi^3}\triangle_0(\xi) i\xi \int_{\Gamma_i} \widehat{\triangle_{k_1}u}(\xi_1,\tau_1)\widehat{\triangle_{k_2}v}(\xi_2,\tau_2)d\tau'.$$

First we consider the estimate of A_1. It follows from Proposition 5.4 and Proposition 5.3 (b) that

$$\|\mathscr{F}^{-1}(A_1)\|_{L_x^2L_t^\infty} \lesssim \left\|\langle\tau'-\xi^3\rangle^{-1}\varphi_0(\xi)\xi\int_{A_1} \widehat{\triangle_{k_1}u}(\xi_1,\tau_1)\widehat{\triangle_{k_2}v}(\xi_2,\tau_2)\right\|_{X_0}.$$

Since in Γ_1 one has $|\xi| \lesssim 2^{-2k_1}$, then we get

$$\begin{aligned}
&\left\|\langle\tau'-\xi^3\rangle^{-1}\varphi_0(\xi) i\xi\int_{\Gamma_1} \widehat{\triangle_{k_1}u}(\xi_1,\tau_1)\widehat{P_{k_2}v}(\xi_2,\tau_2)\right\|_{X_0}\\
&\lesssim \sum_{k_3\leqslant -2k_1+10}\sum_{j_3\geqslant 0} 2^{-j_3/2}2^{k_3} \sum_{j_1\geqslant 0, j_2\geqslant 0} \|1_{D_{k_3,j_3}}\cdot(u_{k_1,j_1} * v_{k_2,j_2})\|_{L^2}.
\end{aligned}$$

Using (5.37) we get

$$\begin{aligned}
\|\mathscr{F}^{-1}(A_1)\|_{L_x^2L_t^\infty} \lesssim& \sum_{k_3\leqslant -2k_1+10}\sum_{j_i\geqslant 0} 2^{-j_3/2}2^{k_3}2^{j_{min}/2}2^{k_3/2}\|u_{k_1,j_1}\|_{L^2}\|v_{k_2,j_2}\|_{L^2}\\
\lesssim& 2^{-3k_1}\|\widehat{\triangle_{k_1}u}\|_{X_{k_1}}\|\widehat{\triangle_{k_2}u}\|_{X_{k_2}},
\end{aligned}$$

which suffices for the bound of A_1.

Next we consider A_3. As for A_1, it follows from Proposition 5.4 and Proposition 5.3 (b) that

$$\begin{aligned}
\|\mathscr{F}^{-1}(A_3)\|_{L_x^2L_t^\infty} \lesssim& \left\|\langle\tau'-\xi^3\rangle^{-1}\varphi_0(\xi)\xi\int_{A_3\cup A_4} \widehat{\triangle_{k_1}u}(\xi_1,\tau_1)\widehat{\triangle_{k_2}v}(\xi_2,\tau_2)\right\|_{X_0}\\
\lesssim& \sum_{k_3\leqslant 0}\sum_{j_3\geqslant 0} 2^{-j_3/2}2^{k_3}\sum_{j_1,j_2\geqslant 0}\|1_{D_{k_3,j_3}}\cdot(u_{k_1,j_1} * v_{k_2,j_2})\|_{L^2}.
\end{aligned}$$

We may assume $j_3 \leqslant 10k_1$ and $|\tau_1 - \xi_1^3| \gtrsim 3|\xi\xi_1\xi_2|$. It follows from Lemma 5.7 (b) that

$$\begin{aligned}\|\mathscr{F}^{-1}(A_3)\|_{L_x^2 L_t^\infty} \lesssim & \sum_{k_3 \leqslant 0} \sum_{j_1 \geqslant k_3+2k_1-10, j_2, j_3 \geqslant 0} 2^{k_3} 2^{j_2/2} 2^{-k_1} \|u_{k_1,j_1}\|_{L^2} \|v_{k_2,j_2}\|_{L^2} \\ \lesssim & k_1 2^{-2k_1} \|\widehat{\Delta_{k_1} u}\|_{X_{k_1}} \|\widehat{\Delta_{k_2} u}\|_{X_{k_2}},\end{aligned}$$

which gives the bound for A_3. Similarly, we can control A_4.

Now we consider A_2, which is the main term. Simple calculations show that

$$\begin{aligned}\mathscr{F}_t^{-1}(A_2) =& \psi(t) \int_0^t e^{i(t-s)\xi^3} \varphi_0(\xi) i\xi \int_{\mathbb{R}^2} e^{is(\tau_1+\tau_2)} \\ & \times \int_{\xi=\xi_1+\xi_2} u_{k_1}(\xi_1, \tau_1) v_{k_2}(\xi_2, \tau_2)\, d\tau_1 d\tau_2 ds\end{aligned}$$

where

$$\begin{aligned}u_{k_1}(\xi_1, \tau_1) =& \varphi_{k_1}(\xi_1) 1_{\{|\tau_1 - \xi_1^3| \ll 3\cdot 2^{2k_1}|\xi|\}} \widehat{u}(\xi_1, \tau_1), \\ v_{k_2}(\xi_2, \tau_2) =& \varphi_{k_2}(\xi_2) 1_{\{|\tau_2 - \xi_2^3| \ll 3\cdot 2^{2k_1}|\xi|\}} \widehat{v}(\xi_2, \tau_2).\end{aligned}$$

By change of variable $\tau_1' = \tau_1 - \xi_1^3$, $\tau_2' = \tau_2 - \xi_2^3$, we get

$$\begin{aligned}\mathscr{F}_t^{-1}(A_2) =& \psi(t) e^{it\xi^3} \varphi_0(\xi) i\xi \int_0^t e^{-is\xi^3} \int_{\mathbb{R}^2} e^{is(\tau_1+\tau_2)} \\ & \times \int_{\xi=\xi_1+\xi_2} e^{is\xi_1^3} u_{k_1}(\xi_1, \tau_1 + \xi_1^3) e^{is\xi_2^3} v_{k_2}(\xi_2, \tau_2 + \xi_2^3)\, d\tau_1 d\tau_2 ds.\end{aligned}$$

By interchange of the integral we get it is equal to

$$\begin{aligned}\psi(t) e^{it\xi^3} \varphi_0(\xi) \xi \int_{\mathbb{R}^2} e^{it(\tau_1+\tau_2)} \int_{\xi=\xi_1+\xi_2} & \frac{e^{it(\xi_1^3+\xi_2^3-\xi^3)} - e^{-it(\tau_1+\tau_2)}}{\tau_1+\tau_2-\xi^3+\xi_1^3+\xi_2^3} \\ & \times u_{k_1}(\xi_1, \tau_1+\xi_1^3) v_{k_2}(\xi_2, \tau_2+\xi_2^3)\, d\tau_1 d\tau_2 \\ & := \mathscr{F}_t^{-1}(II_1) - \mathscr{F}_t^{-1}(II_2).\end{aligned}$$

For the term II_2,

$$\begin{aligned}& \mathscr{F}_t^{-1}(II_2) \\ =& \int_{\mathbb{R}^2} \psi(t) e^{it\xi^3} \varphi_0(\xi) \xi \int_{\xi=\xi_1+\xi_2} \frac{u_{k_1}(\xi_1, \tau_1+\xi_1^3) v_{k_2}(\xi_2, \tau_2+\xi_2^3)}{\tau_1+\tau_2-\xi^3+\xi_1^3+\xi_2^3}\, d\tau_1 d\tau_2.\end{aligned}$$

Since in the integral area $|\tau_1+\tau_2-\xi^3+\xi_1^3+\xi_2^3| \sim |\xi\xi_1\xi_2|$, then from Theorem 5.2 we get

$$\|\mathscr{F}^{-1}(II_2)\|_{L_x^2 L_t^\infty}$$

$$\lesssim \int_{\mathbb{R}^2} \left\| \int_{\xi=\xi_1+\xi_2} \xi \frac{u_{k_1}(\xi_1, \tau_1+\xi_1^3) v_{k_2}(\xi_2, \tau_2+\xi_2^3)}{\tau_1+\tau_2-\xi^3+\xi_1^3+\xi_2^3} \right\|_{L^2_\xi} d\tau_1 d\tau_2$$
$$\lesssim 2^{-\frac{3k_1}{2}} \|\widehat{\Delta_{k_1} u}\|_{X_{k_1}} \|\widehat{\Delta_{k_2} u}\|_{X_{k_2}}.$$

To prove 5.11, it suffices to prove

$$\|\mathscr{F}^{-1}(II_1)\|_{L^2_x L^\infty_t} \lesssim 2^{-3k_1/2} \|\widehat{P_{k_1} u}\|_{X_{k_1}} \|\widehat{P_{k_2} u}\|_{X_{k_2}}.$$

Compare II_1 with II_1' defined as following:

$$\begin{aligned}\mathscr{F}_t^{-1}(II_1') =& \psi(t) e^{it\xi^3} \varphi_0(\xi) \xi \int_{\mathbb{R}^2} e^{it(\tau_1+\tau_2)} \int_{\xi=\xi_1+\xi_2} \frac{e^{it(\xi_1^3+\xi_2^3-\xi^3)}}{-\xi^3+\xi_1^3+\xi_2^3} \\ & \times \; u_{k_1}(\xi_1, \tau_1+\xi_1^3) v_{k_2}(\xi_2, \tau_2+\xi_2^3) \; d\tau_1 d\tau_2.\end{aligned}$$

For II_1' we have

$$\begin{aligned}\mathscr{F}_t^{-1}(II_1') =& \int_{\mathbb{R}^2} \psi(t) \varphi_0(\xi) e^{it(\tau_1+\tau_2)} 1_{\{|\xi| \gg |\tau_1| 2^{-2k_1}\}} 1_{\{|\xi| \gg |\tau_2| 2^{-2k_1}\}} \\ & \times \int_{\xi=\xi_1+\xi_2} \frac{e^{it(\xi_1^3+\xi_2^3)}}{-3\xi_1\xi_2} \mathscr{F}(f_{\tau_1})(\xi_1) \mathscr{F}(g_{\tau_2})(\xi_2) \; d\tau_1 d\tau_2,\end{aligned}$$

where for $\tau_1, \tau_2 \in \mathbb{R}$ we denote

$$\mathscr{F}(f_{\tau_1})(\xi) = \widehat{\Delta_{k_1} u}(\xi, \tau_1+\xi^3), \;\; \mathscr{F}(g_{\tau_2})(\xi) = \widehat{\Delta_{k_2} v}(\xi, \tau_2+\xi^3).$$

Since (We may need a smooth cut-off $1_{\{|\xi| \gg \lambda\}}$) $\forall \, \lambda > 0$

$$\|\mathscr{F}_x^{-1} 1_{\{|\xi| \gg \lambda\}} \mathscr{F}_x u\|_{L^2_x L^\infty_t} \lesssim \|u\|_{L^2_x L^\infty_t},$$

From Theorem 5.2 we get

$$\begin{aligned}\|\mathscr{F}^{-1}(II_1')\|_{L^2_x L^\infty_t} \lesssim & \int_{\mathbb{R}^2} \|S(t)\partial_x^{-1} f_{\tau_1} S(t) \partial_x^{-1} f_{\tau_2}\|_{L^2_x L^\infty_t} d\tau_1 d\tau_2 \\ \lesssim & \int_{\mathbb{R}^2} \|S(t)\partial_x^{-1} f_{\tau_1})\|_{L^4_x L^\infty_t} \|S(t)\partial_x^{-1} f_{\tau_2}\|_{L^4_x L^\infty_t} d\tau_1 d\tau_2 \\ \lesssim & 2^{-\frac{3k_1}{2}} \|\widehat{\Delta_{k_1} u}\|_{X_{k_1}} \|\widehat{\Delta_{k_2} u}\|_{X_{k_2}},\end{aligned}$$

which gives the bound of II_1'.

To finish the proof of Proposition 5.11, it remains to prove

$$\|\mathscr{F}^{-1}(II_1 - II_1')\|_{L^2_x L^\infty_t} \lesssim 2^{-3k_1/2} \|\widehat{P_{k_1} u}\|_{X_{k_1}} \|\widehat{P_{k_2} u}\|_{X_{k_2}}.$$

Since in the integral area $|\tau_i| \ll 2^{2k_1}|\xi|$, $i = 1, 2$, thus in the hyperplane

$$\frac{1}{\tau_1+\tau_2-\xi^3+\xi_1^3+\xi_2^3} - \frac{1}{-\xi^3+\xi_1^3+\xi_2^3} = \sum_{n=1}^{\infty} \frac{1}{3\xi\xi_1\xi_2} \left(\frac{\tau_1+\tau_2}{3\xi\xi_1\xi_2}\right)^n.$$

then we get

$$\begin{aligned}
&\mathscr{F}_t^{-1}(II_1 - II_1') \\
&= \psi(t)\varphi_0(\xi)\xi \int_{\mathbb{R}^2} e^{it(\tau_1+\tau_2)} \int_{\xi=\xi_1+\xi_2} \sum_{n=1}^{\infty} \frac{1}{3\xi\xi_1\xi_2}\left(\frac{\tau_1+\tau_2}{3\xi\xi_1\xi_2}\right)^n \\
&\quad \times e^{it(\xi_1^3+\xi_2^3)} u_{k_1}(\xi_1, \tau_1+\xi_1^3) v_{k_2}(\xi_2, \tau_2+\xi_2^3)\, d\tau_1 d\tau_2.
\end{aligned}$$

Decomposing dyadically on the low frequency, $\{\chi_k(\xi)\}_{k=-\infty}^{\infty}$ denote homogeneous decomposition, we get

$$\begin{aligned}
&\mathscr{F}_t^{-1}(II_1 - II_1') \\
&= \sum_{n=1}^{\infty} \int_{\mathbb{R}^2} e^{it(\tau_1+\tau_2)} \sum_{2^{k_3} \gg 2^{-2k_1}\max(|\tau_1|,|\tau_2|)} \psi(t) e^{it(\xi_1^3+\xi_2^3)} \chi_{k_3}(\xi) \\
&\quad \times \int_{\xi=\xi_1+\xi_2} \left(\frac{\tau_1+\tau_2}{3\xi\xi_1\xi_2}\right)^n \frac{u_{k_1}(\xi_1, \tau_1+\xi_1^3) v_{k_2}(\xi_2, \tau_2+\xi_2^3)}{3\xi_1\xi_2}\, d\tau_1 d\tau_2.
\end{aligned}$$

We rewrite it into the following form

$$\begin{aligned}
&\mathscr{F}_t^{-1}(II_1 - II_1') \\
&= \sum_{n=1}^{\infty} \int_{\mathbb{R}^2} e^{it(\tau_1+\tau_2)} \sum_{2^{k_3} \gg 2^{-2k_1}\max(|\tau_1|,|\tau_2|)} \psi(t) e^{it(\xi_1^3+\xi_2^3)} \chi_{k_3}(\xi)(\xi/2^{k_3})^{-n} \\
&\quad \times 2^{-nk_3} \int_{\xi=\xi_1+\xi_2} \left(\frac{\tau_1+\tau_2}{3\xi_1\xi_2}\right)^n \frac{u_{k_1}(\xi_1, \tau_1+\xi_1^3) v_{k_2}(\xi_2, \tau_2+\xi_2^3)}{3\xi_1\xi_2}\, d\tau_1 d\tau_2.
\end{aligned}$$

Since $\chi_{k_3}(\xi)(\xi/2^{k_3})^{-n}$ is the multiplier for the $L_x^2 L_t^\infty$, then as for II_1' we get

$$\begin{aligned}
&\|\mathscr{F}^{-1}(II_1 - II_1')\|_{L_x^2 L_t^\infty} \\
&\lesssim \sum_{n=1}^{\infty} \int_{\mathbb{R}^2} \sum_{2^{k_3} \gg 2^{-2k_1}\max(|\tau_1|,|\tau_2|)} C^n |\tau_1+\tau_2|^n 2^{-nk_3} \\
&\quad \times \left\| \psi(t) e^{it(\xi_1^3+\xi_2^3)} \int_{\xi=\xi_1+\xi_2} \frac{\mathscr{F}(f_{\tau_1})(\xi_1)\mathscr{F}(g_{\tau_2})(\xi_2)}{3\xi_1^{n+1}\xi_2^{n+1}} \right\|_{L_x^2 L_t^\infty} d\tau_1 d\tau_2.
\end{aligned}$$

By Theorem 5.2 and summing on k_3 we get for some $M \gg 1$

$$\begin{aligned}
&\|\mathscr{F}^{-1}(II_1 - II_1')\|_{L_x^2 L_t^\infty} \\
&\lesssim \sum_{n=1}^{\infty} \int_{\mathbb{R}^2} \sum_{2^{k_3} \gg 2^{-2k_1}\max(|\tau_1|,|\tau_2|)} C^n |\tau_1+\tau_2|^n 2^{-nk_3} 2^{-2nk_1} \\
&\quad \times 2^{-3k_1/2} \|\mathscr{F}(f_{\tau_1})\|_{L^2} \|\mathscr{F}(g_{\tau_2})\|_{L^2} d\tau_1 d\tau_2
\end{aligned}$$

$$\lesssim \sum_{n=1}^{\infty}(C/M)^n \int_{\mathbb{R}^2} 2^{-3k_1/2}\|\mathscr{F}(f_{\tau_1})\|_{L^2}\|\mathscr{F}(g_{\tau_2})\|_{L^2}d\tau_1 d\tau_2$$
$$\lesssim 2^{-3k_1/2}\|\widehat{P_{k_1}u}\|_{X_{k_1}}\|\widehat{P_{k_2}u}\|_{X_{k_2}}.$$

Therefore, we complete the proof of the proposition. □

Inspired by the proposition, then we may use the structure $L_x^2 L_t^\infty$ for the low frequency part. It remains to control the high-low interaction.

Proposition 5.12. *If $k \geqslant 10$, $|k-k_2| \leqslant 5$, then for any $u, v \in \bar{F}^0$*

$$\|\langle\tau-\xi^3\rangle^{-1}\varphi_k(\xi)\xi(\widehat{\triangle_0 u} * \widehat{\triangle_{k_2} v})\|_{X_k} \lesssim \|\triangle_0 u\|_{L_x^2 L_t^\infty}\|\widehat{\triangle_{k_2} v}\|_{X_{k_2}}. \quad (5.73)$$

Proof. We may assume $k=k_2$. From the definition of X_k we get

$$\begin{aligned}&\|\langle\tau-\xi^3\rangle^{-1}\varphi_k(\xi)\xi(\widehat{\triangle_0 u} * \widehat{\triangle_k v})\|_{X_k}\\ &\lesssim 2^k \sum_{j\geqslant 0} 2^{-j/2}\|\widehat{\triangle_0 u} * \widehat{\triangle_{k_2} v}\|_{L^2_{\xi,\tau}}.\end{aligned} \quad (5.74)$$

It follows from Plancherel's equality and Proposition 5.4 that

$$2^k\|\widehat{\triangle_0 u} * \widehat{\triangle_{k_2} v}\|_{L^2_{\xi,\tau}} \lesssim 2^k\|\triangle_0 u\|_{L_x^2 L_t^\infty}\|\triangle_k u\|_{L_x^\infty L_t^2} \lesssim \|\triangle_0 u\|_{L_x^2 L_t^\infty}\|\widehat{\triangle_k v}\|_{X_k},$$

thus, the proposition is proved. □

Next we prove a crucial bilinear estimate which is key to prove Theorem 5.7. For $u, v \in \bar{F}^s$ we define bilinear operator

$$B(u,v) = \psi(t/4)\int_0^t S(t-\tau)\partial_x\big(\psi^2(\tau)u(\tau)\cdot v(\tau)\big)d\tau. \quad (5.75)$$

Consider the following integral equation

$$u = \psi(t)S(t)u_0 + \psi(t/4)\int_0^t S(t-\tau)\partial_x\big(\psi^2(\tau)u(\tau)\cdot u(\tau)\big)d\tau. \quad (5.76)$$

In order to apply the contraction principle, the rest task is to show the boundedness $B: \bar{F}^s \times \bar{F}^s \to \bar{F}^s$.

Proposition 5.13. *Assume $-3/4 \leqslant s \leqslant 0$. Then there exists $C>0$ such that*

$$\|B(u,v)\|_{\bar{F}^s} \leqslant C(\|u\|_{\bar{F}^s}\|v\|_{\bar{F}^{-3/4}} + \|u\|_{\bar{F}^{-3/4}}\|v\|_{\bar{F}^s}) \quad (5.77)$$

hold for all $u,\ v \in \bar{F}^s$.

Proof. From the definition of F^s we have

$$\|B(u,v)\|_{\bar{F}^s}^2 = \|\triangle_0 B(u,v)\|_{\bar{X}_0}^2 + \sum_{k_1\geqslant 1} 2^{2k_1 s}\|\varphi_{k_1}(\xi)\mathscr{F}[B(u,v)]\|_{X_{k_1}}^2. \quad (5.78)$$

First we consider the second term on the right-hand side of (5.78). By applying Littlewood-Paley decomposition to u, v we get

$$\|\varphi_{k_1}(\xi)\mathscr{F}[B(u,v)]\|X_{k_1} \lesssim \sum_{k_2,k_3\geqslant 0} \|\varphi_{k_1}(\xi)\mathscr{F}[B(\triangle_{k_2}(u),\triangle_{k_3}(v))]\|_{X_{k_1}}. \quad (5.79)$$

It follows from Proposition 5.3 (b) that the right-hand side of (5.79) is bounded by

$$\sum_{k_2,k_3\geqslant 0} \|\langle\tau-\xi^3\rangle^{-1}\varphi_{k_1}(\xi)\xi\widehat{\psi(t)\triangle_{k_2}u} * \widehat{\psi(t)\triangle_{k_3}v})\|_{X_{k_1}}. \quad (5.80)$$

From symmetry we may assume $k_2 \leqslant k_3$ in (5.80). It suffices to prove

$$\Big(\sum_{k_1\geqslant 1} 2^{2k_1 s}\Big[\sum_{k_2,k_3\geqslant 0} \|\langle\tau-\xi^3\rangle^{-1}\varphi_{k_1}(\xi)\xi\widehat{\psi(t)\triangle_{k_2}u} * \widehat{\psi(t)\triangle_{k_3}v})\|_{X_{k_1}}\Big]^2\Big)^{1/2}$$
$$\lesssim \|u\|_{\bar{F}^{-3/4}}\|v\|_{\bar{F}^s}. \quad (5.81)$$

If $k_{max} \leqslant 20$, then from Proposition 5.8 we get (5.80) is bounded by

$$\sum_{k_{max}\leqslant 20} \|\triangle_{k_2}u\|_{L_t^\infty L_x^2}\|\triangle_{k_3}v\|_{L_t^\infty L_x^2},$$

which is sufficient to give (5.81), since in this case $\|\triangle_k u\|_{L_t^\infty L_x^2} \lesssim \|\triangle_k u\|_{X_k}$ for $k \geqslant 1$, and $\|\triangle_k u\|_{L_t^\infty L_x^2} \lesssim \|\triangle_k u\|_{\bar{X}_k}$ for $k = 0$. Now we assume $k_{max} \geqslant 20$ in (5.81). There are three cases. If $|k_1 - k_3| \leqslant 5, k_2 \leqslant k_1 - 10$, then we use Proposition 5.5 (a) for $k_2 = 0$, and (b) for $k_2 \geqslant 1$; if $|k_1 - k_3| \leqslant 5, k_1 - 9 \leqslant k_2 \leqslant k_3$, then we use Proposition 5.6; if $|k_2 - k_3| \leqslant 5, 1 \leqslant k_1 \leqslant k_2 - 5$, then we use Proposition 5.9 (b). Therefore, we get (5.81).

It remains to prove

$$\|B(u,v)\|_{\bar{X}_0} \leqslant C(\|u\|_{\bar{F}^s}\|v\|_{\bar{F}^{-3/4}} + \|u\|_{\bar{F}^{-3/4}}\|v\|_{\bar{F}^s}). \quad (5.82)$$

By applying Littlewood-Paley decomposition to u, v we get

$$\|B(u,v)\|_{\bar{X}_0} \leqslant \sum_{k_2,k_3\geqslant 0} \|B(\triangle_{k_2}u,\triangle_{k_3}v)\|_{\bar{X}_0}.$$

If $\max(k_2,k_3) \leqslant 10$, then from Proposition 5.14 and Proposition 5.8 we get

$$\|B(\triangle_{k_2}u,\triangle_{k_3}v)\|_{\bar{X}_0} \lesssim \|\triangle_{k_2}u\|_{L_t^\infty L_x^2}\|\triangle_{k_3}v\|_{L_t^\infty L_x^2},$$

which is sufficient to give (5.82). If $\max(k_2, k_3) \geqslant 10$, then $|k_2 - k_3| \leqslant 5$. It follows from Proposition 5.11 that

$$\begin{aligned}\|B(u,v)\|_{\bar{X}_0} \leqslant & \sum_{|k_2-k_3|\leqslant 5,\ k_2,k_3\geqslant 10} 2^{-3k_2/2}\|\mathscr{F}(\triangle_{k_2}u)\|_{X_{k_2}}\|\mathscr{F}(\triangle_{k_3}v)\|_{X_{k_3}} \\ \lesssim & \|u\|_{\bar{F}^{-3/4}}\|v\|_{\bar{F}^{-3/4}}\end{aligned}$$

which gives (5.82). Therefore, the proposition is proved. □

Similarly, as the proof of 5.3 (a), it is easy to prove the following proposition by Lemma 5.4.

Proposition 5.14. *Assume $s \in \mathbb{R}$ and $\phi \in H^s$. Then there exists $C > 0$ such that*

$$\|\psi(t)S(t)\phi\|_{\bar{F}^s} \leqslant C\|\phi\|_{H^s}.$$

To prove Theorem 5.7, we need the bilinear form contraction principle which is easy to prove. The proof is left to the readers as an exercise.

Lemma 5.8. *Assume $(X, \|\cdot\|)$ is a Banach space with the norm $\|\cdot\|$. Let $B : X \times X \to X$ be a bilinear operator satisfying*

$$\|B(x_1,x_2)\| \leqslant \eta\|x_1\|\|x_2\|, \quad \forall\ x_1, x_2 \in X.$$

Then for all $y \in X$ with $4\eta\|y\| < 1$, the equation $x = y + B(x,x)$ has unique solution $x \in X$ such that $\|x\| < \frac{1}{2\eta}$.

Using Lemma 5.8, Proposition 5.13 and Proposition 5.14, we can prove Theorem 5.7.

5.5 I-method

In the last two sections, we studied the local well-posedness for the KdV equation. In this section, we study the global well-posedness, namely extend the local solution to a global one. KdV equation (5.32) is completely integrable, and hence has infinite conservation laws. With these conservation laws, one can derive some a priori estimates of the solution. For example, if u is a smooth solution to the KdV equation (5.32), then for $k \in N \cup \{0\}$ and $T > 0$

$$\|u(t)\|_{H^k} \leqslant C(T, \|u_0\|_{H^k}),\ \forall\ t \in [-T, T]. \tag{5.83}$$

With these a priori estimates and local well-posedness, one can easily obtain the following: KdV equation is globally well-posed in H^k, where $k \in N \cup$

$\{0\}$. In view of the local well-posedness results, it is a natural question: what about global well-posedness in H^s for $s \in [-3/4, 0)$? For the regularity below L^2, there is no conservation law, and hence the global well-posedness do not hold automatically. *I-method* is an effective method to study this kind of problem. It was introduced by J. Colliander, M. Keel, G. Staffilani, H. Takaoka, and T. Tao [42], inspired by Bourgain's frequency decomposition techniques [17].

We explain briefly the basic ideas of *I-method*. We recall the proof of (5.83) in the case $k = 0, 1$. Multiplying u on both sides of equation (5.32), and then integrating on x, we get

$$\frac{d}{dt}\left(\|u(t)\|_{L^2}^2\right) = 0,$$

which is (5.83) at $k = 0$. For $k = 1$, it is easy to see that

$$\frac{d}{dt}\left(\|u(t)\|_{H^1}^2\right) = 0$$

fails, the reason is the quantity $\|u(t)\|_{H^1}$ does not exploit the structure of the equation. A natural idea is to find some other quantity $Q_1(u(t))$ such that

$$Q_1(u(t)) \sim \|u(t)\|_{H^1}^2, \quad \frac{d}{dt}[Q_1(u(t))] = 0.$$

If such $Q_1(u(t))$ exists, then we get $\forall\, t \in \mathbb{R}$

$$\|u(t)\|_{H^1} \sim \|Q_1(u(t))\| = \|Q_1(u_0)\| \lesssim \|u_0\|_{H^1},$$

which is (5.83) at $k = 1$. Now we know such quantity exists, for instance, we can take

$$Q_1(u(t)) = \int_{\mathbb{R}} u_x^2 - \frac{1}{3}u^3 + Cu^2 dx.$$

Actually, obviously $\frac{d}{dt}[Q(u(t))] = 0$, it suffices to show $Q_1(u(t)) \sim \|u(t)\|_{H^1}^2$. By Gagliardo-Nirenberg's inequality we have

$$\|u\|_{L^3}^3 \lesssim \|u\|_{L^2}^{5/2} \|u_x\|_{L^2}^{1/2}.$$

Then using

$$ab \leqslant \frac{a^p}{p} + \frac{b^q}{q}, 1 < p \leqslant q < \infty, 1/p + 1/q = 1,$$

and taking $q = 4$, we get

$$\|u\|_{L^3}^3 \leqslant \frac{3\|u\|_{L^2}^{10/3}}{4} + \frac{\|u_x\|_{L^2}^2}{4} \leqslant \frac{3\|u_0\|_{L^2}^{4/3}\|u\|_{L^2}^2}{4} + \frac{\|u_x\|_{L^2}^2}{4}.$$

Finally take $C = \|u_0\|_{L^2}^{4/3}$.

From this, for general H^s norm, we also need to find quantity $Q_s(u(t))$, they have similar properties as $Q_1(u(t))$. However, since such quantity does not exist for $s < 0$, then we relax the condition to $Q_s(u(t))$ increase much slower than the norm $\|u\|_{H^s}$ during the evolution. The issues that how to construct $Q_s(u(t))$ and control the increasing rate are systematically studied in 'I-method'.

Given $m : \mathbb{R}^k \to \mathbb{C}$, m is said to be symmetric if

$$m(\xi_1, \cdots, \xi_k) = m(\sigma(\xi_1, \cdots, \xi_k))$$

for all $\sigma \in S_k$, where S_k is the permutation group for k elements. The symmetrization of m is defined by

$$[m]_{sym}(\xi_1, \xi_2, \cdots, \xi_k) = \frac{1}{k!} \sum_{\sigma \in S_k} m(\sigma(\xi_1, \xi_2, \cdots, \xi_k)).$$

For each m, we define a k-linear functional acting on k functions $u_1, \cdots, u_k$(m is said to be k-multiplier) as following

$$\Lambda_k(m; u_1, \cdots, u_k) = \int_{\xi_1 + \cdots + \xi_k = 0} m(\xi_1, \cdots, \xi_k) \widehat{u_1}(\xi_1) \cdots \widehat{u_k}(\xi_k).$$

Usually we apply Λ_k on k functions which are all u. For convenience, $\Lambda_k(m; u, \cdots, u)$ is simply denoted by $\Lambda_k(m)$. From symmetries we see $\Lambda_k(m) = \Lambda_k([m]_{sym})$. By using the KdV equation (5.32) we can get the following proposition.

Proposition 5.15. *Assume u satisfies* (5.32) *and m is a symmetric function. Then*

$$\frac{d\Lambda_k(m)}{dt} = \Lambda_k(mh_k) - \frac{ik}{2} \Lambda_{k+1}\big(m(\xi_1, \cdots, \xi_{k-1}, \xi_k + \xi_{k+1})(\xi_k + \xi_{k+1})\big),$$

where

$$h_k = i(\xi_1^3 + \xi_2^3 + \cdots + \xi_k^3).$$

Next we introduce the I-operator. Assume $m : \mathbb{R} \to \mathbb{R}$ is a real-valued even function, then we define Fourier multiplier operator Iu as following

$$\widehat{Iu}(\xi) = m(\xi)\widehat{u}. \tag{5.84}$$

Define the modified energy $E_I^2(t)$

$$E_I^2(t) = \|Iu\|_{L^2}^2.$$

Since m is real-valued even function and u is real-valued, then by Plancherel's equality we get

$$E_I^2(t) = \Lambda_2(m(\xi_1)m(\xi_2)).$$

It follows from Proposition 5.15 that

$$\frac{d}{dt}E_I^2(t) = \Lambda_2(m(\xi_1)m(\xi_2)h_2) - i\Lambda_3(m(\xi_1)m(\xi_2+\xi_3)(\xi_2+\xi_3)).$$

Since $\xi_1^3+\xi_2^3=0$ for $\xi_1+\xi_2=0$, then the first term vanishes. Symmetrizing the second term we get

$$\frac{d}{dt}E_I^2(t) = \Lambda_3(-i[m(\xi_1)m(\xi_2+\xi_3)(\xi_2+\xi_3)]_{sym}).$$

Denote

$$M_3(\xi_1,\xi_2,\xi_3) = -i[m(\xi_1)m(\xi_2+\xi_3)(\xi_2+\xi_3)]_{sym}.$$

Define a new modified energy

$$E_I^3(t) = E_I^2(t) + \Lambda_3(\sigma_3),$$

where symmetric function σ_3 will be set momentarily. The role of σ_3 is to make a cancelation. It follows from Proposition 5.15 that

$$\frac{d}{dt}E_I^3(t) = \Lambda_3(M_3) + \Lambda_3(\sigma_3 h_3) - \frac{3}{2}i\Lambda_4(\sigma_3(\xi_1,\xi_2,\xi_3+\xi_4)(\xi_3+\xi_4)). \tag{5.85}$$

Take $\sigma_3 = -M_3/h_3$ such that the two trilinear terms in (5.85) cancels. Denote

$$M_4(\xi_1,\xi_2,\xi_3,\xi_4) = -i\frac{3}{2}[\sigma_3(\xi_1,\xi_2,\xi_3+\xi_4)(\xi_3+\xi_4)]_{sym},$$

then we have

$$\frac{d}{dt}E_I^3(t) = \Lambda_4(M_4).$$

Similarly we define a new modified energy

$$E_I^4(t) = E_I^3(t) + \Lambda_4(\sigma_4),$$

where

$$\sigma_4 = -\frac{M_4}{h_4}.$$

It is easy to get

$$\frac{d}{dt}E_I^4(t) = \Lambda_5(M_5)$$

where

$$M_5(\xi_1,\ldots,\xi_5) = -2i[\sigma_4(\xi_1,\xi_2,\xi_3,\xi_4+\xi_5)(\xi_4+\xi_5)]_{sym}.$$

This procedure can be made for a finite times, but we stop here since it suffices for our purposes.

In order to control the norm $\|u\|_{H^s}$, it seems to convenient to take $m(\xi) = \langle\xi\rangle^s$. However, it is difficult to control its increase. Basing on the L^2 conservation, we take m as following: m is a smooth even function which has the form

$$m(\xi) = \begin{cases} 1, & |\xi| < N, \\ N^{-s}|\xi|^s, & |\xi| > 2N. \end{cases} \tag{5.86}$$

It is easy to see that if $N \to \infty$ then $Iu \to u$, and hence the increase rate of the modified energy $E_I^k(t)$ tends to 0 as $N \to \infty$. The basic question is how fast it tends to 0. If m is of the form (5.86), then it is easy to see m^2 satisfies

$$\begin{cases} m^2(\xi) \sim m^2(\xi') \ \ |\xi| \sim |\xi'|, \\ (m^2)'(\xi) = O(\frac{m^2(\xi)}{|\xi|}), \\ (m^2)''(\xi) = O(\frac{m^2(\xi)}{|\xi|^2}). \end{cases} \tag{5.87}$$

Next we need to estimate the multipliers M_3, M_4, M_5. We will use two mean value formulas: if $|\eta|, |\lambda| \ll |\xi|$

$$|a(\xi+\eta) - a(\xi)| \lesssim |\eta| \sup_{|\xi'|\sim|\xi|} |a'(\xi')|, \tag{5.88}$$

and

$$|a(\xi+\eta+\lambda) - a(\xi+\eta) - a(\xi+\lambda) + a(\xi)| \lesssim |\eta||\lambda| \sup_{|\xi'|\sim|\xi|} |a''(\xi')|. \tag{5.89}$$

First we give the estimate of M_3.

Proposition 5.16. *Let m be given by* (5.86). *In the set* $\{\xi_1+\xi_2+\xi_3 = 0, |\xi_i| \sim N_i\}$ *where N_i is dyadic, we have*

$$|M_3(\xi_1,\xi_2,\xi_3)| \lesssim \max(m^2(\xi_1), m^2(\xi_2), m^2(\xi_3)) \min(N_1, N_2, N_3).$$

Proof. From symmetry we may assume $N_1 = N_2 \geqslant N_3$. If $N_3 \gtrsim N_1$, then it follows directly from the definition of m. If $N_3 \ll N_1$, from the mean-value formula we get

$$m^2(\xi_1)\xi_1 - m^2(\xi_1+\xi_3)(\xi_1+\xi_3) + m^2(\xi_3)\xi_3 \leqslant \max(m^2(\xi_1), m^2(\xi_2))N_3.$$

Thus the proposition is proved. □

To estimate M_4, first we extend the multiplier σ_3 from the hyperplane to the entire space.

Proposition 5.17. *If m is of the form (5.86), then for any dyadic numbers $\lambda \leqslant \mu$ there exists an extension of σ_3 from $\{\xi_1+\xi_2+\xi_3=0, |\xi_1| \sim \lambda,\ |\xi_2|,\ |\xi_3| \sim \mu\}$ to the entire space $\{(\xi_1,\xi_2,\xi_3) \in \mathbb{R}^3, |\xi_1| \sim \lambda,\ |\xi_2|,\ |\xi_3| \sim \mu\}$ such that*

$$|\partial_1^{\beta_1}\partial_2^{\beta_2}\partial_3^{\beta_3}\sigma_3(\xi_1,\xi_2,\xi_3)| \leqslant Cm^2(\lambda)\mu^{-2}\lambda^{-\beta_1}\mu^{-\beta_2-\beta_3}, \tag{5.90}$$

where the constant C is independent of λ, μ.

Proof. Since on the hyperplane $\{(\xi_1,\xi_2,\xi_3) : \xi_1+\xi_2+\xi_3=0\}$ we have

$$h_3 = i(\xi_1^3+\xi_2^3+\xi_3^3) = 3i\xi_1\xi_2\xi_3,$$

then $|h| \sim \lambda\mu^2$. From

$$\begin{aligned} M_3(\xi_1,\xi_2,\xi_3) =& -i[m(\xi_1)m(\xi_2+\xi_3)(\xi_2+\xi_3)]_{sym} \\ =& i(m^2(\xi_1)\xi_1+m^2(\xi_2)\xi_2+m^2(\xi_3)\xi_3), \end{aligned}$$

we will discuss case by case. If $\lambda \sim \mu$, then we extend σ_3 as following

$$\sigma_3(\xi_1,\xi_2,\xi_3) = -\frac{i(m^2(\xi_1)\xi_1+m^2(\xi_2)\xi_2+m^2(\xi_3)\xi_3)}{3i\xi_1\xi_2\xi_3},$$

which suffices for the purpose. If $\lambda \ll \mu$, then we extend σ_3 as following

$$\sigma_3(\xi_1,\xi_2,\xi_3) = -\frac{i(m^2(\xi_1)\xi_1+m^2(\xi_2)\xi_2-m^2(\xi_1+\xi_2)(\xi_1+\xi_2))}{3i\xi_1\xi_2\xi_3}.$$

Then (5.90) follows from (5.88) and (5.87). □

Next we give the estimate of σ_4. The proof here was due to [96]. Compared to the proof given in [42], the proof here is more flexible to a general class of dispersive equations. But the ideas are the same.

Proposition 5.18. *Let m be of the form (5.86). Then on the area $|\xi_i| \sim N_i, |\xi_j+\xi_k| \sim N_{jk}$, where N_i, N_{jk} are dyadic numbers, we have*

$$\frac{|M_4(\xi_1,\xi_2,\xi_3,\xi_4)|}{|h_4|} \lesssim \frac{m^2(\min(N_i,N_{jk}))}{(N+N_1)(N+N_2)(N+N_3)(N+N_4)}. \tag{5.91}$$

Proof. From symmetry we may assume $N_1 \geqslant N_2 \geqslant N_3 \geqslant N_4$. Since $\xi_1+\xi_2+\xi_3+\xi_4=0$, then $N_1 \sim N_2$. We may also assume $N_1 \sim N_2 \gtrsim N$, otherwise M_4 vanishes, since $m^2(\xi)=1$ if $|\xi| \leqslant N$, and if $\max(N_{12},N_{13},N_{14}) \ll N_1$, then $\xi_3 \approx -\xi_1$, $\xi_4 \approx -\xi_1$, which contradicts to $\xi_1+\xi_2+\xi_3+\xi_4=0$. The right-hand side of (5.91) is actually

$$\frac{m^2(\min(N_i, N_{jk}))}{{N_1}^2(N+N_3)(N+N_4)}. \tag{5.92}$$

Since on the hyperplane $\xi_1+\xi_2+\xi_3+\xi_4=0$

$$h_4 = \xi_1^3+\xi_2^3+\xi_3^3+\xi_4^3 = 3(\xi_1+\xi_2)(\xi_1+\xi_3)(\xi_1+\xi_4),$$

thus we get

$$\begin{aligned}
&CM_4(\xi_1,\xi_2,\xi_3,\xi_4)\\
&= [\sigma_3(\xi_1,\xi_2,\xi_3+\xi_4)(\xi_3+\xi_4)]_{sym}\\
&= \sigma_3(\xi_1,\xi_2,\xi_3+\xi_4)(\xi_3+\xi_4)+\sigma_3(\xi_1,\xi_3,\xi_2+\xi_4)(\xi_2+\xi_4)\\
&\quad +\sigma_3(\xi_1,\xi_4,\xi_2+\xi_3)(\xi_2+\xi_3)+\sigma_3(\xi_2,\xi_3,\xi_1+\xi_4)(\xi_1+\xi_4)\\
&\quad +\sigma_3(\xi_2,\xi_4,\xi_1+\xi_3)(\xi_1+\xi_3)+\sigma_3(\xi_3,\xi_4,\xi_1+\xi_2)(\xi_1+\xi_2)\\
&= [\sigma_3(\xi_1,\xi_2,\xi_3+\xi_4)-\sigma_3(-\xi_3,-\xi_4,\xi_3+\xi_4)](\xi_3+\xi_4)\\
&\quad +[\sigma_3(\xi_1,\xi_3,\xi_2+\xi_4)-\sigma_3(-\xi_2,-\xi_4,\xi_2+\xi_4)](\xi_2+\xi_4)\\
&\quad +[\sigma_3(\xi_1,\xi_4,\xi_2+\xi_3)-\sigma_3(-\xi_2,-\xi_3,\xi_2+\xi_3)](\xi_2+\xi_3)\\
&:= I+II+III.
\end{aligned} \tag{5.93}$$

We will get (5.91) by studying different cases.

Case 1. $|N_4| \gtrsim \frac{N}{2}$.

Case 1a. $N_{12}, N_{13}, N_{14} \gtrsim N_1$.

For this case, it follows directly from (5.90) that

$$\frac{|M_4(\xi_1,\xi_2,\xi_3,\xi_4)|}{|h_4|} \lesssim \frac{m^2(N_4)}{N_1N_2N_3N_4}. \tag{5.94}$$

Case 1b. $N_{12} \ll N_1$, $N_{13} \gtrsim N_1$, $N_{14} \gtrsim N_1$.

First we consider the contribution of I. From (5.90) we get

$$\frac{|I|}{|h_4|} \lesssim \frac{m^2(\min(N_4, N_{12}))}{N_1N_2N_3N_4}. \tag{5.95}$$

For the contribution of II, if $N_{12} \gtrsim N_3$, then we use (5.88) and (5.90), else if $N_{12} \ll N_3$, then we use (5.88)(5.90). Thus we get

$$\frac{II}{h_4} \lesssim \frac{m^2(N_4)}{N_1N_1N_1N_3}. \tag{5.96}$$

From symmetry, the contribution of III is identical to that of II.

Case 1c. $N_{12} \ll N_1$, $N_{13} \ll N_1$, $N_{14} \gtrsim N_1$.

Since $N_{12} \ll N_1$, $N_{13} \ll N_1$, then $N_1 \sim N_2 \sim N_3 \sim N_4$.

First we consider the contribution of I. Since

$$\begin{aligned}I =&[\sigma_3(\xi_1,\xi_2,\xi_3+\xi_4)-\sigma_3(-\xi_3,\xi_2,\xi_3+\xi_4)](\xi_3+\xi_4)\\&+[\sigma_3(-\xi_3,\xi_2,\xi_3+\xi_4)-\sigma_3(-\xi_3,-\xi_4,\xi_3+\xi_4)](\xi_3+\xi_4)\\=&I_1+I_2.\end{aligned}$$

Using (5.90) and (5.88) we get

$$\frac{I}{|h_4|}\lesssim\frac{I_1}{|h_4|}+\frac{I_2}{|h_4|}\lesssim\frac{m^2(N_{12})}{N_1^4}.$$

The contribution of II is identical to that of I.

Finally we consider III. Since

$$\begin{aligned}III =&1/2[\sigma_3(\xi_1,\xi_4,\xi_2+\xi_3)-\sigma_3(-\xi_2,-\xi_3,\xi_2+\xi_3)\\&-\sigma_3(-\xi_3,-\xi_2,\xi_2+\xi_3)+\sigma_3(\xi_4,\xi_1,\xi_2+\xi_3)](\xi_2+\xi_3),\end{aligned}$$

then using (5.89) four times we get

$$\frac{III}{|h_4|}\lesssim\frac{m^2(N_1)}{N_1^4}.$$

Case 1d. $N_{12}\ll N_1$, $N_{13}\gtrsim N_1$, $N_{14}\ll N_1$.

This case is identical to Case 1c.

Case 2. $N_4\ll N/2$.

It is obvious that in this case $m^2(\min(N_i,N_{jk}))=1$, and $N_{13}\sim|\xi_1+\xi_3|=|\xi_2+\xi_4|\sim N_1$.

Case 2a. $N_1/4>N_{12}\gtrsim N/2$.

Since $N_4\ll N/2$ and $|\xi_3+\xi_4|=|\xi_1+\xi_2|\gtrsim N/2$, then $N_3\gtrsim N/2$. From $|h_4|\sim N_{12}N_1^2$, we control the six terms in (5.93) respectively and get

$$\frac{|M_4|}{|h_4|}\lesssim\frac{1}{N_1^2N_3N}. \tag{5.97}$$

Case 2b. $N_{12}\ll N/2$.

Since $N_{12}=N_{34}\ll N/2$ and $N_4\ll N/2$, then $N_3\ll N/2$, and $N_{13}\sim N_{14}\sim N_1$.

For the contribution of I, since $N_3,N_4,N_{34}\ll N/2$, then from the definition of m we have $\sigma_3(-\xi_3,-\xi_4,\xi_3+\xi_4)=0$. Using (5.90) we get

$$\frac{|I|}{|h_4|}\lesssim\frac{|\sigma_3(\xi_1,\xi_2,\xi_3+\xi_4)|}{N_1^2}\lesssim\frac{1}{N_1^4}. \tag{5.98}$$

Next we consider the contribution of II and III. We can not deal with the two terms separately, but need to exploit a cancelation between the two terms. Rewrite $II+III$ as following

$$II+III=[\sigma_3(\xi_1,\xi_3,\xi_2+\xi_4)-\sigma_3(-\xi_2,-\xi_4,\xi_2+\xi_4)](\xi_2+\xi_4)$$

$$
\begin{aligned}
&+[\sigma_3(\xi_1,\xi_4,\xi_2+\xi_3)-\sigma_3(-\xi_2,-\xi_3,\xi_2+\xi_3)](\xi_2+\xi_3)\\
=&[\sigma_3(\xi_1,\xi_3,\xi_2+\xi_4)-\sigma_3(-\xi_2,-\xi_4,\xi_2+\xi_4)]\xi_4\\
&+[\sigma_3(\xi_1,\xi_4,\xi_2+\xi_3)-\sigma_3(-\xi_2,-\xi_3,\xi_2+\xi_3)]\xi_3\\
&+[\sigma_3(\xi_1,\xi_3,\xi_2+\xi_4)-\sigma_3(-\xi_2,-\xi_4,\xi_2+\xi_4)\\
&\quad+\sigma_3(\xi_1,\xi_4,\xi_2+\xi_3)-\sigma_3(-\xi_2,-\xi_3,\xi_2+\xi_3)]\xi_2\\
=&J_1+J_2+J_3.
\end{aligned}
\tag{5.99}
$$

For J_1, since

$$
\frac{|J_1|}{|h_4|}\leqslant\frac{|[\sigma_3(\xi_1,\xi_3,\xi_2+\xi_4)-\sigma_3(-\xi_2,-\xi_4,\xi_2+\xi_4)]\xi_4|}{|h_4|}
\tag{5.100}
$$

if $N_{12}\ll N_3$ (in this case $N_3\sim N_4$), then use (5.88) twice, otherwise use (5.88) and (5.90), thus we get

$$
\frac{|J_1|}{|h_4|}\lesssim\frac{1}{N_1^4}.
$$

J_2 is identical to J_1. Now consider J_3. First we assume $N_{12}\gtrsim N_3$. From the symmetry of σ_3 we get

$$
\begin{aligned}
J_3=&[\sigma_3(\xi_1,\xi_3,\xi_2+\xi_4)-\sigma_3(-\xi_2-\xi_3,\xi_3,\xi_2)\\
&+\sigma_3(\xi_1,\xi_4,\xi_2+\xi_3)-\sigma_3(-\xi_2-\xi_4,\xi_4,\xi_2)]\xi_2.
\end{aligned}
$$

From (5.88) and that $N_{12}\gtrsim N_3$ we get

$$
\frac{|J_3|}{|h_4|}\lesssim\frac{1}{N_1^4}.
$$

If $N_{12}\ll N_3$, then $N_3\sim N_4$. Rewrite J_3 as following

$$
\begin{aligned}
J_3=&[\sigma_3(-\xi_2,\xi_3,\xi_2+\xi_4)-\sigma_3(-\xi_2,-\xi_4,\xi_2+\xi_4)\\
&+\sigma_3(\xi_1,\xi_4,\xi_2+\xi_3)-\sigma_3(\xi_1,-\xi_3,\xi_2+\xi_3)]\xi_2\\
&+[\sigma_3(\xi_1,\xi_3,\xi_2+\xi_4)-\sigma_3(-\xi_2,\xi_3,\xi_2+\xi_4)\\
&+\sigma_3(\xi_1,-\xi_3,\xi_2+\xi_3)-\sigma_3(-\xi_2,-\xi_3,\xi_2+\xi_3)]\xi_2\\
=&J_{31}+J_{32}.
\end{aligned}
$$

It follows from (5.88) that

$$
\frac{|J_{32}|}{|h_4|}\lesssim\frac{1}{N_1^4}.
\tag{5.101}
$$

It suffices to control J_{31}. Since in this case $m^2(\xi_3)=m^2(\xi_4)=1$, then we get

$$
J_{31}=[\sigma_3(-\xi_2,\xi_3,\xi_2+\xi_4)-\sigma_3(\xi_1,-\xi_3,\xi_2+\xi_3)
$$

$$
\begin{aligned}
&- \sigma_3(-\xi_2, -\xi_4, \xi_2+\xi_4) + \sigma_3(\xi_1, \xi_4, \xi_2+\xi_3)]\xi_2 \\
&= \frac{-m^2(\xi_2)\xi_2 + \xi_3 + m^2(\xi_2+\xi_4)(\xi_2+\xi_4)}{-\xi_2\xi_3(\xi_2+\xi_4)}\xi_2 \\
&\quad - \frac{-m^2(\xi_2)\xi_2 - \xi_4 + m^2(\xi_2+\xi_4)(\xi_2+\xi_4)}{\xi_2\xi_4(\xi_2+\xi_4)}\xi_2 \\
&\quad + \frac{m^2(\xi_1)\xi_1 + \xi_4 + m^2(\xi_2+\xi_3)(\xi_2+\xi_3)}{\xi_1\xi_4(\xi_2+\xi_3)}\xi_2 \\
&\quad - \frac{m^2(\xi_1)\xi_1 - \xi_3 + m^2(\xi_2+\xi_3)(\xi_2+\xi_3)}{-\xi_1\xi_3(\xi_2+\xi_3)}\xi_2.
\end{aligned}
$$

Noting that there is a cancelation, then we get

$$
\begin{aligned}
J_{31} = &- \frac{\xi_3+\xi_4}{\xi_3\xi_4}\frac{-m^2(\xi_2)\xi_2 + m^2(\xi_2+\xi_4)(\xi_2+\xi_4)}{\xi_2(\xi_2+\xi_4)}\xi_2 \\
&+ \frac{\xi_3+\xi_4}{\xi_3\xi_4}\frac{m^2(\xi_1)\xi_1 + m^2(\xi_2+\xi_3)(\xi_2+\xi_3)}{\xi_1(\xi_2+\xi_3)}\xi_2.
\end{aligned} \tag{5.102}
$$

Rewrite (5.102) as following

$$
\begin{aligned}
&- \frac{-m^2(\xi_2)\xi_2 + m^2(\xi_2+\xi_4)(\xi_2+\xi_4) + m^2(\xi_1)\xi_1 + m^2(\xi_2+\xi_3)(\xi_2+\xi_3)}{\xi_2(\xi_2+\xi_4)}\xi_2 \\
&\times \frac{\xi_3+\xi_4}{\xi_3\xi_4} + \frac{\xi_3+\xi_4}{\xi_3\xi_4}[m^2(\xi_1)\xi_1 + m^2(\xi_2+\xi_3)(\xi_2+\xi_3)] \\
&\times \left[\frac{1}{\xi_1(\xi_2+\xi_3)} + \frac{1}{\xi_2(\xi_2+\xi_4)}\right]\xi_2.
\end{aligned}
$$

Thus, for the first term we use (5.89), and (5.88) for the second term, finally we get

$$
\frac{|J_{31}|}{|h_4|} \lesssim \frac{1}{N_1^4},
$$

Therefore, the proposition is proved. □

The estimate of M_5 follows immediately from the estimate of σ_4.

Proposition 5.19. *If m is of the form* (5.86), *then*

$$
|M_5(\xi_1, \ldots, \xi_5)| \lesssim \left[\frac{m^2(N_{*45})N_{45}}{(N+N_1)(N+N_2)(N+N_3)(N+N_{45})}\right]_{sym},
$$

where

$$
N_{*45} = \min(N_1, N_2, N_3, N_{45}, N_{12}, N_{13}, N_{23}).
$$

Next we will prove two properties. One is to show the modified energy $E_I^4(t)$ increase slowly, the other is to prove it is close to $E_I^2(t)$. To control the increase rate of $E_I^4(t)$, it suffices to control its derivative

$$\frac{d}{dt}E_I^4(t) = \Lambda_5(M_5).$$

Proposition 5.20. *Assume $I \subset \mathbb{R}$ and $|I| \lesssim 1$. Let $0 \leqslant k_1 \leqslant \ldots \leqslant k_5$ and $k_4 \geqslant 10$. Then*

$$\left| \int_I \int \prod_{i=1}^{5} P_{k_i}(w_i)(x,t)dxdt \right| \lesssim 2^{\frac{5(k_1+k_2+k_3)}{12}} 2^{-k_4-k_5} \prod_{j=1}^{5} \|\widehat{\triangle_{k_j}(w_j)}\|_{X_{k_j}}, \tag{5.103}$$

where on the right-hand side if $k_j = 0$ then replace X_{k_j} by $\bar{X}_0$.

Proof. It follows from Hölder inequality that the left-hand side of (5.103) is bounded by

$$\prod_{i=1}^{3} \|\triangle_{k_i}(w_i)\|_{L_x^3 L_{t\in I}^\infty} \cdot \|\triangle_{k_4}(w_4)\|_{L_x^\infty L_t^2} \cdot \|\triangle_{k_5}(w_5)\|_{L_x^\infty L_t^2}.$$

For the term $\|\triangle_{k_4}(w_4)\|_{L_x^\infty L_t^2}$ and $\|\triangle_{k_5}(w_5)\|_{L_x^\infty L_t^2}$, we use Proposition 5.4.

For the term $\|\triangle_{k_i}(w_i)\|_{L_x^3 L_{t\in I}^\infty}$, we use interpolation between $\|\triangle_{k_i}(w_i)\|_{L_x^2 L_{t\in I}^\infty}$ and $\|\triangle_{k_i}(w_i)\|_{L_x^4 L_{t\in I}^\infty}$, then use Proposition 5.4. □

Proposition 5.21. *Assume $\delta \lesssim 1$. Let m be given by (5.86) with $s = -3/4$. Then*

$$\left| \int_0^\delta \Lambda_5(M_5; u_1, \cdots, u_5)dt \right| \lesssim N^{-15/4} \prod_{j=1}^{5} \|I(u_j)\|_{\bar{F}^0(\delta)}. \tag{5.104}$$

Proof. It follows from Proposition 5.19 that the left-hand side of (5.104) is bounded by

$$\sum_{k_i \geqslant 0} \bigg| \int_0^\delta \Lambda_5 \Big(\frac{N_{45} m^2(N_{*45})}{(N+N_1)(N+N_2)(N+N_3)(N+N_{45}) \prod_{i=1}^5 m(N_i)};$$
$$P_{k_1}u_1, \cdots, P_{k_5}u_5 \Big) dt \bigg| \lesssim N^{-\frac{15}{4}} \prod_{i=1}^{5} \|u_j\|_{\bar{F}^0(\delta)},$$

where $N_i = 2^{k_i}$. Cancel $\frac{N_{*45}}{(N+N_{45})} \leqslant 1$ and consider only the worst case $m^2(N_{*45}) = 1$. it suffices to prove

$$\textstyle\sum_{k_1,\cdots,k_5 \geqslant 0} \left| \int_0^\delta \Lambda_5 \left(\prod_{i=1}^3 \frac{1}{(N+N_i)m(N_i)} \frac{1}{m(N_4)} \frac{1}{m(N_5)}; P_{k_1}u_1, \cdots, P_{k_5}u_5 \right) dt \right|$$

$$\lesssim N^{-\frac{15}{4}} \prod_{i=1}^{5} \|u_j\|_{\bar{F}^0(\delta)}.$$

From symmetry we may assume $N_1 \geqslant N_2 \geqslant N_3\ N_4 \geqslant N_5$ and then at least two of N_i satisfy $N_i \gtrsim N$. Fix the extension $\widetilde{u}_i$ of u_i such that $\|\widetilde{u}_i\|_{\bar{F}^0} \lesssim 2\|u_i\|_{\bar{F}^0(\delta)}$, still denoted by u_i.

From (5.86) and $s = -3/4$, it is easy to see $\frac{1}{(N+N_i)m(N_i)} \lesssim N^{-3/4}\langle N_i\rangle^{-1/4}$ and

$$\frac{1}{m(N_4)m(N_5)} \lesssim N^{-3/2} N_4^{3/4} N_5^{3/4}.$$

Hence we need to control

$$N^{-\frac{15}{4}} \sum_{k_i} \int_0^\delta \Lambda_5 \left(\prod_{i=1}^{3} \langle N_i\rangle^{-1/4} N_4^{3/4} N_5^{3/4}; u_1, \cdots, u_5 \right) dt. \tag{5.105}$$

If $N_2 \sim N_1 \gtrsim N$, $N_4 \lesssim N_2$, then consider the worst case $N_1 \geqslant N_2 \geqslant N_4 \geqslant N_5 \geqslant N_3$. We get from (5.103) that

$$(5.105) \lesssim N^{-\frac{15}{4}} \sum_{N_i} \langle N_1\rangle^{-5/4} \langle N_2\rangle^{-5/4} \langle N_3\rangle^{1/6} N_4^{7/6} N_5^{7/6} \prod_{i=1}^{5} \|\widehat{P_{k_i} u}\|_{X_{k_i}}$$

$$\lesssim N^{-\frac{15}{4}} \prod_{j=1}^{5} \|Iu_j\|_{\bar{F}^0(\delta)}. \tag{5.106}$$

The rest cases $N_4 \sim N_5 \gtrsim N$, $N_1 \lesssim N_5$ or $N_1 \sim N_4 \gtrsim N$ can be handled similarly. □

The following proposition shows that $E_I^4(t)$ is very close to $E_I^2(t)$.

Proposition 5.22. *Assume I the Fourier multiplier with symbol m given by (5.86) and $s = -3/4$. Then*

$$|E_I^4(t) - E_I^2(t)| \lesssim \|Iu(t)\|_{L^2}^3 + \|Iu(t)\|_{L^2}^4.$$

Proof. Since $E_I^4(t) = E_I^2 + \Lambda_3(\sigma_3) + \Lambda_4(\sigma_4)$, it suffices to prove

$$|\Lambda_3(\sigma_3; u_1, u_2, u_3)| \lesssim \prod_{j=1}^{3} \|Iu_j(t)\|_{L^2}, \tag{5.107}$$

and

$$|\Lambda_4(\sigma_4; u_1, u_2, u_3, u_4)| \lesssim \prod_{j=1}^{4} \|Iu_j(t)\|_{L^2}. \tag{5.108}$$

We may assume $\widehat{u_j}$ is non-negative. To prove (5.107), it suffices to prove

$$\left|\Lambda_3\left(\frac{m^2(\xi_1)\xi_1+m^2(\xi_2)\xi_2+m^2(\xi_3)\xi_3}{\xi_1\xi_2\xi_3 m(\xi_1)m(\xi_2)m(\xi_3)};u_1,u_2,u_3\right)\right|\lesssim\prod_{i=1}^{3}\|u_j\|_2. \quad (5.109)$$

Making dyadic frequency decomposition, we get the left-hand side of (5.109) is bounded by

$$\sum_{k_i\geqslant 0}\left|\Lambda_3\left(\frac{m^2(\xi_1)\xi_1+m^2(\xi_2)\xi_2+m^2(\xi_3)\xi_3}{\xi_1\xi_2\xi_3 m(\xi_1)m(\xi_2)m(\xi_3)};\triangle_{k_1}u_1,\triangle_{k_2}u_2,\triangle_{k_3}u_3\right)\right|. \quad (5.110)$$

Denote $N_i=2^{k_i}$. From symmetry we may assume $N_1\geqslant N_2\geqslant N_3$, and hence $N_1\sim N_2\gtrsim N$.

Case 1. $N_3\ll N$.

Now $m_3(N_3)=1$, hence we have

$$\begin{aligned}(5.112)&\lesssim\sum_{k_i\geqslant 0}\left|\Lambda_3\left(\frac{N^sN^s}{N_1^{1+s}N_1^{1+s}};\triangle_{k_1}u_1,\triangle_{k_2}u_2,\triangle_{k_3}u_3\right)\right|\\&\lesssim\sum_{k_i\geqslant 0}\left|\Lambda_3\left(N_1^{-1/4}N_2^{-1/4};\triangle_{k_1}u_1,\triangle_{k_2}u_2,\triangle_{k_3}u_3\right)\right|.\end{aligned}$$

It suffices to prove

$$\sum_{k_i\geqslant 0}\int_{\xi_1+\xi_2+\xi_3=0,|\xi_i|\sim N_i}N_1^{-1/2}\prod_{i=1}^{3}\eta_{k_i}(\xi_i)\widehat{u_i}(\xi_i)\lesssim\prod_{i=1}^{3}\|u_i\|_{L^2}.$$

Define $v_i(x)$, whose Fourier transform is

$$\widehat{v_i}(\xi)\doteq N_i^{-1/6}\widehat{u_i}(\xi)\chi_{\{|\xi|\sim N_i\}}(\xi).$$

From Sobolev's embedding theorem we get $\|v_i\|_{L^3}\lesssim\|u_i\|_{L^2}$, thus use Hölder's inequality we get

$$\begin{aligned}&\sum_{k_i\geqslant 0}\int_{\xi_1+\xi_2+\xi_3=0,|\xi_i|\sim N_i}N_1^{-1/2}\prod_{i=1}^{3}\eta_{k_i}(\xi_i)\widehat{u_i}(\xi_i)\\&\lesssim\sum_{k_i\geqslant 0}N_1^{-1/6}N_3^{1/6}\prod_{i=1}^{3}\|v_i\|_{L^3}\lesssim\prod_{i=1}^{3}\|u_i\|_{L^2}.\end{aligned}$$

Case 2. $N_3\gtrsim N$. It is easy to see that

$$(5.112)\lesssim\sum_{k_i\geqslant 0}\left|\Lambda_3\left(\frac{N_3^{-3/4}N^{-3/4}}{N_1^{1/2}};P_{k_1}u_1,P_{k_2}u_2,P_{k_3}u_3\right)\right|\lesssim\prod_{i=1}^{3}\|u_i\|_{L^2}.$$

Thus (5.107) is proved.

Next show (5.108). It suffices to prove

$$\left|\Lambda_4\left(\frac{\sigma_4}{m(\xi_1)m(\xi_2)m(\xi_3)m(\xi_4)}; u_1, u_2, u_3, u_4\right)\right| \lesssim \prod_{i=1}^{4} \|u_j\|_2. \quad (5.111)$$

By dyadic frequency decomposition, we get the left-hand side of (5.111) is bounded by

$$\sum_{k_i \geqslant 0} \left|\Lambda_4\left(\frac{\sigma_4}{m(\xi_1)m(\xi_2)m(\xi_3)m(\xi_4)}; \triangle_{k_1}u_1, \triangle_{k_2}u_2, \triangle_{k_3}u_3, \triangle_{k_4}u_4\right)\right|. \quad (5.112)$$

Denote $N_i = 2^{k_i}$. From symmetry we may assume $N_1 \geqslant N_2 \geqslant N_3 \geqslant N_4$, and hence $N_1 \sim N_2 \gtrsim N$. In the integration area

$$\left|\frac{\sigma_4}{m(\xi_1)m(\xi_2)m(\xi_3)m(\xi_4)}\right| \lesssim \frac{1}{\prod_{i=1}^4 (N+N_i)m(N_i)} \lesssim \frac{N^{-3}}{\prod_{i=1}^4 N_i^{1/4}}.$$

We get from Hölder's inequality that

$$\begin{aligned}(5.112) &\lesssim \sum_{k_i \geqslant 0} \frac{N^{-3}}{\prod_{i=1}^4 N_i^{1/4}} \|\triangle_{k_1}u_1\|_{L^2} \|\triangle_{k_2}u_2\|_{L^2} \|\triangle_{k_3}u_3\|_{L^\infty} \|\triangle_{k_4}u_4\|_{L^\infty} \\ &\lesssim \prod_{i=1}^{4} \|u_j\|_2.\end{aligned}$$

Thus we proved the proposition. □

Now we are ready to extend the local solution to a global one. First we need a variant local well-posedness which can be proved similarly.

Proposition 5.23. *Let $-3/4 \leqslant s \leqslant 0$. Assume ϕ satisfies $\|I\phi\|_{L^2(\mathbb{R})} \leqslant 2\epsilon_0 \ll 1$. Then there exists a unique solution to* (5.32) *on* $[-1,1]$ *such that*

$$\|Iu\|_{\bar{F}^0(1)} \leqslant C\epsilon_0, \quad (5.113)$$

where C is independent of N.

For any given $u_0 \in H^{-3/4}$ and time $T > 0$, our purpose is to construct a solution on $[0,T]$. If u is a solution to KdV equation with initial data u_0, then for any $\lambda > 0$, $u_\lambda(x,t) = \lambda^{-2}u(x/\lambda, t/\lambda^3)$ is also a solution to KdV equation with initial data $u_{0,\lambda} = \lambda^{-2}u_0(x/\lambda)$. Simple calculations show that

$$\|Iu_{0,\lambda}\|_{L^2} \lesssim \lambda^{-\frac{3}{2}-s} N^{-s} \|u_0\|_{H^s}.$$

For fixed N(N will be determined later), we take $\lambda \sim N^{-\frac{2s}{3+2s}}$ such that

$$\lambda^{-\frac{3}{2}-s} N^{-s} \|\phi\|_{H^s} = \epsilon_0 < 1.$$

For simplicity of notations, we still denote u_λ by u, $u_{0,\lambda}$ by u_0, and assume $\|Iu_0\|_{L^2} \leqslant \epsilon_0$. The purpose is then to construct solutions on $[0, \lambda^3 T]$. By Proposition 5.23 we get a solution on $[0,1]$, and we need to extend the solution. It suffices to control the modified energy $E_I^2(t) = \|Iu\|_{L^2}^2$.

First we control $E_I^2(t)$ for $t \in [0,1]$, we will prove $E_I^2(t) < 4\epsilon_0^2$. By standard bootstrap, we may assume $E_I^2(t) < 5\epsilon_0^2$. Then from Proposition 5.22 we get

$$E_I^4(0) = E_I^2(0) + O(\epsilon_0^3)$$

and

$$E_I^4(t) = E_I^2(t) + O(\epsilon_0^3).$$

From Proposition 5.21 we get for all $t \in [0,1]$

$$E_I^4(t) \leqslant E_I^4(0) + C\epsilon_0^5 N^{-15/4}.$$

Thus

$$\begin{aligned}\|Iu(1)\|_{L^2}^2 = E_I^4(1) + O(\epsilon_0^3) \leqslant & E_I^4(0) + C\epsilon_0^5 N^{-15/4} + O(\epsilon_0^3) \\ = & \epsilon_0^2 + C\epsilon_0^5 N^{-15/4} + O(\epsilon_0^3) < 4\epsilon_0^2.\end{aligned}$$

Thus the solution u can be extended to $t \in [0,2]$. Iterating this procedure M steps, then we get for $t \in [0, M+1]$

$$E_I^4(t) \leqslant E_I^4(0) + CM\epsilon_0^5 N^{-15/4}$$

as long as $MN^{-15/4} \lesssim 1$. Thus we get

$$E_I^2(M) = E_I^4(t) + O(\epsilon_0^3) = \epsilon_0^2 + O(\epsilon_0^3) + CM\epsilon_0^5 N^{-15/4} < 4\epsilon_0^2,$$

hence the solution can be extended to $t \in [0, N^{15/4}]$. Take $N(T)$ sufficiently large such that

$$N^{15/4} > \lambda^3 T \sim N^3 T.$$

Therefore, the u is extended to $[0, \lambda^3 T]$.

In the end of this section, we prove some properties of the global solution. By the scaling we get

$$\sup_{t\in[0,T]} \|u(t)\|_{H^s} \sim \lambda^{3/2+s} \sup_{t\in[0,\lambda^3 T]} \|u_\lambda(t)\|_{H^s} \leqslant \lambda^{3/2+s} \sup_{t\in[0,\lambda^3 T]} \|Iu_\lambda(t)\|_{L^2},$$

$$\|I\phi_\lambda\|_{L^2} \lesssim N^{-s} \|\phi_\lambda\|_{H^s} \sim N^{-s} \lambda^{-3/2-s} \|\phi\|_{H^s}.$$

From the proof we have

$$\sup_{t\in[0,\lambda^3T]} \|Iu_\lambda(t)\|_{L^2} \lesssim \|I\phi_\lambda\|_{L^2},$$

and hence

$$\sup_{t\in[0,T]} \|u(t)\|_{H^s} \lesssim N^{-s}\|\phi\|_{H^s}.$$

Take λ such that $\|I\phi_\lambda\|_{L^2} \sim \epsilon_0 \ll 1$, thus we get

$$\lambda = \lambda(N, \epsilon_0, \|\phi\|_{H^s}) \sim \left(\frac{\|\phi\|_{H^s}}{\epsilon_0}\right)^{\frac{2}{3+2s}} N^{-\frac{2s}{3+2s}}.$$

Take N such that $N^{\frac{15}{4}} > \lambda^3 T \sim_{c\|\phi\|_{H^s},\epsilon_0} N^{-\frac{6s}{3+2s}}T$, then $N \sim T^{4/3}$. Thus the global solution $u(x,t)$ satisfies

$$\|u(t)\|_{H^{-3/4}} \lesssim (1+|t|)\|\phi\|_{H^{-3/4}}.$$

For the case $s \in (-3/4, 0]$, we left the proof to the readers.

5.6 Schrödinger equation with derivative

In above several sections, we use the Bourgain's spaces to prove some bilinear estimates with respect to KdV equation. However, in some cases, for example: local well-posedness of the KdV equation in $H^{-3/4}$, it is not enough to get the local result directly using the Bourgain's spaces. In this section, we consider the following Cauchy problem of the Schrödinger equation with derivative such that the reader can understand more the method of the Bourgain's spaces

$$\begin{cases} iu_t + u_{xx} = i\lambda(|u|^2u)_x, \ (x,t) \in \mathbb{R}^2 \\ u(x,0) = u_0(x). \end{cases} \tag{5.114}$$

From Section 5.1 in this chapter, it follows that dispersion relation of (5.114) $\phi(\xi) = -\xi^2$. Thus we can obtain the following the definition of the norm of the Bourgain's spaces $X^{s,b}$ with respect to Schrödinger equation with derivative:

$$\|u\|_{X^{s,b}} = \|\langle\xi\rangle^2\langle\tau+\xi^2\rangle^b\widehat{u}(\xi,\tau)\|_{L^2}.$$

In order to prove the well-posedness of the Cauchy problem (5.114), from the standard method, it suffices to show

$$\|\partial_x(u\bar{v}w)\|_{X^{s,b-1}} \lesssim \|u\|_{X^{s,b}}\|v\|_{X^{s,b}}\|w\|_{X^{s,b}}. \tag{5.115}$$

But the above inequality is invalid for any $s, b \in \mathbb{R}$, one can refer to [91].

In this section, we will use gauge transform to improve the derivative nonlinearity, one can refer to [105; 184; 213]. The result in this section was first given by Takaoka[213]. Via the gauge transformation which was first used by Hayashi [102] (see also [106])

$$v(x,t) = \mathcal{G}_\lambda(u)(x,t) = e^{-i\lambda \int_{-\infty}^{x} |u(y,t)|^2 dy} u(x,t), \tag{5.116}$$

(5.114) is formally rewritten as the Cauchy problem

$$\begin{cases} i\partial_t v + \partial_x^2 v = -i\lambda v^2 \partial_x \bar{v} - \frac{|\lambda|^2}{2} |v|^4 v, \\ v(x,0) = v_0(x), \end{cases} \tag{5.117}$$

where $v_0(x) = e^{-i\lambda \int_{-\infty}^{x} |u_0(y)|^2 dy} u_0(x)$. The corresponding inverse gauge transformation,

$$u(x,t) = e^{i\lambda \int_{-\infty}^{x} |v(y,t)|^2 dy} v(x,t) = \mathcal{G}_{-\lambda}(v).$$

Since the map $G_\lambda$is a bicontinuous map from H^s to H^s, we can show that the global well-posedness of (5.114) in H^s is equivalent to that of (5.117). in the following, we only consider the Cauchy problem (5.117).

Notice that the nonlinear term of the original equation $i\lambda(|u|^2 u)_x = 2i\lambda|u|^2 u_x + i\lambda u^2 \bar{u}_x$, via the gauge transformation, the derivative in the nonlinearity $|u|^2 u_x$ has been replaced by the quintic nonlinearity $|v|^4 v$. The Strichartz estimate can control the nonlinearity $|v|^4 v$ easy. But, at first sight, there still exists a derivative in the nonlinearity $u^2 \bar{u}_x$ in equation (5.117). However a derivative of the complex conjugate of the solution u can be handled while a derivative of u cannot since $\omega(\xi) = -\xi^2$ is a even function.

From the definition of Bourgain's spaces, it follows that for complex conjugate of u

$$\|\bar{u}\|_{X^{s,b}} = \|\langle\xi\rangle^s \langle\tau - \xi^2\rangle^b \mathcal{F}u\|_{L^2_\xi L^2_\tau}, \; \|u\|_{X^{s,b}} = \|\langle\xi\rangle^s \langle\tau + \xi^2\rangle^b \mathcal{F}u\|_{L^2_\xi L^2_\tau}.$$

For the nonlinearity$u^2\bar{u}_x$,we can prove that following estimate holds for $s \geqslant 1/2$ and $b > 1/2$

$$\|uv\bar{w}_x\|_{X_{s,b-1}} \lesssim \|u\|_{X_{s,b}} \|v\|_{X_{s,b}} \|w\|_{X_{s,b}}. \tag{5.118}$$

But for $|u|^2 u_x$, the above inequality is invalid.

Now we mainly estimate the nonlinear term $v^2 \partial_x \bar{v}$. Similarly with the KdV equation, we also need some space-time estimates (Strichartz estimate, local smoothing effect and maximal function estimate) of Schrödinger eqution, the proofs of these estimates are similar with one of the Benjamin-Ono equation (the dispersion relation $\omega(\xi) = -|\xi|\xi$) in Section 5.2.

Lemma 5.9. *The group of Schrödinger equation $S(t) = \mathscr{F}_x^{-1} e^{it\xi^2} \mathscr{F}_x$ satisfies :*

$$\|S(t)v_0\|_{L_x^6 L_t^6} \lesssim \|v_0\|_{L^2},$$
$$\|D^{1/2}S(t)v_0\|_{L_x^\infty L_t^2} \lesssim \|v_0\|_{L^2},$$
$$\|D^{-1/4}S(t)v_0\|_{L_x^4 L_t^\infty} \lesssim \|v_0\|_{L^2}.$$

Using Lemma 5.3 and Lemma 5.9, we can easily obtain the following lemma:

Lemma 5.10. *Let $u \in X^{0,b}$ with $1/2 < b < 1$. Then*

$$\|u\|_{L_x^6 L_t^6} \lesssim \|u\|_{X^{0,b}},$$
$$\|D^{1/2}u\|_{L_x^\infty L_t^2} \lesssim \|u\|_{X^{0,b}},$$
$$\|D^{-1/4}u\|_{L_x^4 L_t^\infty} \lesssim \|u\|_{X^{0,b}}.$$

By Lemma 5.10, we can obtain the following trilinear estimates.

Theorem 5.8. *Assume $s \geqslant 1/2$, $1/2 < b < 2/3$ and $b' > 1/2$. Then*

$$\|v_1 v_2 \partial_x \bar{v}_3\|_{X_{s,b-1}} \leqslant C\|v_1\|_{X_{s_1,b'}} \|v_2\|_{X_{s_2,b'}} \|v_3\|_{X_{s_3,b'}}. \tag{5.119}$$

Proof. By duality and the Plancherel identity, it suffices to show that

$$\int_{\Gamma_4} \frac{\langle\xi_4\rangle^s \langle\tau_4 - \xi_4^2\rangle^{b-1} \xi_3 \prod_{i=1}^4 f_i(\xi_i, \tau_i)}{\langle\tau_1 + \xi_1^2\rangle^{b'} \langle\tau_2 + \xi_2^2\rangle^{b'} \langle\tau_3 - \xi_3^2\rangle^{b'} \prod_{j=1}^3 \langle\xi_i\rangle^s} \lesssim \prod_{i=1}^4 \|f_i\|_{L^2}, \tag{5.120}$$

for all $f_i \in L^2(\mathbb{R}^2) > 0$, $i = 1, 2, 3, 4$; where the hyperplane Γ_4 is defined by

$$\Gamma_4 = \{(\xi, \tau) \in \mathbb{R}^4 \times \mathbb{R}^4 : \xi_1 + \xi_2 + \xi_3 + \xi_4 = 0, \tau_1 + \tau_2 + \tau_3 + \tau_4 = 0\}$$

which we endow with the standard measure: $\theta_i = (\xi_i, \tau_i)$

$$\int_{\Gamma_4} h(\theta_1, \theta_2, \theta_3, \theta_4) = \int_{\mathbb{R}^6} h(\theta_1, \theta_2, \theta_3, -\theta_1 - \theta_2 - \theta_3) d\theta_1 d\theta_2 d\theta_3.$$

From Plancherel identity, it follows that

$$\int_{\Gamma_4} \prod_{i=1}^4 f_i(\xi_i, \tau_i) = \int_{\mathbb{R}^2} \prod_{i=1}^4 \mathscr{F}^{-1}(f_i)(x,t) dx dt. \tag{5.121}$$

By symmetry it suffices to estimate the integral in the domain $|\xi_1| \leqslant |\xi_2|$. We define $|\xi|_{max}$, $|\xi|_{sub}$, $|\xi|_{thd}$, $|\xi|_{min}$ be the maximum, second largest, third largest and minimum of $|\xi_1|$, $|\xi_2|$, $|\xi_3|$, $|\xi_4|$; and define $\sigma_j^\pm = \tau_j \pm \xi_j^2$, $j = 1, 2, 3, 4$. Let

$$F_i = \mathscr{F}^{-1}\left(\frac{f_i}{\langle \tau + \xi^2 \rangle^{b'}}\right), i = 1, 2;\ F_k = \mathscr{F}^{-1}\left(\frac{f_k}{\langle \tau - \xi^2 \rangle^{b'}}\right), k = 3, 4.$$

$$K(\xi_1, \xi_2, \xi_3, \xi_4) = \frac{|\xi_3|\langle \xi_4 \rangle^s}{\langle \xi_1 \rangle^s \langle \xi_2 \rangle^s \langle \xi_3 \rangle^s}.$$

In order to obtain the boundedness of the integral in the left side of (5.120), we split the domain of integration in several pieces as in [91].

Case 1. If $\max(|\xi_1|, |\xi_2|, |\xi_3|, |\xi_4|) \leqslant 10$, then

$$K(\xi_1, \xi_2, \xi_3, \xi_4) \leqslant C.$$

By Lemma 5.10 and (5.121), the left side of (5.120) restricted to this domain is bounded by

$$\int_{\Gamma_4} \frac{\prod_{i=1}^4 f_i(\xi_i, \tau_i)}{\langle \tau_1 + \xi_1^2 \rangle^{b'} \langle \tau_2 + \xi_2^2 \rangle^{b'} \langle \tau_3 - \xi_3^2 \rangle^{b'}} \lesssim \|f_4\|_2 \prod_{i=1}^3 \|F_i\|_{L^6} \lesssim \prod_{i=1}^4 \|f_i\|_{L^2}.$$

Case 2. If $|\xi|_{max} \gg 1$ and $|\xi|_{thd} \ll |\xi|_{sub}$, then from $\xi_1 + \xi_2 + \xi_3 + \xi_4 = 0$ it follows that $|\xi|_{max} \sim |\xi|_{sub}$. We split this case into the following several subcases.

Subcase 2a. If $|\xi_2| \ll |\xi_3| \sim |\xi_4|$, then recall the following identity on the hyperplane Γ_4

$$h(\xi_1, \xi_2, \xi_3, \xi_4) = \sigma_1^+ + \sigma_2^+ + \sigma_3^- + \sigma_4^- = 2(\xi_1 + \xi_3)(\xi_2 + \xi_3). \quad (5.122)$$

This implies that

$$\max(|\sigma_1^+|, |\sigma_2^+|, |\sigma_3^-|, |\sigma_4^-|) \gtrsim |\xi_3|^2.$$

By symmetry, we only consider $|\sigma_4^-| = \max(|\sigma_1^+|, |\sigma_2^+|, |\sigma_3^-|, |\sigma_4^-|)$ and $|\sigma_2^+| = \max(|\sigma_1^+|, |\sigma_2^+|, |\sigma_3^-|, |\sigma_4^-|)$.

If $|\sigma_4^-| = \max(|\sigma_1^+|, |\sigma_2^+|, |\sigma_3^-|, |\sigma_4^-|)$, then for $b \leqslant 3/4$ and $s \geqslant 1/4$, we have

$$\frac{K(\xi_1, \xi_2, \xi_3, \xi_4)}{|\sigma_4^-|^{1-b}} \lesssim \frac{|\xi_3||\xi_3|^{2(b-1)}}{\langle \xi_1 \rangle^s \langle \xi_2 \rangle^s} \lesssim \frac{|\xi_3|^{1/2}}{\langle \xi_1 \rangle^s \langle \xi_2 \rangle^s}.$$

By Lemma 5.10, (5.121) and Hölder inequality, the left side of (5.120) restricted to this domain is bounded by

$$\int_{\Gamma_4} \frac{|\xi_3||\xi_3|^{1/2} \prod_{i=1}^4 f_i(\xi_i, \tau_i)}{\prod_{i=1}^2 \langle \xi_i \rangle^s \langle \tau_1 + \xi_1^2 \rangle^{b'} \langle \tau_2 + \xi_2^2 \rangle^{b'} \langle \tau_3 - \xi_3^2 \rangle^{b'}}$$

$$\lesssim \|J^{-s}F_1\|_{L^4_x L^\infty_t} \|J^{-s}F_2\|_{L^4_x L^\infty_t} \|f_4\|_{L^2_{x,t}} \|\Lambda^{1/2}F_3\|_{L^\infty_x L^2_t} \lesssim \prod_{i=1}^{4} \|f_i\|_{L^2}.$$

If $|\sigma_2^+| = \max(|\sigma_1^+|, |\sigma_2^+|, |\sigma_3^-|, |\sigma_4^-|)$, then from $\langle\sigma_4^-\rangle^{1-b}\langle\sigma_2^+\rangle^{b'} \gtrsim \langle\sigma_4^-\rangle^{b'}\langle\sigma_2^+\rangle^{1-b}$, similarly with above, for $b \leqslant 3/4$ and $s \geqslant 1/4$, the left side of (5.120) restricted to this domain is bounded by

$$\int_{\Gamma_4} \frac{|\xi_3||\xi_3|^{2(b-1)} \prod_{i=1}^4 f_i(\xi_i, \tau_i)}{\prod_{i=1}^2 \langle\xi_i\rangle^s \langle\tau_1 + \xi_1^2\rangle^{b'} \langle\tau_3 - \xi_3^2\rangle^{b'} \langle\tau_4 - \xi_4^2\rangle^{b'}}$$

$$\lesssim \|J^{-s}F_1\|_{L^4_x L^\infty_t} \|F_4\|_{L^4_x L^\infty_t} \|f_2\|_{L^2_{x,t}} \|\Lambda^{1/2}F_3\|_{L^\infty_x L^2_t} \lesssim \prod_{i=1}^{4} \|f_i\|_{L^2},$$

which follows from Lemma 5.10, (5.121) and Hölder inequality.

Subcase 2b. Assume $|\xi_2| \sim |\xi_3| \sim |\xi|_{max}$ or $|\xi_2| \sim |\xi_4| \sim |\xi|_{max}$. By symmetry, we can assume $|\xi_2| \sim |\xi_3| \sim |\xi|_{max}$.

If $|\xi_2 + \xi_3| \leqslant 1$ or $|\xi_4| \lesssim |\xi_1|$, then $\langle\xi_1\rangle \sim \langle\xi_4\rangle$. For $s \geqslant 1/2$, we have

$$K(\xi_1, \xi_2, \xi_3, \xi_4) \leqslant C.$$

Similarly with Case 1, the left side of (5.120) restricted to this domain is bounded by

$$\int_{\Gamma_4} \frac{\prod_{i=1}^4 f_i(\xi_i, \tau_i)}{\langle\tau_1 + \xi_1^2\rangle^{b'} \langle\tau_2 + \xi_2^2\rangle^{b'} \langle\tau_3 - \xi_3^2\rangle^{b'}} \lesssim \prod_{i=1}^{4} \|f_i\|_{L^2}.$$

If $|\xi_2 + \xi_3| \geqslant 1$ and $|\xi_4| \gg |\xi_1|$, then

$$\max(|\sigma_1^+|, |\sigma_2^+|, |\sigma_3^-|, |\sigma_4^-|) \gtrsim |\xi_1 + \xi_3||\xi_2 + \xi_3| \sim \langle\xi_3\rangle\langle\xi_4\rangle.$$

By symmetry, we only consider $|\sigma_4^-| = \max(|\sigma_1^+|, |\sigma_2^+|, |\sigma_3^-|, |\sigma_4^-|)$. Then for $s \geqslant 1/2$ and $b \leqslant 3/4$

$$\frac{K(\xi_1, \xi_2, \xi_3, \xi_4)}{|\sigma_4^-|^{1-b}} \lesssim \frac{|\xi_3|^{1/2}}{\langle\xi_1\rangle^s \langle\xi_3\rangle^{1-b} \langle\xi_4\rangle^{1-b}} \lesssim \frac{|\xi_3|^{1/2}}{\langle\xi_1\rangle^{1/4}\langle\xi_2\rangle^{1/4}}.$$

Similarly with Subcase 2a, we can obtain the result.

Subcase 2c. If $|\xi_1| \sim |\xi_3| \sim |\xi|_{max}$ or $|\xi_1| \sim |\xi_4| \sim |\xi|_{max}$, then similarly with Subcase 2b, we can obtain the result.

Subcase 2d. If $|\xi_1| \sim |\xi_2| \sim |\xi|_{max}$, then $\max(|\sigma_1^+|, |\sigma_2^+|, |\sigma_3^-|, |\sigma_4^-|) \gtrsim |\xi_3|^2$, similarly with Subcase 2a, we can obtain the result.

Case 3. If $|\xi|_{min} \ll |\xi|_{thd} \sim |\xi|_{sub} \sim |\xi|_{max}$, then we have

$$\max(|\sigma_1^+|, |\sigma_2^+|, |\sigma_3^-|, |\sigma_4^-|) \gtrsim |\xi_3|^2.$$

Similarly with Subcase 2a, we can obtain the result.

Case 4. If $|\xi|_{min} \sim |\xi|_{thd} \sim |\xi|_{sub} \sim |\xi|_{max}$, similarly with Case 1, we can obtain the result.

This completes the proof of theorem. □

Next we estimate the nonlinearity $|u|^4u$. From the above standard argument, it suffices to show that

$$\|u_1u_2u_3\bar{u}_4\bar{u}_5\|_{X^{s,b-1}}\lesssim\prod_{i=1}^{5}\|u_i\|_{X^{s,b}}. \tag{5.123}$$

Notice that there exists the condition $s\geqslant 1/2$ when we estimate the nonlinearity $u^2\bar{u}_x$, so we do not need to obtain the sharp index of s for (5.123), the index $s\geqslant 1/2$ is enough for the local well-posedness. But the author believe that the sharp index of s for (5.123) is $s\geqslant 0$.

Lemma 5.11. *Let $s\in\mathbb{R}$, $1/2<b\leqslant 1$. Then*

$$\|\psi(t/T)u\|_{X^{s,b-1}}\lesssim\|u\|_{L_T^{2/(3-2b)}H_x^s}.$$

Proof. Without loss of generality, we assume $s=0$. Using the definition of $X^{s,b}$, imbedding inequality $L_t^{2/(3-2b)}\hookrightarrow H_t^{b-1}$ and Minkowski's inequality, we have

$$\begin{aligned}\|\psi(t/T)u\|_{X^{0,b-1}}=&\big\|\|e^{-it\omega(\xi)}\psi(t/T)(\mathscr{F}_xu)(\xi,t)\|_{H_t^{b-1}}\big\|_{L_\xi^2}\\ \lesssim&\big\|\|e^{-it\omega(\xi)}\psi(t/T)(\mathscr{F}_xu)(\xi,t)\|_{L_t^{2/(3-2b)}}\big\|_{L_\xi^2}\\ \lesssim&\|u\|_{L_T^{2/(3-2b)}L_x^2}.\end{aligned}$$

This completes the proof of lemma. □

Theorem 5.9. *Let $s\geqslant 1/2$ and $1/2<b\leqslant 1$. Then there exists some $\theta>0$ such that*

$$\|\psi(t/T)u_1u_2u_3\bar{u}_4\bar{u}_5\|_{X^{s,b-1}}\lesssim T^\theta\prod_{i=1}^{5}\|u_i\|_{X^{s,b}}. \tag{5.124}$$

Proof. Assume $u_1=\cdots=u_5=u$, it suffices to show that

$$\|\psi(t/T)|u|^4u\|_{X^{s,b-1}}\lesssim T^\theta\|u\|_{X^{s,b}}^5. \tag{5.125}$$

Using Lemma 5.11, Lemma 5.10 and the Leibniz rule for fractional derivatives, we have

$$\begin{aligned}\|\psi(t/T)|u|^4u\|_{X^{s,b-1}}\lesssim&\||u|^4u\|_{L_T^{2/(3-2b)}H_x^s}\lesssim\|u^4\|_{L_T^\infty L_x^2}\|u\|_{L_T^{2/(3-2b)}H_\infty^s}\\ \lesssim&T^\theta\|u\|_{L_T^\infty L_x^8}^4\|u\|_{L_T^4H_\infty^s}\lesssim T^\theta\|u\|_{X^{s,b}}^5.\end{aligned}$$

This completes the proof. □

Using the proof of Theorem 5.5 in Section 5.3, we have

Proposition 5.24. *Let* $1/2 \leqslant s < 1$. *For* $v_0 \in H^s$, *then there exists* $b > 1/2$ *and* $T = T(\|v_0\|_{H^{1/2}}) > 0$ *such that the Cauchy problem* (5.117) *admits a unique local solution* $v \in X_T^{s,b}$.

For the original equation (5.114), we have

Theorem 5.10. *Let* $\lambda \in \mathbb{R}$ *and* $s \geqslant 1/2$. *For* $u_0 \in H^s(\mathbb{R})$, *then there exists* $b > 1/2$ *and* $T = T(\|u_0\|_{H^{1/2}}) > 0$ *such that the Cauchy problem* (5.114) *admits a unique local solution*

$$u \in C([-T,T] : H^s(\mathbb{R})), \quad \mathcal{G}_\lambda(u) \in X_T^{s,b}.$$

Moreover, given $t \in (0,T)$, *the map* $u_0 \to u(t)$ *is Lipschitz continuous from* H^s *to* $C(0,T;H^s)$.

Proof. For $u_0 \in H^s$ and $1/2 \leqslant s < 1$, we define $v_0 \in H^s$

$$v_0(x) = e^{-i\lambda/2\int_{-\infty}^{x}|u_0(y)|^2dy}u_0(x).$$

Assume $v_0^{(n)}$ is a sequence in H^∞ satisfying $v_0^{(n)} \to v_0$ in H^s. Let v_n be a solution of the Cauchy problem (5.117) obtained in Proposition 5.24. Define

$$\begin{aligned} u_n(x,t) =& \mathcal{G}_{-\lambda}(v_n)(x,t), \\ u_0^{(n)}(x) =& e^{i\lambda/2\int_{-\infty}^{x}|v_0^{(n)}(y)|^2dy}v_0^{(n)}(x). \end{aligned}$$

From the definition of u_n, it follows that

$$\|u_n\|_{L_T^\infty L_x^2} = \|v_n\|_{L_T^\infty L_x^2}.$$

Using the Leibniz rule for fractional derivatives and Sovolev inequality, for $1/2 \leqslant s < 1$, we have

$$\begin{aligned} \|D^s u_n\|_{L_T^\infty L_x^2} \lesssim & \|D^s v_n\|_{L_T^\infty L_x^2} + \left\| D^s\left(e^{i\lambda/2\int_{-\infty}^{x}|v_0^{(n)}(y)|^2dy}\right)v_n\right\|_{L_T^\infty L_x^2} \\ \lesssim & \|D^s v_n\|_{L_T^\infty L_x^2} + \left\| D^s\left(e^{i\lambda/2\int_{-\infty}^{x}|v_0^{(n)}(y)|^2dy}\right)\right\|_{L_T^\infty L_x^p} \|v_n\|_{L_T^\infty L_x^q} \\ \lesssim & \|D^s v_n\|_{L_T^\infty L_x^2} + \|v_n\|^2_{L_T^\infty L_x^{2p_1}} \|v_n\|_{L_T^\infty H_x^{1/2}} \\ \lesssim & (1 + \|v_n\|^2_{L_T^\infty H_x^s})\|v_n\|_{L_T^\infty H_x^s}, \end{aligned}$$

where $2 < p < \infty$, $1/p + 1/q = 1/2$ and $1/p_1 = 1/p + 1 - s$. Similarly with above, we have

$$\|u_n - u_m\|_{L_T^\infty H_x^s} \lesssim (1 + \|v_n\|_{L_T^\infty H_x^s} + \|v_m\|_{L_T^\infty H_x^s})^2 \|v_n - v_m\|_{L_T^\infty H_x^s}.$$

It suffices to show that the map $\mathcal{G}_\lambda$ is Lipschitz continuous from H^s to L^p for any $2 \leqslant p < \infty$. In fact, we have

$$\begin{aligned}&\left\| e^{i\lambda/2\int_{-\infty}^x |f(y)|^2 dy} f(x) - e^{i\lambda/2\int_{-\infty}^x |g(y)|^2 dy} g(x) \right\|_{L^p} \\ &\lesssim \|f-g\|_{H^s} + \left\| e^{i\lambda/2\int_{-\infty}^x |f(y)|^2 dy} - e^{i\lambda/2\int_{-\infty}^x |g(y)|^2 dy} \right\|_{L^\infty} (\|f\|_{H^s} + \|g\|_{H^s}).\end{aligned}$$

Using the fact that $|f(x)-f(y)| \leqslant \|f'\|_{L^1}$, we have

$$\begin{aligned}&\left\| e^{i\lambda/2\int_{-\infty}^x |f(y)|^2 dy} - e^{i\lambda/2\int_{-\infty}^x |g(y)|^2 dy} \right\|_{L^\infty} \\ \lesssim &\left\| e^{i\lambda/2\int_{-\infty}^x (|f(y)|^2 - |g(y)|^2) dy} - 1 \right\|_{L^\infty} \\ \lesssim &\| |f|^2 - |g|^2 \|_{L^1} \lesssim \|f-g\|_{H^s} (\|f\|_{H^s} + \|g\|_{H^s}).\end{aligned}$$

Let $n \to \infty$, we can show that u is a solution of the Cauchy problem (5.114). This completes the proof of theorem. □

5.7 Some other dispersive equations

Except for KdV equation and Schrödinger equation, there exist some other important dispersive equations which have been considered by lots of mathematicer recently. In this section, the authors give some methods for some other dispersive equations such that the readers can understand more the method of Bourgain's spaces.

From the bilinear estimates of KdV equation and trilinear estimates of Schrödinger equation with derivative, it follows that we can use not only the method of Tao's $[k; Z]$ multiplier [216], but also the method of the local smoothing effects and maximal function estimates [213] to consider the well-posedness of dispersive equations. We take the bilinear estimates of KdV equation for example to show the difference of the two methods. First, we can use the method of Tao's $[k; Z]$ multiplier to prove Theorem 5.3. That is

$$\|\partial_x(u_1 u_2)\|_{X^{s,b-1}} \leqslant C\|u_1\|_{X^{s,b'}} \|u_2\|_{X^{s,b'}}, \ s > -3/4, b, b' > 1/2. \quad (5.126)$$

However, (5.126) holds with the condition $s > -5/8$ using only Strichartz estimate, the local smoothing effects and the maximal function estimates of KdV equation, one can refer to [128]. If $-5/8 \geqslant s > -3/4$, Kenig, Ponce and Vega [132] used the following inequalities to obtain (5.126).

$$\int_{-\infty}^{\infty} \frac{dx}{(1+|x-\alpha|)^{2b}(1+|x-\beta|)^{2b}} \leqslant \frac{C}{(1+|\alpha-\beta|)^{2b}}, \ b > \frac{1}{2} \quad (5.127)$$

$$\int_{-\infty}^{\infty} \frac{dx}{(1+|x|)^{2b}\sqrt{a-x}} \leqslant \frac{C}{(1+|a|)^{\frac{1}{2}}},\ b > \frac{1}{2} \tag{5.128}$$

$$\int_{-\infty}^{\infty} \frac{dx}{(1+|x-\alpha|)^{2(1-b)}(1+|x-\beta|)^{2b}} \leqslant \frac{C}{(1+|\alpha-\beta|)^{2(1-b)}},\ b > \frac{1}{2} \tag{5.129}$$

$$\int_{|x|\leqslant\beta} \frac{dx}{(1+|x|)^{2(1-b)}\sqrt{a-x}} \leqslant \frac{C(1+\beta)^{2(b-\frac{1}{2})}}{(1+|a|)^{\frac{1}{2}}},\ b > \frac{1}{2}. \tag{5.130}$$

As in Section 5.3 of this chapter, we first define $\xi = \xi_1 + \xi_2,\quad \tau = \tau_1 + \tau_2$ $\sigma = \tau - \xi^3,\quad \sigma_1 = \tau_1 - \xi_1^3,\quad \sigma_2 = \tau_2 - \xi_2^3$. We split the the hyperplane $\{(\xi,\xi_1,\xi_2)\times(\tau,\tau_1,\tau_2)\in\mathbb{R}^3\times\mathbb{R}^3 : \xi = \xi_1+\xi_2, \tau = \tau_1+\tau_2\}$ into suitable several parts. For the case: $|\xi_1| \sim |\xi_2| \gg |\xi|$ and $|\sigma| \geqslant |\sigma_1|, |\sigma_2|$, we only use the inequality (5.127); for other cases, similarly with the paper [128], we use the Strichartz estimates, the local smoothing effects and the maximal function estimates, thus we can also obtain (5.126). This can simplify the proof of the paper [132], the readers can try it. From this, it follows that the interaction: *high* $\times$ *high* $\to$ *low* is worst when one consider low regularity solution in Sobolev spaces of negative indice. Moreover, corresponding to the inequality (5.127), there exists the following inequality, which is also important when one consider the local well-posedness of the dispersive equations

$$\int_{\mathbb{R}} \frac{dx}{\langle x-\alpha\rangle^{2b}\langle x-\beta\rangle^{2b}} \leqslant \frac{C}{\langle\alpha-\beta\rangle^{4b-1}},\ \frac{1}{4} < b < \frac{1}{2}. \tag{5.131}$$

Next we take the fourth-order nonlinear Schrödinger equation for example, and show how to consider the well-posedness for the dispersive equation with complicated phase function (the dispersion relations of KdV equation $\phi(\xi) = \xi^3$ and Schrödinger equation $\phi(\xi) = |\xi|^2$ do not have non-zero singular points, that is, the solutions of the phase function $\phi(\xi) = 0$, $\phi'(\xi) = 0$ and $\phi''(\xi) = 0$ are only zero; the complicated phase functions are generally assumed to have non-zero singular points). The Cauchy problem for the fourth-order Schrödinger equation which describes the motion of the vortex filament as follows:

$$i\partial_t u + \partial_x^2 u + \nu\partial_x^4 u = F(u,\bar{u},\partial_x u,\partial_x\bar{u},\partial_x^2 u,\partial_x^2\bar{u}),\quad (x,t)\in\mathbb{R}\times\mathbb{R}, \tag{5.132}$$

where ν is a non-zero real constant, the nonlinear term F is given by

$$\begin{aligned} &F(u,\bar{u},\partial_x u,\partial_x\bar{u},\partial_x^2 u,\partial_x^2\bar{u}) \\ &= -\frac{1}{2}|u|^2 u + \lambda_1|u|^4 u + \lambda_2(\partial_x u)^2\bar{u} + \lambda_3|\partial_x u|^2 u + \lambda_4 u^2\partial_x^2\bar{u} + \lambda_5|u|^2\partial_x^2 u \end{aligned}$$

with $\lambda_1 = -\frac{3\mu}{4}$, $\lambda_2 = -2\mu+\frac{\nu}{2}$, $\lambda_3 = -4\mu-\nu$, $\lambda_5 = -2\mu+\nu$, where μ is also a real constant. The equation (5.132) describes the three-dimensional motion of an isolated vortex filament, which is embedded in inviscid incompressible fluid fulfilled in an infinite region. Note that the dispersion relation of the equation (5.132) $\phi(\xi) = \nu\xi^4 - \xi^2$ has non-zero singular points. we can not directly use the local smoothing effects and the maximal function estimates of Section 5.2 in this chapter to consider it. So we need use the following Fourier restriction operators

$$P^N f = \int_{|\xi|\geqslant N} e^{ix\xi}\hat{f}(\xi)d\xi, \quad P_N f = \int_{|\xi|\leqslant N} e^{ix\xi}\hat{f}(\xi)d\xi, \ \forall N > 0, \quad (5.133)$$

to eliminate the singularity of the phase function. Using the Fourier restriction operators and the results of Section 5.2, we have

$$\|D_x^{\frac{3}{2}} P^{2a} S(t)\varphi\|_{L_x^\infty L_t^2} \leqslant C\|\varphi\|_{L^2}, \quad (5.134)$$

$$\|D_x^{-\frac{1}{4}} P^a S(t)\varphi\|_{L_x^4 L_t^\infty} \leqslant C\|\varphi\|_{L^2}, \quad (5.135)$$

$$\left(\int_{-\infty}^{\infty} \sup_{[-T,T]} |P^a S(t)u_0|^2 dx\right)^{1/2} \leqslant C_{T,s}\|u_0\|_{H^s}, \ s > 1, \quad (5.136)$$

where a depends on ν. Then using (5.134), (5.135), (5.136) and the Strichartz estimates, and the Bourgain's spaces, we can obtain the local well-posedness of the Cauchy problem (5.132). One can refer to [109; 110].

For some other dispersive equation, in order to obtain the well-posedness, we can not directly use the Bourgain's spaces $X^{s,b}$ to construct the contraction mapping. So we need new method to consider them. In this book, we do not introduce them in details, one can refer to Guo's thesis [93]. For example: the generalized Benjamin-Ono (BO) equation:

$$\partial_t u + \mathcal{H}\partial_{xx} u = \mu\partial_x(u^k), \quad u(x,0) = u_0(x), \quad (5.137)$$

where $k \in \mathbb{Z}$, $\mu \in \mathbb{R}$, and $\mathcal{H}$ is the Hilbert transform $\mathcal{H}(f)(x) = \frac{1}{\pi}p.v.\int_{\mathbb{R}} \frac{f(y)}{x-y}dy$. From $\widehat{\mathcal{H}f} = -i\mathrm{sgn}(\xi)\widehat{f}(\xi)$, it follows that the dispersion relation of BO equation $\omega(\xi) = -|\xi|\xi$. This model was first introduced by Benjamin [10] and Ono [182], and it describe one-dimensional internal waves in deep water. Unlike Schrödinger equation, if initial data u_0 is a real number, then the solution u of (5.137) is also a real number. The real and complex BO equation are different.

Compared with KdV equation, BO equation has weaker dispersive effect. If $k = 2$, the Cauchy problem of BO equation (5.137) is badly behaved with respect to Picard iterative methods in standard Sobolev spaces. But for the definition of some weaker well-posedness (for example: the solution mapping is only continuous), some local results of the real BO equation were obtained. For now, the best local result was obtained by Kenig and Ionescu[113], who obtained local well-posedness in L^2. Their method based on the gauge transform and the Besov-type Bourgain's spaces $X^{s,b}$. If $k = 3$, the best result of the well-posedness was obtained by Kenig and Takaoka [136], who used the dyadic gauge transform to prove that the Cauchy problem is globally well-posed in $H^{1/2}$. Moreover,they showed that the index $s = 1/2$ is sharp in the sense: the solution map $u_0 \to u$ as mapping from H^s to $C([-T,T]; H^s)$ is no longer uniformly continuous when $s < 1/2$. Recently, in the paper [91], the author followed the idea in [113], used the contraction mapping to obtain the global well-posedness in $H^{1/2}$ with small data in L^2 norm, here the author did not use the gauge transform. So this method can apply to complex BO equation.

KdV equation and BO equation can be viewed as the special cases of the following dispersion generalized BO equation:

$$\partial_t u + |\partial_x|^{1+\alpha} \partial_x u = \mu \partial_x (u^k), \quad u(x,0) = u_0(x), \tag{5.138}$$

where $0 \leqslant \alpha \leqslant 1$, $u : \mathbb{R}^2 \to \mathbb{R}$ and $|\partial_x|$ is the Fourier multiplier operator with symbol $|\xi|$. This model is very interesting from mathematical view, it let us understand the relation between the dispersion effect and the well-posedness. From the intensity of dispersion effect, it lies between KdV equation and BO equation, for example: if $0 < \alpha < 1$, then it is similar with BO equation, H^s assumption on the initial data is insufficient for the local well-posedness via Picard iteration by showing the solution mapping fails to be C^2 smooth from H^s to $C([-T,T]; H^s)$ at the origin for any s, one can refer to [167]. For the methods to obtained the well-posedness for (5.138), the method of refined energy was used in [127], the method of short time $X^{s,b}$ in [115] was used in [92], and a para-differential renormalization technique was used in [108].

Notice that the spatial dimension of the dispersive equations considered above in this chapter is one. How about the application of the Bourgain's spaces to dispersive equations in higher dimension? For example: the Bourgain's spaces can be applied to the higher dimensional Schrödinger equation. Compared with one dimension case, cases in higher dimension are more delicate. For instance, a quadratic non-linear Schrödinger equation($n = 1, 2$)

$$iu_t + \Delta u = N(u, \bar{u}), \; u(x,0) = u_0(x),$$

where $(x,t) \in \mathbb{R}^n \times \mathbb{R}$, $N(u, \bar{u}) = c_1|u|^2 + c_2 u^2 + c_3 \bar{u}^2$. Using only the Strichartz estimates, one can obtain the well-posedness in H^s with $s \geqslant 0$.

However, by the Bourgain's spaces, it was showed that it is well-posed in H^s with $s > -3/4(c_1 = 0)$ and H^s with $s > -1/4(c_1 \neq 0)$ respectively, one can refer to [131; 41]. These results can be improved, one can refer to [9; 142]. For other nonlinearity $|u|^p u$ to non-linear Schrödinger equation, one can not obtain the better result of low regularity solution using Bourgain's spaces, but one can obtain the better result of global well-posedness using I-method, one can refer to [45]. Finally, we introduce KP equation, which can be viewed as two dimensional KdV equation. The KP equation is also an important shallow water model. The Cauchy problem of KP-I equation as follows:

$$\partial_t u + \partial_x^3 u - \partial_x^{-1}\partial_y^2 u + \partial_x(u^2/2) = 0; \; u(x,y,0) = \phi(x,y), \tag{5.139}$$

where $u(x,y,t) : \mathbb{R}^3 \to \mathbb{R}$. The KP-I equation and The KP-II equation, in which the sign of the term $\partial_x^{-1}\partial_y^2 u$ in (5.139) is $+$ instead of $-$, arise in physical contexts as models for the propagation of dispersive long waves with weak transverse effects. The KP-II equation is well understood from the point of view of well-posedness, one can refer to [16; 214; 212; 97; 98]. For the KP-I equation, Molinet, Saut and Tzvetkov [169] showed that it is badly behaved with respect to Picard iterative methods in standard Sobolev spaces H^{s_1,s_2} with any $s_1, s_2 \in \mathbb{R}$. On the positive side, it is known that the KP-I initial value problem is globally well-posed in the second energy spaces [122; 168]; these global well-posedness results rely on refined energy methods. Recently, Ionescu, Kenig and Tataru [115] obtained the global well-posedness in the natural energy space $\mathbb{E}^1 = \{\phi \in L^2(\mathbb{R}^2), \partial_x\phi \in L^2(\mathbb{R}^2), \partial_x^{-1}\partial_y\phi \in L^2(\mathbb{R}^2)\}$, they introduce a new method, which can be looked as the blend of Bourgain's spaces method and energy method. Recently, the authors in [95] remove the condition $\partial_x^{-1}\partial_y\phi \in L^2(\mathbb{R}^2)$. Compared with the result of KP-II equation, the global well-posedness of KP-I equation in L^2 is a open problem.

Chapter 6

Frequency-uniform decomposition techniques

W. Orlicz had a small apartment and he once applied to the city administration for a bigger one. The answer of an employee was: "Your apartment is really small but we cannot accept your claim since we know that you have your own spaces!" —— Lech Maligranda[2]

In this chapter we study the NLS and the NLKG by using the frequency-uniform decomposition techniques, see [243; 245; 246; 247].

Let Q_k be the unit cube with the center at k, $\{Q_k\}_{k\in\mathbb{Z}^n}$ constitutes a decomposition of $\mathbb{R}^n$. Such a kind of decomposition goes back to the work of N. Wiener [252], and we say that it is the Wiener decomposition of $\mathbb{R}^n$. We can roughly write

$$\Box_k \sim \mathscr{F}^{-1}\chi_{Q_k}\mathscr{F}, \quad k \in \mathbb{Z}^n, \tag{6.1}$$

where χ_E denotes the characteristic function on the set E. Since Q_k is a translation of Q_0, $\Box_k$ $(k \in \mathbb{Z}^n)$ have the same localized structures in the frequency space, which are said to be the *frequency-uniform decomposition operators*. Similar to Besov spaces, one can use $\{\Box_k\}_{k\in\mathbb{Z}^n}$ and $\ell^q(L^p)$ to generate a class of function spaces, so called modulation spaces for which the norm is defined by

$$\|f\|_{M^s_{p,q}} = \left(\sum_{k\in\mathbb{Z}^n} \langle k\rangle^{sq} \|\Box_k f\|_p^q\right)^{1/q}.$$

6.1 Why does the frequency-uniform decomposition work

Comparing with the dyadic decomposition, the frequency-uniform decomposition has at least two advantages for the Schrödinger semi-group:

[2]W. Orlicz (1903-1990), Polish mathematician and he is known for his Orlicz spaces, L. Maligranda is one of his students, see http://www.sm.luth.se/~lech.

(1) $\Box_k e^{it\Delta} : L^{p'} \to L^p$ satisfies a uniform truncated decay;
(2) $\Box_k e^{it\Delta}$ is uniformly bounded in L^p.

It is known that $S(t) = e^{it\Delta} : L^p \to L^p$ if and only if $p = 2$. This is one of the main reasons that we can not solve NLS in L^p ($p \neq 2$). However, if we consider the frequency-uniform decomposition, there holds

$$\|\Box_k S(t) f\|_p \lesssim (1+|t|)^{n|1/2-1/p|} \|\Box_k f\|_p. \tag{6.2}$$

Taking the summation to the above inequality over all $k \in \mathbb{Z}^n$, we can get the relevant estimate in modulation spaces, which enable us to solve NLS in modulation spaces $M^0_{p,1}$, $1 \leqslant p \leqslant \infty$.

Now we give a comparison between frequency-uniform and dyadic decompositions. Recalling that

$$\triangle_k \sim \mathscr{F}^{-1} \chi_{\{|\xi| \sim 2^k\}} \mathscr{F}$$

and noticing that $|\{\xi : |\xi| \sim 2^k\}| = O(2^{nk})$ and $|Q_k| = 1$, we see that their Bernstein's estimates are quite different:

$$\|\triangle_k f\|_q \lesssim 2^{n(1/p-1/q)k} \|\triangle_k f\|_p, \quad \|\Box_k f\|_q \lesssim \|\Box_k f\|_p, \quad p \leqslant q.$$

We find that there is no regularity increasement for the Bernstein estimate of $\Box_k$, which leads to the Schrödinger semi-group $S(t) = e^{it\Delta}$ satisfies the following truncated decay estimate,

$$\|\Box_k S(t) f\|_p \lesssim (1+|t|)^{-n(1/2-1/p)} \|\Box_k f\|_{p'}, \quad p \geqslant 2,\ 1/p + 1/p' = 1. \tag{6.3}$$

Comparing it with the classical $L^{p'} \to L^p$ estimate

$$\|\triangle_k S(t) f\|_p \lesssim |t|^{-n(1/2-1/p)} \|\triangle_k f\|_{p'}, \tag{6.4}$$

we see that the singularity at $t = 0$ disappears in (6.3).

Recalling that in the classical Strichartz inequalities, in order to handle the singularity of $|t|^{-n(1/2-1/p)}$ at $t = 0$, we need condition $n(1/2-1/p) \leqslant 1$, which arises from the Hardy-Littlewood-Sobolev inequality and it is an essential condition. To solve NLS with the nonlinearity $|u|^\sigma u$, one need to improve the spatial regularity at least at the $\dot{H}^{n/2-2/\sigma}$ level for $\sigma \geqslant 4/n$.

Considering the Strichartz inequalities with the frequency-uniform decomposition, due to the truncated decay, we can remove the condition $n(1/2 - 1/p) \leqslant 1$. As a result, in solving NLS with the nonlinearity $|u|^\sigma u$, it is not necessary to improve the spatial derivative regularity if σ is large.

The frequency-uniform decomposition is more delicate than the dyadic decomposition and it is easier to handle the derivatives in the nonlinearity, say for $k = (k_1, ..., k_n)$ with $k_1 \gg 1$ and $j \gg 1$,

$$\|\partial_{x_1} \Box_k f\|_p \sim k_1 \|\Box_k f\|_p, \quad \|\partial_{x_1} \triangle_j f\|_p \lesssim 2^j \|\triangle_j f\|_p.$$

Using the frequency-uniform decomposition techniques, the initial data should belong to modulation spaces $M^s_{p,q}$. The modulation space $M^s_{p,q}$ coincides with the Sobolev space H^s in the case $p = q = 2$. Recently, this technique was also developed to the Navier-Stokes equation and the dissipative nonlinear electrohydrodynamic system [247; 116; 56].

6.2 Frequency-uniform decomposition, modulation spaces

Roughly speaking, dyadic decomposition operators combined with function spaces $\ell^q(L^p)$ generate Besov spaces, frequency-uniform decomposition operators joint with function spaces $\ell^q(L^p)$ produce modulation spaces. The modulation space was introduced by Feichtinger [73] in 1983, from the history point of view, it was defined by the short-time Fourier transform[3]. During the past twenty years, the frequency-uniform decomposition had not been attached the importance to applications and it is even not mentioned in Gröchenig's book [86]. However, from PDE point of view, the combination of frequency-uniform decomposition operators and Banach function spaces $\ell^q(X(\mathbb{R}^n))$[4] seems to be important in making nonlinear estimates, which contains an automatic decomposition on high-low frequencies.

We now give an exact definition on frequency-uniform decomposition operators. Since χ_{Q_k} can not make differential operations, one needs to replace χ_{Q_k} in (6.1) by a smooth cut-off function. Let $\rho \in \mathscr{S}(\mathbb{R}^n)$, $\rho : \mathbb{R}^n \to [0,1]$ be a smooth function verifying $\rho(\xi) = 1$ for $|\xi|_\infty \leqslant 1/2$ and $\rho(\xi) = 0$ for $|\xi|_\infty \geqslant 1$[5]. Let ρ_k be a translation of ρ,

$$\rho_k(\xi) = \rho(\xi - k), \quad k \in \mathbb{Z}^n. \tag{6.6}$$

[3]Let $g \in \mathscr{S}(\mathbb{R}^n)$,

$$V_g f(x,\omega) = \int_{\mathbb{R}^n} e^{-it\omega} \overline{g(t-x)} f(t) dt.$$

$V_g f$ is said to be the short-time Fourier transform of f. The norm on modulation spaces is given by

$$\|f\|^\circ_{M^s_{p,q}} = \left(\int_{\mathbb{R}^n} \left(\int_{\mathbb{R}^n} |V_g f(x,\omega)|^p dx \right)^{q/p} \langle\omega\rangle^{sq} d\omega \right)^{1/q}. \tag{6.5}$$

One can prove that $\|\cdot\|^\circ_{M^s_{p,q}}$ and $\|\cdot\|_{M^s_{p,q}}$ are equivalent norms; cf. [73] for a proof on modulation spaces defined in an Abel group and [246] for a straightforward proof.

[4]X is a Banach function space defined in $\mathbb{R}^n$.

[5]For $\xi = (\xi_1, ..., \xi_n)$, $|\xi|_\infty := \max_{i=1,...,n} |\xi_i|$.

We see that $\rho_k(\xi) = 1$ in Q_k and so, $\sum_{k\in\mathbb{Z}^n} \rho_k(\xi) \geqslant 1$ for all $\xi \in \mathbb{R}^n$. Denote

$$\sigma_k(\xi) = \rho_k(\xi) \left(\sum_{k\in\mathbb{Z}^n} \rho_k(\xi) \right)^{-1}, \quad k \in \mathbb{Z}^n. \tag{6.7}$$

Then we have

$$\begin{cases} |\sigma_k(\xi)| \geqslant c, \quad \forall\, \xi \in Q_k, \\ \operatorname{supp} \sigma_k \subset \{\xi : |\xi - k|_\infty \leqslant 1\}, \\ \sum_{k\in\mathbb{Z}^n} \sigma_k(\xi) \equiv 1, \quad \forall\, \xi \in \mathbb{R}^n, \\ |D^\alpha \sigma_k(\xi)| \leqslant C_{|\alpha|}, \quad \forall\, \xi \in \mathbb{R}^n,\ \alpha \in (\mathbb{N} \cup \{0\})^n. \end{cases} \tag{6.8}$$

Hence, the set

$$\Upsilon_n = \{\{\sigma_k\}_{k\in\mathbb{Z}^n} : \{\sigma_k\}_{k\in\mathbb{Z}^n} \text{ satisfies (6.8)}\} \tag{6.9}$$

is non-void. If there is no confusion, in the sequel we will write $\Upsilon = \Upsilon_n$. Let $\{\sigma_k\}_{k\in\mathbb{Z}^n} \in \Upsilon$. Denote

$$\square_k := \mathscr{F}^{-1} \sigma_k \mathscr{F}, \quad k \in \mathbb{Z}^n. \tag{6.10}$$

$\{\square_k\}_{k\in\mathbb{Z}^n}$ are said to be frequency-uniform decomposition operators. For $k \in \mathbb{Z}^n$, we denote $|k| = |k_1| + ... + |k_n|$, $\langle k \rangle = 1 + |k|$. Let $s \in \mathbb{R}$, $0 < p, q \leqslant \infty$,

$$M^s_{p,q}(\mathbb{R}^n) = \left\{ f \in \mathscr{S}'(\mathbb{R}^n) : \|f\|_{M^s_{p,q}} = \left(\sum_{k\in\mathbb{Z}^n} \langle k \rangle^{sq} \|\square_k f\|_p^q \right)^{1/q} < \infty \right\}. \tag{6.11}$$

For simplicity, we write $M^0_{p,q} = M_{p,q}$. $M^s_{p,q}$ is said to be the modulation space.

6.2.1 *Basic properties on modulation spaces*

As indicated in Proposition 1.16, if we consider a function f with a compact support set in the frequency space, then we can compare $\|f\|_p$ with $\|f\|_q$, which is one of the advantages of the frequency localizations. The following is a refinement of Proposition 1.16.

Lemma 6.1. *Let Ω be a compact subset of $\mathbb{R}^n$*, $\operatorname{diam} \Omega < 2R$, $0 < p \leqslant q \leqslant \infty$. *Then there exists a constant $C > 0$ which only depends on p, q and R such that*

$$\|f\|_q \leqslant C \|f\|_p, \quad \forall\, f \in L^p_\Omega, \tag{6.12}$$

where $L^p_\Omega = \{f \in L^p : \operatorname{supp} \hat{f} \subset \Omega\}$.

Proof. We emphasize that the constant in (6.12) is independent of the position of Ω. Let $\psi \in \mathscr{S}(\mathbb{R}^n)$ satisfy supp $\widehat{\psi} \subset B(0, 2R)$ and $\widehat{\psi}|_{B(0,R)} = 1$. Due to diam $\Omega < 2R$, we can take ξ_0 satisfting $\Omega \subset B(\xi_0, R)$. For any $f \in \mathscr{S}(\mathbb{R}^n) \cap L^p_\Omega$, we have $\widehat{f} = \widehat{f} \cdot \widehat{\psi}(\cdot - \xi_0)$, which implies that

$$f(x) = c \int_{\mathbb{R}^n} f(x-y) e^{\mathrm{i}\xi_0 y} \psi(y) dy.$$

If $1 \leqslant p \leqslant \infty$, using Young's inequality, we can get the conclusion. If $0 < p < 1$,

$$\|f\|_\infty \leqslant C_\psi \left(\sup_y |f(x-y)|^{1-p} \right) \int_{\mathbb{R}^n} |f(x-y)|^p dy,$$

we obtain that for $q = \infty$ and $0 < p < 1$, the result holds. For general $0 < p \leqslant q < \infty$, by Hölder's inequality, L^q norm can be controlled by L^p and L^∞ norms, from which we get the result, as desired. □

Lemma 6.2 (L^p_Ω-multiplier). *Let $\Omega \subset \mathbb{R}^n$ be a compact subset, $0 < r \leqslant \infty$ and $\sigma_r = n(1/(r \wedge 1) - 1/2)$. If $s > \sigma_r$, then there exists a $C > 0$ such that*

$$\|\mathscr{F}^{-1} \varphi \mathscr{F} f\|_r \leqslant C \|\varphi\|_{H^s} \|f\|_r \tag{6.13}$$

holds for all $f \in L^r_\Omega$ and $\varphi \in H^s$.

Proof. If $r \geqslant 1$, in view of Bernstein's multiplier estimates, we have the result, see Proposition 1.11. If $r < 1$, the proof is similar to the case $r > 1$, cf. [224]. □

Proposition 6.1 (Completeness). *Let $0 < p, q \leqslant \infty$ and $s \in \mathbb{R}$.*

(1) *$M^s_{p,q}$ is a (quasi-) Banach space. Moreover, if $1 \leqslant p, q \leqslant \infty$, then $M^s_{p,q}$ is a Banach space.*
(2) *$\mathscr{S}(\mathbb{R}^n) \subset M^s_{p,q} \subset \mathscr{S}'(\mathbb{R}^n)$.*
(3) *Let $0 < p, q < \infty$, then $\mathscr{S}(\mathbb{R}^n)$ is dense in $M^s_{p,q}$.*

Proof. Analogous to Besov spaces, we can prove the consequence and the details are omitted. □

Proposition 6.2 (Equivalent norm). *Let $\{\sigma_k\}_{k \in \mathbb{Z}^n}$, $\{\varphi_k\}_{k \in \mathbb{Z}^n} \in \Upsilon$. Then $\{\sigma_k\}_{k \in \mathbb{Z}^n}$ and $\{\varphi_k\}_{k \in \mathbb{Z}^n}$ generate equivalent norms on $M^s_{p,q}$.*

Proof. We have the translation identity,

$$(\mathscr{F}^{-1}m\mathscr{F}f)(x) = e^{ixk}\left[\mathscr{F}^{-1}m(\cdot+k)\mathscr{F}(e^{-iky}f(y))\right](x).$$

For convenience, we denote

$$\square_k^\sigma := \mathscr{F}^{-1}\sigma_k\mathscr{F}, \quad \square_k^\varphi := \mathscr{F}^{-1}\varphi_k\mathscr{F}.$$

Noticing the almost orthogonality of $\square_k^\sigma$

$$\square_k^\sigma = \sum_{|\ell|_\infty\leqslant 1} \square_k^\sigma\square_{k+\ell}^\varphi, \tag{6.14}$$

we have

$$\|\square_k^\sigma f\|_p \leqslant \sum_{|\ell|_\infty\leqslant 1} \|\square_{k+\ell}^\varphi(\square_k^\sigma f)\|_p.$$

Using the multiplier estimate

$$\|\square_{k+\ell}^\varphi(\square_k^\sigma f)\|_p \lesssim \|\sigma_k\|_{H^s}\|\square_{k+\ell}^\varphi f\|_p, \quad s > n(1/(1\wedge p)-1/2),$$

and $\|\sigma_k\|_{H^s} \leqslant C$, we immediately have

$$\|\square_k^\sigma f\|_p \lesssim \sum_{|\ell|_\infty\leqslant 1} \|\square_{k+\ell}^\varphi f\|_p.$$

So, $\|f\|_{M^s_{p,q}}^{\{\sigma_k\}} \leqslant \|f\|_{M^s_{p,q}}^{\{\varphi_k\}}$. $\square$

Proposition 6.2 indicates that one can choose $\{\sigma_k\}_{k\in\mathbb{Z}^n} \in \Upsilon_n$ according to our requirement. In applications of PDE, it is convenient to us the following $\{\sigma_k\}_{k\in\mathbb{Z}^n} \in \Upsilon_n$. Let $\{\eta_k\}_{k\in\mathbb{Z}} \in \Upsilon_1$, we denote

$$\sigma_k(\xi) := \eta_{k_1}(\xi_1)...\eta_{k_n}(\xi_n), \tag{6.15}$$

then we have $\{\sigma_k\}_{k\in\mathbb{Z}^n} \in \Upsilon_n$. the above $\sigma_k(\xi)$ realizes the separation of different variables.

Proposition 6.3 (Embedding). *Let $s_1, s_2 \in \mathbb{R}$ and $0 < p_1, p_2, q_1, q_2 \leqslant \infty$.*

(1) *If $s_2 \leqslant s_1$, $p_1 \leqslant p_2$ and $q_1 \leqslant q_2$, then $M^{s_1}_{p_1,q_1} \subset M^{s_2}_{p_2,q_2}$.*
(2) *If $q_2 < q_1$ and $s_1 - s_2 > n/q_2 - n/q_1$, then $M^{s_1}_{p,q_1} \subset M^{s_2}_{p,q_2}$.*

Proof. Recall that

$$\|\square_k f\|_{p_2} \leqslant \sum_{|\ell|_\infty\leqslant 1} \|\mathscr{F}^{-1}\sigma_k\mathscr{F}(\square_{k+\ell}f)\|_{p_2}.$$

Using Lemma 6.1, we have

$$\|\Box_k f\|_{p_2} \lesssim \sum_{|\ell|_\infty \leqslant 1} \|\mathscr{F}^{-1}\sigma_k \mathscr{F}(\Box_{k+\ell} f)\|_{p_1} \lesssim \sum_{|\ell|_\infty \leqslant 1} \|\Box_{k+\ell} f\|_{p_1}.$$

In view of $\ell^{q_1} \subset \ell^{q_2}$, we can get the conclusion of (1).

Now we prove (2). Using Hölder's inequality,

$$\begin{aligned}\|f\|_{M^{s_2}_{p,q_2}} &= \left(\sum_{k\in\mathbb{Z}^n} \langle k\rangle^{s_2 q_2} \|\Box_k f\|_p^{q_2}\right)^{1/q_2} \\ &\leqslant \|f\|_{M^{s_1}_{p,q_1}} \left(\sum_{k\in\mathbb{Z}^n} \langle k\rangle^{(s_2-s_1)q_1q_2/(q_1-q_2)}\right)^{(q_1-q_2)/q_1q_2}. \end{aligned} \tag{6.16}$$

Noticing that

$$\sum_{k\in\mathbb{Z}^n} \langle k\rangle^{(s_2-s_1)q_1q_2/(q_1-q_2)} \lesssim \sum_{i=0}^{\infty} \langle i\rangle^{n-1+(s_2-s_1)q_1q_2/(q_1-q_2)}, \tag{6.17}$$

$s_1 - s_2 > n/q_2 - n/q_1$ implies that the right hand side of (6.17) is a convergent series. It follows that (2) holds. □

Proposition 6.4 (Dual space). *Let $s \in \mathbb{R}$ and $0 < p, q < \infty$. If $p \geqslant 1$, we denote $1/p + 1/p' = 1$; If $0 < p < 1$, we write $p' = \infty$. Then*

$$(M^s_{p,q})^* = M^{-s}_{p',q'}. \tag{6.18}$$

If $p \geqslant 1$, Proposition 6.4 is similar to that of Besov spaces, however, if $0 < p < 1$, the result is quite different from that of Besov spaces. The details of the proof of Proposition 6.4 can be found in [246] by following the proof of the relevant result in Besov spaces.

Remark 6.1. If $p, q \in [1, \infty]$, Propositions 6.1 and 6.4 were obtained by Feichtinger [73]. In [246; 247], the cases $0 < p < 1$ and $0 < q < 1$ were considered. The proofs of the results in this section are due to [246; 247].

Soon after the work [247], Kobayashi [148] independently defined $M_{p,q}$ for all $0 < p, q \leqslant \infty$ and obtained Proposition 6.1. Almost at the same time as [246], Kobayashi [149] discussed the dual space of $M_{p,q}$ and obtained partial results of Proposition 6.4: if $0 < p < 1$ or $1 < q < \infty$, he obtained $M_{p',q'} \subset (M_{p,q})^* \subset M_{\infty,\infty}$. For the other cases, he showed $(M_{p,q})^* = M_{p',q'}$. Recently, by using the molecular decomposition techniques of modulation spaces, Kobayashi and Sawano [150] reconsidered the dual space of $M^s_{p,q}$ and they also obtained the result of Proposition 6.4. It is worth to mention that Triebel [225] introduced a class of generalized modulation spaces for all indices $0 < p, q \leqslant \infty$, however, those spaces have no complete norms, which seems harder to use in the study of PDEs.

6.3 Inclusions between Besov and modulation spaces

From the definitions, we see that Besov spaces and modulation spaces are rather similar, both of them are the combinations of frequency decomposition operators and function spaces $\ell^q(L^p)$. In fact, we have the following inclusion results.

Theorem 6.1 (Embedding). *Let $0 < p, q \leqslant \infty$ and $s_1, s_2 \in \mathbb{R}$. We have the following results.*

(1) $B^{s_1}_{p,q} \subset M^{s_2}_{p,q}$ *if and only if* $s_1 \geqslant s_2 + \tau(p,q)$, *where*

$$\tau(p,q) = \max\left\{0,\ n\left(\frac{1}{q} - \frac{1}{p}\right),\ n\left(\frac{1}{q} + \frac{1}{p} - 1\right)\right\};$$

(2) $M^{s_1}_{p,q} \subset B^{s_2}_{p,q}$ *if and only if* $s_1 \geqslant s_2 + \sigma(p,q)$, *where*

$$\sigma(p,q) = \max\left\{0,\ n\left(\frac{1}{p} - \frac{1}{q}\right),\ n\left(1 - \frac{1}{p} - \frac{1}{q}\right)\right\}.$$

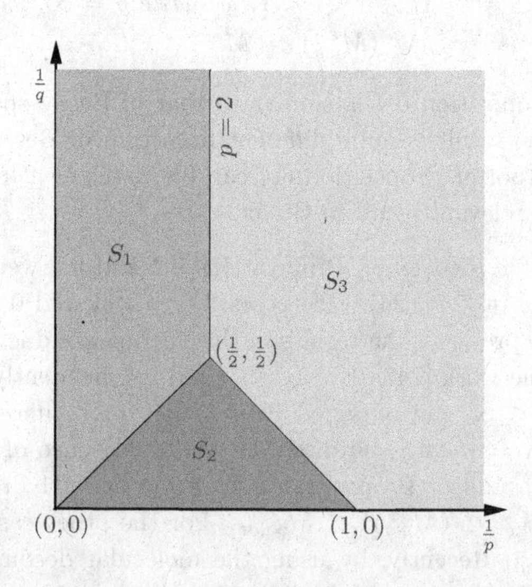

Fig. 6.1 The distribution of $\tau(p,q)$ $\mathbb{R}^2_+$: $\tau(p,q) = n(\frac{1}{q} - \frac{1}{p})$ in S_1; $\tau(p,q) = 0$ in S_2; $\tau(p,q) = n(\frac{1}{p} + \frac{1}{q} - 1)$ in S_3.

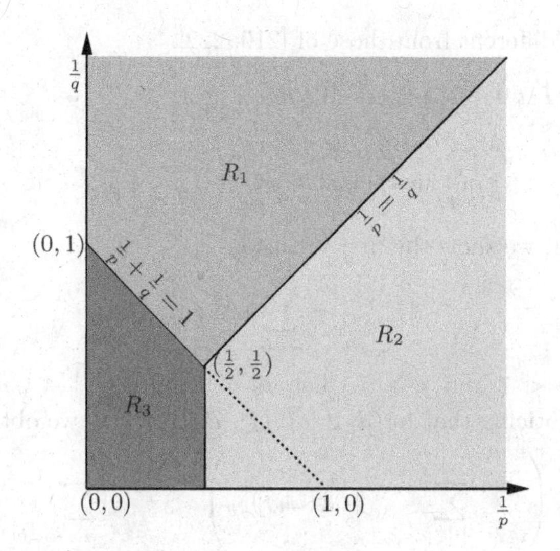

Fig. 6.2 The distribution of $\sigma(p, q)$: $\sigma(p, q) = 0$ in R_1; $\sigma(p, q) = n(\frac{1}{p} - \frac{1}{q})$ in R_2; $\sigma(p, q) = n(1 - \frac{1}{p} - \frac{1}{q})$ in R_3.

The inclusions between Besov and modulation spaces in the cases $(1/p, 1/q) \in [0, 1]^2$ were first discussed by Gröbner [85] and he has never published his results. When $(1/p, 1/q)$ is in the vertices of the square $[0, 1]^2$, Gröbner's results are optimal. Afterwards, Toft [222] obtained the sufficiency of Theorem 6.1 in the cases $(1/p, 1/q) \in [0, 1]^2$. Sugimoto and Tomita [210] showed the necessity of the first inclusion of Theorem 6.1 in the cases $(1/p, 1/q) \in [0, 1]^2$, and by duality they obtained the second inclusion is also sharp if $(1/p, 1/q) \in [0, 1]^2$ and $p, q \neq \infty$. Sugimoto and Tomita's idea is to use Feichtinger's norm and the dilation property of modulation spaces. In [245; 246; 247] the authors proved the conclusions of Theorem 6.1 by using frequency-uniform decomposition techniques.

Corollary 6.1. *We have the following inclusions.*

$$B^{s+n/2}_{2,1} \subset M^s_{2,1} \subset B^s_{2,1}, \quad B^{s+n}_{\infty,1} \subset M^s_{\infty,1} \subset B^s_{\infty,1}.$$

The above embedding theorem is of importance for the study of nonlinear PDEs and we give the details of the proof. We can first prove Theorem 6.1 for some special p, q and then by the interpolation to carry out the other cases. The following proof is a collection of [85; 247; 246;

245], which is different from those of [210; 222].

Lemma 6.3. *Let* $0 < p, q \leqslant \infty$. *We have*

$$M_{p,q} \subset B^0_{p,q}, \quad \forall\, q \leqslant p \wedge 1,$$
$$M^{n(1/(p\wedge 1)-1/q)}_{p,q} \subset B^0_{p,q}, \quad \forall\, q \geqslant 1 \wedge p.$$

Proof. First, we show the first inclusion

$$\|f\|^q_{B^0_{p,q}} = \sum_{k=0}^{\infty} \|\triangle_k f\|^q_p. \tag{6.19}$$

Case 1. $q < 1$ and $p \geqslant 1$. Let $a_k = \max(0, 2^{k-1} - \sqrt{n})$ and $b_k = 2^{k+1} + \sqrt{n}$. Noticing that for $|i| \notin [a_k, b_k]$, $\triangle_k \Box_i f = 0$, we obtain that

$$\|\triangle_k f\|^q_p \leqslant \left(\sum_{i\in\mathbb{Z}^n,\ |i|\in[a_k,b_k]} \|\triangle_k(\Box_i f)\|_p \right)^q \lesssim \sum_{i\in\mathbb{Z}^n,\ |i|\in[a_k,b_k]} \|\Box_i f\|^q_p. \tag{6.20}$$

From (6.20) we deduce the result.

Case 2. $q < 1$ and $p < 1$. From $p < 1$ and $q/p \leqslant 1$ it follows that

$$\begin{aligned} \|\triangle_k f\|^q_p &\leqslant \left(\sum_{i\in\mathbb{Z}^n,\ |i|\in[a_k,b_k]} \int |\triangle_k(\Box_i f)(x)|^p dx \right)^{q/p} \\ &\leqslant \sum_{i\in\mathbb{Z}^n,\ |i|\in[a_k,b_k]} \|\triangle_k(\Box_i f)\|^q_p. \end{aligned} \tag{6.21}$$

In view of the multiplier estimate on L^p_Ω, $\triangle_k \Box_i : L^p_{B(i,\sqrt{n})} \to L^p_{B(i,\sqrt{n})}$. By (6.21) we get the conclusion.

Next, we show the second inclusion. We consider the following two cases.

Case 1. $p \geqslant 1$. Denote

$$\Lambda_0 = \{k \in \mathbb{Z}^n : B(k, \sqrt{n}) \cap \{\xi : |\xi| \in [0, 2)\} \neq \emptyset\}, \tag{6.22}$$
$$\Lambda_j = \{k \in \mathbb{Z}^n : B(k, \sqrt{n}) \cap \{\xi : |\xi| \in [2^{j-1}, 2^{j+1})\} \neq \emptyset\}, \quad j \geqslant 1. \tag{6.23}$$

In view of the multiplier estimate on L^p_Ω,

$$\|\triangle_j f\|_p \lesssim \sum_{k\in\Lambda_j} \sum_{|\ell|_\infty \leqslant 1} \|\triangle_j \Box_{k+\ell} \Box_k f\|_p \lesssim \sum_{k\in\Lambda_j} \|\Box_k f\|_p. \tag{6.24}$$

Since $q \geqslant p \wedge 1$, we have

$$(a_1 + ... + a_m)^q \leqslant m^{q-1}(a_1^q + ... + a_m^q). \tag{6.25}$$

For $k \in \Lambda_j$, one has that $|k| \sim 2^j$ and Λ_j overlaps at most $O(2^{nj})$ cubes, which follows that

$$
\begin{aligned}
\|f\|^q_{B^0_{p,q}} &\lesssim \sum_{j=0}^{\infty} \left(\sum_{k\in\Lambda_j} \|\Box_k f\|_p \right)^q \\
&\lesssim \sum_{j=0}^{\infty} 2^{jn(q-1)} \sum_{k\in\Lambda_j} \|\Box_k f\|^q_p \\
&\lesssim \sum_{j=0}^{\infty} \sum_{k\in\Lambda_j} \langle k\rangle^{n(q-1)} \|\Box_k f\|^q_p \\
&\lesssim \|f\|^q_{M^{n(1-1/q)}_{p,q}}. \qquad (6.26)
\end{aligned}
$$

Case 2. $p < 1$. In view of $q/p > 1$, we have

$$(a_1 + ... + a_m)^{q/p} \leqslant m^{q/p-1}(a_1^{q/p} + ... + a_m^{q/p}). \qquad (6.27)$$

By the multiplier estimate on L^p_Ω,

$$
\begin{aligned}
\|\triangle_j f\|^q_p &\lesssim \left(\sum_{k\in\Lambda_j} \sum_{\ell\in\Lambda} \int_{\mathbb{R}^n} |\triangle_j \Box_{k+\ell} \Box_k f|^p dx \right)^{q/p} \\
&\lesssim 2^{jn(q/p-1)} \sum_{k\in\Lambda_j} \|\Box_k f\|^q_p. \qquad (6.28)
\end{aligned}
$$

Similar to Case 1, we have the conclusion, as desired. □

Lemma 6.4. *We have*

$$B^{n(1/q-1/2)}_{2,q} \subset M_{2,q}, \quad \forall\, 0 < q \leqslant 2.$$

Proof. By Plancherel's identity,

$$\|f\|^q_{M_{2,q}} \sim \sum_{k\in\mathbb{Z}^n} \|\chi_{Q_k} \widehat{f}\|^q_2. \qquad (6.29)$$

Denote $\Lambda_j = \{k \in \mathbb{Z}^n : |k| \in [2^{j-1}, 2^j)\}$. Let $L \gg 1$. One has that

$$\|f\|^q_{M_{2,q}} \lesssim \sum_{|k|\leqslant 2^L} \|\chi_{Q_k} \widehat{f}\|^q_2 + \sum_{j=L}^{\infty} \sum_{k\in\Lambda_j} \|\chi_{Q_k} \widehat{f}\|^q_2. \qquad (6.30)$$

It is easy to see that Λ_j has at most $O(2^{nj})$ elements. So,

$$\|f\|^q_{M_{2,q}} \lesssim \|f\|^q_2 + \sum_{j=L}^{\infty} 2^{jn(1-q/2)} \left\| \sum_{k\in\Lambda_j} \chi_{Q_k} \widehat{f} \right\|^q_2$$

$$\lesssim \|f\|_2^q + \sum_{j=L}^{\infty} 2^{jn(1-q/2)} \|\chi_{|\cdot|\in(2^{j-2},2^{j+1})} \widehat{f}\|_2^q$$

$$\lesssim \|f\|^q_{B^{n(1/q-1/2)}_{2,q}}, \tag{6.31}$$

which is the result, as desired. □

Lemma 6.5. *Let* $1 \leqslant p, q \leqslant \infty$. *We have*

$$M^{\sigma(p,q)}_{p,q} \subset B^0_{p,q}, \ \ \sigma(p,q) = \max\left(0,\ n\left(\frac{1}{p \wedge p'} - \frac{1}{q}\right)\right). \tag{6.32}$$

Proof. By the dual versions of Lemmas 6.3 and 6.4, we have

$$M^n_{p,\infty} \subset B^0_{p,\infty}, \ \ 1 \leqslant p \leqslant \infty, \tag{6.33}$$

$$M^{n(1/2-1/q)}_{2,q} \subset B^0_{2,q}, \ \ 2 \leqslant q \leqslant \infty. \tag{6.34}$$

Taking $p = 1, \infty$ and $q = \infty$, we have

$$M^n_{1,\infty} \subset B^0_{1,\infty}, \ \ M^{n/2}_{2,\infty} \subset B^0_{2,\infty}, \ \ M^n_{\infty,\infty} \subset B^0_{\infty,\infty}. \tag{6.35}$$

By the complex interpolation on (6.35) (see Appendix),

$$M^{\max(n/p,\, n/p')}_{p,\infty} \subset B^0_{p,\infty}, \ \ 1 \leqslant p \leqslant \infty. \tag{6.36}$$

Lemma 6.3 also implies that

$$M_{p,1} \subset B^0_{p,1}, \ \ 1 \leqslant p \leqslant \infty. \tag{6.37}$$

Recall that

$$M_{2,2} = B^0_{2,2}. \tag{6.38}$$

Making a complex interpolation on (6.36)–(6.38), we obtain the result, as desirted. □

When $0 < p < 1$, we need the following multiplier estimate (cf. Peetre [192] and Triebel [224]).

Proposition 6.5. *Let* $\Omega \subset \mathbb{R}^n$ *be a compact set, and* $0 < p \leqslant 1$. *Then*

$$\|\mathscr{F}^{-1} M \mathscr{F} f\|_p \lesssim \|M\|_{B^{n(1/p-1/2)}_{2,p}} \|f\|_p$$

holds for all $f \in L^p_\Omega$ *and* $M \in B^{n(1/p-1/2)}_{2,p}$.

Corollary 6.2. *Let* $b > 0$ *and* $0 < p \leqslant 1$. *Then*

$$\|\mathscr{F}^{-1} M \mathscr{F} f\|_p \leqslant C \|M(b\,\cdot)\|_{B^{n(1/p-1/2)}_{2,p}} \|f\|_p$$

holds for all $f \in L^p_{B(0,b)}$ *and* $M \in B^{n(1/p-1/2)}_{2,p}$, *where* $C > 0$ *is independent of* $b > 0$.

Proof. If $f \in L^p_{B(0,b)}$, then $f(b^{-1}\cdot) \in L^p_{B(0,1)}$. In view of

$$(\mathscr{F}^{-1}M\mathscr{F}f)(x) = [\mathscr{F}^{-1}M(b\,\cdot)\widehat{f(b^{-1}\cdot)}](bx),$$

it follows from Proposition 6.5 that Corollary 6.2 holds. □

Lemma 6.6. *Let $0 < p \leqslant \infty$. Then we have*

$$B^{n(1/(p\wedge 1)-1)}_{p,\infty} \subset M_{p,\infty}. \tag{6.39}$$

Proof. First, we consider the case $0 < p < 1$. By Corollary 6.2, for any $|k| \gg 1$, $|k| \in [2^{j-1}, 2^j)$,

$$\begin{aligned} \|\Box_k f\|_p &= \left\| \mathscr{F}^{-1}\sigma_k \sum_{\ell=-4}^{4} \varphi_{j+\ell}\mathscr{F}f \right\|_p \\ &\lesssim \|\sigma_k(2^{j+5}\,\cdot)\|_{B^{n(1/p-1/2)}_{2,p}} \sum_{\ell=-4}^{4} \|\triangle_{j+\ell}f\|_p . \end{aligned} \tag{6.40}$$

Using the scaling in Besov spaces (cf. [224])

$$\|g(\lambda\,\cdot)\|_{B^s_{p,q}} \leqslant \lambda^{s-n/p}\|g\|_{B^s_{p,q}}, \quad \lambda \gtrsim 1,$$

we obtain that[6]

$$\|\sigma_k(2^{j+5}\,\cdot)\|_{B^{n(1/p-1/2)}_{2,p}} \lesssim 2^{jn(1/p-1)}\|\sigma_k\|_{B^{n(1/p-1/2)}_{2,p}} \lesssim 2^{jn(1/p-1)}. \tag{6.41}$$

Inserting (6.41) into (6.40), we immediately have

$$\|\Box_k f\|_p \lesssim 2^{jn(1/p-1)} \sum_{\ell=-4}^{4} \|\triangle_{j+\ell}f\|_p . \tag{6.42}$$

By (6.42), we get (6.39).

Now we consider the case $p > 1$. By Young's inequality, for $|k| \gg 1$,

$$\begin{aligned} \|\Box_k f\|_p &= \left\| \mathscr{F}^{-1}\sigma_k \sum_{\ell=-4}^{4} \varphi_{j+\ell}\mathscr{F}f \right\|_p \\ &\lesssim \|\mathscr{F}^{-1}\sigma_k\|_1 \sum_{\ell=-4}^{4} \|\triangle_{j+\ell}f\|_p \lesssim \sum_{\ell=-4}^{4} \|\triangle_{j+\ell}f\|_p . \end{aligned} \tag{6.43}$$

This implies the result. □

[6]We can assume that $\sigma_k = \sigma_0(\cdot - k)$ in the definition of $M^s_{p,q}$.

Proof. [Proof of Theorem 6.1] (Sufficiency) First, we show that $M^{s+\sigma(p,q)}_{p,q} \subset B^s_{p,q}$. By Lemma 6.5, we see that the conclusion holds if $1 \leqslant p, q \leqslant \infty$. By Lemma 6.3 we have the result if $0 < p \leqslant 1$ or $0 < q < 1$.

Next, we prove that $B^{s+\tau(p,q)}_{p,q} \subset M^s_{p,q}$. Denote $\mathbb{R}^2_+ = \{(1/p, 1/q) : 1/p,\ 1/q \geqslant 0\}$ and (see Fig. 6.1),

$$\begin{aligned} S_1 &= \{(1/p,\ 1/q) \in \mathbb{R}^2_+ : 1/q \geqslant 1/p,\ 1/p \leqslant 1/2\}; \\ S_2 &= \{(1/p,\ 1/q) \in \mathbb{R}^2_+ : 1/q \leqslant 1/p,\ 1/p + 1/q \leqslant 1\}; \\ S_3 &= \mathbb{R}^2_+ \setminus (S_1 \cup S_2). \end{aligned}$$

We first consider the case $(1/p, 1/q) \in S_3$. We have $\tau(p,q) = n(1/p + 1/q - 1)$. Take (p_0, q_0) and (p_1, q_1) satisfying

$$\begin{aligned} &\frac{1}{p_0} = \frac{1}{p} + \frac{1}{q}, \quad \frac{1}{q_0} = 0; \\ &\frac{1}{p_1} = \frac{1}{2}, \quad \frac{1}{q_1} = \frac{1}{p} + \frac{1}{q} - \frac{1}{2}. \end{aligned}$$

Let $\theta = \frac{1}{q}(\frac{1}{p} + \frac{1}{q} - \frac{1}{2})^{-1}$, we have

$$\begin{aligned} &\frac{1}{p} = \frac{1-\theta}{p_0} + \frac{\theta}{p_1}, \quad \frac{1}{q} = \frac{1-\theta}{q_0} + \frac{\theta}{q_1}; \\ &\frac{1}{p} + \frac{1}{q} - 1 = (1-\theta)\left(\frac{1}{p_0} - 1\right) + \left(\frac{1}{q_1} - \frac{1}{2}\right)\theta. \end{aligned}$$

By Lemmas 6.4 and 6.6,

$$B^{n(1/q_1 - 1/2)}_{2,q_1} \subset M_{2,q_1}, \quad B^{n(1/p_0 - 1)}_{p_0,\infty} \subset M_{p_0,\infty}.$$

A complex interpolation yields

$$B^{n(1/p+1/q-1)}_{p,q} \subset M_{p,q}.$$

Secondly, we consider the case $(1/p, 1/q) \in S_1$. If $(1/p, 1/q) \in \dot{S}_1$ ($\dot{S}_1$ denotes the set of all inner points of S_1), then $(1/p, 1/q)$ can be lying in the segment by connecting $(1/\infty, 1/\infty)$ and a point $(1/p_1, 1/q_1)$ in the line $\{(1/p, 1/q) : p = 2, q < 2\}$. By a complex interpolation, we have

$$B^{n(1/q-1/p)}_{p,q} \subset M_{p,q}. \tag{6.44}$$

Thirdly, if $(1/p, 1/q) \in S_2$, the dual version of the first inclusion implies the result, as desired.

(Necessity) We need to show that for any $0 < \eta \ll 1$,

$$B^{\tau(p,q)-\eta}_{p,q} \not\subset M_{p,q}. \tag{6.45}$$

Case 1.1. $(1/p, 1/q) \in S_3$. Let $f = \mathscr{F}^{-1}\varphi_j$, $j \gg 1$. We have

$$\|f\|^q_{B^{\tau(p,q)-\eta}_{p,q}} = \sum_{\ell=-1}^{1} 2^{q(\tau(p,q)-\eta)(j+\ell)} \|\mathscr{F}^{-1}\varphi_{j+\ell}\varphi_j\|^q_p$$
$$\lesssim 2^{q(n/q-\eta)j}. \tag{6.46}$$

Assume without loss of generality that

$$\varphi_j(\xi) = 1, \quad \text{if } \xi \in D_j := \left\{\xi : \frac{5}{4} \cdot 2^{j-1} \leqslant |\xi| \leqslant \frac{3}{4} \cdot 2^{j+1}\right\}.$$

Noticing that

$$\Lambda_j = \{k \in \mathbb{Z}^n : B(k, \sqrt{n}) \subset D_j\}$$

contains at least $O(2^{jn})$ elements, we have

$$\|f\|^q_{M_{p,q}} = \sum_{k\in\mathbb{Z}^n} \|\Box_k \mathscr{F}^{-1}\varphi_j\|^q_p \geqslant \sum_{k\in\Lambda_j} \|\mathscr{F}^{-1}\sigma_k\varphi_j\|^q_p \gtrsim 2^{nj}. \tag{6.47}$$

By (6.46) and (6.47),

$$\|f\|_{M_{p,q}} \gtrsim 2^{\eta j} \|f\|_{B^{\tau(p,q)-\eta}_{p,q}},$$

which implies that (6.45) holds.

Case 1.2. $(1/p, 1/q) \in S_2$. We consider the case $q = \infty$. Taking $k(j) = (2^j, 0, ..., 0)$ and $f = \mathscr{F}^{-1}\sigma_{k(j)}$, we see that

$$\|f\|_{M_{p,\infty}} \gtrsim 1 \gtrsim 2^{\eta j} \|f\|_{B^{-\eta}_{p,\infty}}.$$

If $q < \infty$, we show that

$$M_{p,q} \not\subset B^{\varepsilon}_{p,r} \cup B^{\varepsilon}_{\infty,\infty}. \tag{6.48}$$

Let $f \in \mathscr{S}(\mathbb{R}^n)$ be a Schwartz function satisfying $\mathrm{supp}\widehat{f} \subset \{\xi : |\xi_i| < 1/2,\ i = 1, ..., n\}$. Let $N \gg 1$, $0 < \varepsilon \ll 1$,

$$k(j) = (2^{Nj}, 0, ..., 0) \in \mathbb{Z}^n, \tag{6.49}$$

$$\widehat{F}(\xi) = \sum_{j=1}^{\infty} 2^{-\varepsilon Nj} \widehat{f}(\xi - k(j)). \tag{6.50}$$

Noticing that $\mathrm{supp}\widehat{F} \subset \cup_{j=1}^{\infty} Q_{k(j)}$, we have

$$\Box_k F = 0 \text{ if } k \neq k(j) + \ell, \quad |\ell|_\infty \leqslant 1. \tag{6.51}$$

Since $N \gg 1$, we see that

$$\|F\|^q_{M_{p,q}} \leqslant \sum_{j=1}^{\infty} \sum_{|\ell|_\infty \leqslant 1} \|\Box_{k(j)+\ell} F\|^q_p$$

$$\leqslant \sum_{j=1}^{\infty} \sum_{|\ell|_\infty \leqslant 1} 2^{-\varepsilon Nqj} \|\mathscr{F}^{-1}\sigma_{k(j)+\ell}\widehat{f}(\cdot - k(j))\|_p^q$$

$$= \sum_{j=1}^{\infty} \sum_{|\ell|_\infty \leqslant 1} 2^{-\varepsilon Nqj} \|\Box_\ell f\|_p^q \lesssim 1, \tag{6.52}$$

which leads to $F \in M_{p,q}$. On the other hand, letting $s > \varepsilon$, we have

$$\|F\|_{B^s_{p,r}}^r \geqslant \sum_{j=1}^{\infty} 2^{srj} \| \triangle_j F\|_p^r$$
$$\gtrsim \sum_{j=1}^{\infty} 2^{srNj} \| \triangle_{Nj} F\|_p^r$$
$$\gtrsim \sum_{j=1}^{\infty} 2^{(s-\varepsilon)rNj} \| \mathscr{F}^{-1}\varphi_{Nj}\chi_{Q_{k(j)}}\widehat{f}(\cdot - k(j))\|_p^r, \tag{6.53}$$

where $\varphi_j = \varphi(2^{-j}\cdot)$. We can assume that $\varphi(\xi) = 1$ if $|\xi| \in [3/4, 5/4]$. Hence,

$$\mathscr{F}^{-1}\varphi_{Nj}\chi_{Q_{k(j)}}\widehat{f}(\cdot - k(j)) = \mathscr{F}^{-1}\chi_{Q_{k(j)}}\widehat{f}(\cdot - k(j)). \tag{6.54}$$

By (6.53) and (6.54),

$$\|F\|_{B^s_{p,r}}^r \gtrsim \sum_{j=1}^{\infty} 2^{(s-\varepsilon)rNj} = \infty. \tag{6.55}$$

The above discussion also implies that $F \notin B^s_{\infty,\infty}$.

Case 1.3. $(1/p, 1/q) \in S_1$. We can assume that $\sigma_k(\xi) = 0$ if $\xi \notin \tilde{Q}_k := \{\xi : |\xi_i - k_i| \leqslant 5/8,\ 1 \leqslant i \leqslant n\}$, and the dyadic decomposition function sequence satisfying $\varphi_j(\xi) = 1$ if $\xi \in D_j (= \{\xi : \frac{5}{4} \cdot 2^{j-1} \leqslant |\xi| \leqslant \frac{3}{4} \cdot 2^{j+1}\})$. Let

$$A_j = \{k \in \mathbb{Z}^n : \tilde{Q}_k \subset D_j\}, \quad j \gg 1. \tag{6.56}$$

It is easy to see that A_j has at most $O(2^{nj})$ elements. Let $f \in \mathscr{S}(\mathbb{R}^n)$ be a radial Schwartz function satisfying $\mathrm{supp}\widehat{f} \subset B(0, 1/8)$,

$$g(x) = \sum_{k \in A_j} e^{\mathrm{i}xk}(\tau_k f)(x), \quad \tau_k f = f(\cdot - k). \tag{6.57}$$

Taking notice of $\mathrm{supp}\,\widehat{\tau_k f} \subset B(0, 1/8)$, we see that $\mathrm{supp}\,\tau_k(\widehat{\tau_k f}) \cap \mathrm{supp}\,\sigma_\ell = \emptyset$, $k \neq \ell$. So,

$$\|g\|_{M_{p,q}} \geqslant \left(\sum_{k \in A_j} \|\mathscr{F}^{-1}\sigma_k \mathscr{F} g\|_p^q \right)^{1/q}$$

$$= \left(\sum_{k \in A_j} \|\mathscr{F}^{-1}\sigma_0(\widehat{\tau_k f})\|_p^q \right)^{1/q} \gtrsim 2^{jn/q}. \tag{6.58}$$

On the other hand, $\operatorname{supp} \hat{g} \subset \{\xi : 2^{j-1} \leqslant |\xi| \leqslant 2^{j+1}\}$. Hence,

$$\|g\|_{B_{p,q}^{n(1/q-1/p)-\eta}} \leqslant \left(\sum_{\ell=-1}^{1} 2^{nj(1-q/p)-\eta jq} \|\mathscr{F}^{-1}\varphi_{j+\ell}\mathscr{F}g\|_p^q \right)^{1/q}. \tag{6.59}$$

Using the multiplier estimates and Hölder's inequality, we have

$$\|\mathscr{F}^{-1}\varphi_{j+\ell}\mathscr{F}g\|_p \lesssim \|g\|_p \leqslant \|g\|_\infty^{1-2/p}\|g\|_2^{2/p}. \tag{6.60}$$

By Plancherel's identity,

$$\|g\|_2 = \|\hat{g}\|_2 = \left(\int_{\mathbb{R}^n} \sum_{k \in A_j} |\tau_k(e^{-ik\xi}\hat{f}(\xi))|^2 d\xi \right)^{1/2} \lesssim 2^{nj/2}. \tag{6.61}$$

We can further assume that $f(x) = f(|x|)$ is a decreasing function on $|x|$. In view of $f \in \mathscr{S}(\mathbb{R}^n)$, we have

$$|f(x-k)| \lesssim (1+|x-k|)^{-N}, \quad N \gg 1. \tag{6.62}$$

Denote

$$B_0 = \{k \in A_j : |x-k| \leqslant 2\}, \quad B_i = \{k \in A_j : 2^i < |x-k| \leqslant 2^{i+1}\}.$$

B_i contains at most $O(2^{n\,i})$ elements. It follows from (6.62) that

$$|g(x)| \leqslant \sum_{i \geqslant 0} \sum_{k \in B_i} |f(x-k)| \lesssim f(0) + \sum_{i \geqslant 1} 2^{ni}|f(2^i)| \lesssim \sum_{i \geqslant 0} 2^{(n-N)i} \lesssim 1. \tag{6.63}$$

Collecting (6.60), (6.61) and (6.63), we have

$$\|\mathscr{F}^{-1}\varphi_{j+\ell}\mathscr{F}g\|_p \lesssim 2^{nj/p}. \tag{6.64}$$

Inserting (6.64) into (6.59) and using (6.58), we immediately have

$$\|g\|_{B_{p,q}^{n(1/q-1/p)-\eta}} \lesssim 2^{nj/q-\eta j} \lesssim 2^{-\eta j}\|g\|_{M_{p,q}}. \tag{6.65}$$

This implies the conclusion.

Now we show the necessity of the second inclusion. It suffices to show

$$M_{p,q}^{\sigma(p,q)-\eta} \not\subset B_{p,q}^0, \quad \forall \eta > 0. \tag{6.66}$$

Notice that (see Fig. 6.2)

$$R_1 = \{(1/p, 1/q) \in \mathbb{R}_+^2 : 1/q \geqslant 1/p,\ 1/p + 1/q \geqslant 1\},$$

$$R_2 = \{(1/p,\, 1/q) \in \mathbb{R}^2_+ : 1/q \leqslant 1/p,\ 1/p \geqslant 1/2\},$$
$$R_3 = \mathbb{R}^2_+ \setminus (R_1 \cup R_2).$$

We consider the following three cases.

Case 2.1. $(1/p,\, 1/q) \in R_1$. We have discussed it in Case 1.1.

Case 2.2. $(1/p,\, 1/q) \in R_3$. Assume that there exists an $\eta > 0$ such that $M^{n(1-1/p-1/q)-\eta}_{p,q} \subset B^0_{p,q}$, we will get a contradiction. If $1 \leqslant p, q < \infty$, by duality, $B^0_{p',q'} \subset M^{-n(1/p'+1/q'-1)+\eta}_{p',q'}$, which contradicts the first inclusion. If $p = \infty$ or $q = \infty$, one can use the same way as in Case 2.1 to get the result. In fact, letting $f = \mathscr{F}^{-1}\varphi_j$, we have

$$\|f\|_{B^0_{p,q}} \geqslant \|\mathscr{F}^{-1}\varphi_j\varphi_j\|_p \gtrsim 2^{nj(1-1/p)}. \tag{6.67}$$

On the other hand,

$$\|f\|_{M^{n/p'}_{p,\infty}} \leqslant \sup_k \langle k\rangle^{n(1-1/p)} \|\mathscr{F}^{-1}\sigma_k\varphi_j\|_p \lesssim 2^{nj(1-1/p)}, \tag{6.68}$$

$$\|f\|_{M^{n/q'}_{\infty,q}} \leqslant \left(\sum_{|k|\in[2^{j-1},2^{j+1}]} 2^{nj(q-1)} \|\mathscr{F}^{-1}\sigma_k\varphi_j\|^q_\infty \right)^{1/q} \lesssim 2^{nj}. \tag{6.69}$$

From (6.67)–(6.69) it follows that (6.66) holds in the cases $p = \infty$ or $q = \infty$.

Case 2.3. $(1/p,\, 1/q) \in R_2$. Take $f \in \mathscr{S}(\mathbb{R}^n)$ satisfying $f(0) = 1$, $\operatorname{supp} \hat{f} \subset Q_0$. Choose $0 < a \ll 1$ (which will be fixed in (6.75) below). Denote $f_a(x) = f(x/a)$. We easily see that $\operatorname{supp} \hat{f_a} \subset Q_{0,a} := \{\xi : |\xi_i| \leqslant 1/2a,\ 1 \leqslant i \leqslant n\}$. Recall that $D_j = \{\xi : \frac{5}{4} \cdot 2^{j-1} \leqslant |\xi| \leqslant \frac{3}{4} \cdot 2^{j+1}\}$ contains at most $O(a^n 2^{jn})$ pairwise disjoint cubes $Q_{k(i),a} := k(i) + Q_{0,a}$, $i = 1, ..., O(a^n 2^{jn})$. We write $A_j = \{k(i) : i = 1, ..., O(a^n 2^{jn})\}$,

$$g(x) = \sum_{k\in A_j} e^{\mathrm{i}xk} (\tau_k f_a)(x). \tag{6.70}$$

For any $N \gg 1$,

$$|f(x)| \leqslant C_N (1 + |x|)^{-N}, \tag{6.71}$$

it follows that

$$|f_a(x)| \leqslant C_N a^N |x|^{-N}. \tag{6.72}$$

By the continuity of $f(x)$ and $f(0) = 1$, we deduce that there exists a $\varrho > 0$ such that[7]

$$|f_a(x)| > 1/2, \quad x \in B(0, \varrho). \tag{6.73}$$

[7] In fact, ϱ can be chosen as $\varrho = a\varrho_0$, $\varrho_0 > 0$ is independent of a.

In view of (6.70) and (6.73), we get, for any $x \in B(k(i), \varrho)$, that

$$\begin{aligned} |g(x)| &\geqslant |f_a(x-k(i))| - \sum_{k\in A_j\setminus\{k(i)\}} |f_a(x-k)| \\ &\geqslant \frac{1}{2} - \sum_{k\in A_j\setminus\{k(i)\}} |f_a(x-k)|. \end{aligned} \tag{6.74}$$

Denote $A_{j,\ell} := \{k \in A_j : 2^\ell \leqslant |k-k(i)| < 2^{\ell+1}\}$. We can further assume that $f(x)$ is a decreasing function on $|x|$. Since $A_{j,\ell}$ has at most $O(a^n 2^{\ell n})$ elements, we have for any $x \in B(k(i), \varrho)$,

$$\begin{aligned} \sum_{k\in A_j\setminus\{k(i)\}} |f_a(x-k)| &\leqslant \sum_{\ell\geqslant 1}\sum_{k\in A_{j,\ell}} |f_a(x-k)| \\ &\leqslant C\sum_{\ell\geqslant 1} a^n 2^{n\ell} |f_a(2^\ell-\varrho)| \\ &\lesssim \sum_{\ell\geqslant 1} C_N a^{n+N} 2^{(n-N)\ell} \leqslant 1/4, \end{aligned} \tag{6.75}$$

where $N \geqslant n+1$ and $CC_N a^{n+N} \leqslant 1/4$. Hence, it follows from (6.74) and (6.75) that

$$|g(x)| \geqslant 1/4, \quad x \in B(k(i), \varrho). \tag{6.76}$$

By (6.76), we have

$$\|g(x)\|_p \geqslant \left\| \sum_{i=1}^{O(a^n 2^{nj})} \frac{1}{4}\chi_{B(k(i),\varrho)} \right\|_p \gtrsim (a\varrho)^{n/p} 2^{nj/p}, \tag{6.77}$$

where ϱ and a are independent of $j \gg 1$. We can assume that $\varphi_j(\xi) = 1$ for $\xi \in D_j$. Since $\operatorname{supp}\hat{g} \subset D_j$, we have $\mathscr{F}^{-1}\varphi_j\mathscr{F}g = g$. Hence, (6.77) implies that

$$\|g\|_{B^0_{p,q}} \geqslant \|\mathscr{F}^{-1}\varphi_j\mathscr{F}g\|_p \gtrsim (a\varrho)^{n/p} 2^{nj/p}. \tag{6.78}$$

On the other hand,

$$\begin{aligned} \|g\|_{M^{n(1/p-1/q)}_{p,q}} &= \left(\sum_{|k|\in[2^{j-1},2^{j+1}]} 2^{nj(q/p-1)} \|\mathscr{F}^{-1}\sigma_k\mathscr{F}g\|_p^q \right)^{1/q} \\ &\leqslant 2^{nj/p} \sup_{|k|\in[2^{j-1},2^{j+1}]} \|\mathscr{F}^{-1}\sigma_k\mathscr{F}g\|_p. \end{aligned} \tag{6.79}$$

Since $\operatorname{supp}\sigma_k$ overlaps at most finite many $\operatorname{supp}\tau_\ell(\widehat{\tau_\ell f_a})$, using multiplier estimates, we have

$$\|\mathscr{F}^{-1}\sigma_k\mathscr{F}g\|_p \lesssim \|f\|_p. \tag{6.80}$$

Hence, (6.79) and (6.80) imply that

$$\|g\|_{M^{n(1/p-1/q)}_{p,q}} \lesssim 2^{nj/p}. \tag{6.81}$$

By (6.78) and (6.81) we immediately have (6.66). We finish the proof of Theorem 6.1. □

6.4 NLS and NLKG in modulation spaces

As indicated in §6.1, the dispersive semi-group combined with the frequency-uniform decomposition operator has some advantages and we discuss them in this section. The results of this section can be found in [11; 12; 50; 247; 246].

6.4.1 *Schrödinger and Klein-Gordon semigroup in modulation spaces*

Let $S(t) = e^{\mathrm{i}t\triangle}$ denote the Schrödinger semi-group. In [247], Wang, Zhao and Guo obtained the uniform boundedness for the Ginzburg-Landau semi-group $L(t) = e^{(a+\mathrm{i})t\triangle}$ $(a > 0)$ in modulation spaces and their proof is also adapted to the Schrödinger semi-group ($a = 0$ in $L(t)$). First, we show that $\Box_k S(t) : L^p \to L^p$ is uniformly bounded on $k \in \mathbb{Z}^n$ by following the proof in [247].

$$\begin{aligned}\|\Box_k S(t) f\|_p &\leqslant \sum_{|\ell|_\infty \leqslant 1} \|\mathscr{F}^{-1} \sigma_{k+\ell} e^{\mathrm{i}t|\xi|^2} \sigma_k \hat{f}\|_p \\ &\leqslant \sum_{|\ell|_\infty \leqslant 1} \|\mathscr{F}^{-1}(\sigma_{k+\ell} e^{\mathrm{i}t|\xi|^2})\|_1 \|\Box_k f\|_p. \end{aligned} \tag{6.82}$$

Hence, it suffices to estimate $\|\mathscr{F}^{-1}(\sigma_k e^{\mathrm{i}t|\xi|^2})\|_1$. Using the multiplier estimate, we have

$$\begin{aligned}\|\mathscr{F}^{-1}(\sigma_k e^{\mathrm{i}t|\xi|^2})\|_1 &= \|\mathscr{F}^{-1}(\sigma_0 e^{\mathrm{i}t|\xi|^2})\|_1 \\ &\lesssim \|\sigma_0\|_2^{1-n/2L} \sum_{|\alpha|=L} \|D^\alpha(\sigma_0 e^{\mathrm{i}t|\xi|^2})\|_2^{n/2L} \\ &\lesssim (1 + |t|^{n/2}). \end{aligned} \tag{6.83}$$

Noticing that

$$\|\Box_k S(t) f\|_2 = \|\Box_k f\|_2, \tag{6.84}$$

by a complex interpolation, we immediately have

$$\|\Box_k S(t) f\|_p \lesssim (1 + |t|)^{n|1/2-1/p|} \|\Box_k f\|_p. \tag{6.85}$$

Proposition 6.6 (Uniform boundedness of $S(t)$ in $M^s_{p,q}$). *Let $s \in \mathbb{R}$, $1 \leqslant p \leqslant \infty$ and $0 < q < \infty$. Then we have*

$$\|S(t) f\|_{M^s_{p,q}} \lesssim (1 + |t|)^{n|1/2-1/p|} \|f\|_{M^s_{p,q}}. \tag{6.86}$$

Shortly after the work [247], Proposition 6.6 is independently obtained by Bényi, Gröchenig, Okoudjou and Rogers in [12] and their result contains more general semi-group $e^{it(-\Delta)^\alpha}$ with $\alpha \leqslant 1$, whose proof is based on the short-time frequency analysis technique. Miyachi, Nicola, Riveti, Taracco and Tomita [164] were able to consider the case $\alpha > 1$, Chen, Fan and Sun [38] obtained some refined estimates for $e^{it(-\Delta)^\alpha}$ with any $\alpha > 0$ by using the oscillatory integral estimates in higher spatial dimensions.

Now we consider the truncated decay of $S(t)$; cf. [246]. Let us recall the $L^p - L^{p'}$ estimate of $S(t)$,

$$\|S(t)f\|_p \lesssim |t|^{-n(1/2-1/p)}\|f\|_{p'}, \quad 2 \leqslant p \leqslant \infty, \tag{6.87}$$

where $1/p + 1/p' = 1$. We also have

$$\|\Box_k S(t)f\|_p \lesssim \sum_{|\ell|_\infty \leqslant 1} \|\Box_{k+\ell} f\|_{p'}, \quad 2 \leqslant p \leqslant \infty. \tag{6.88}$$

Indeed, by Young's and Hölder's inequalities, we have

$$\begin{aligned}\|\Box_k S(t)f\|_p &\leqslant \sum_{\ell\in\Lambda} \|\mathscr{F}^{-1}\sigma_k\sigma_{k+\ell}\exp(-\mathrm{i}t|\xi|^2)\mathscr{F}f\|_p.\\ &\leqslant \sum_{\ell\in\Lambda} \|\sigma_{k+\ell}\exp(-\mathrm{i}t|\xi|^2)\sigma_k\mathscr{F}f\|_{p'} \lesssim \|\Box_k f\|_p.\end{aligned}$$

Combing (6.87) with (6.88), one has that

$$\|\Box_k S(t)f\|_p \lesssim (1+|t|)^{-n(1/2-1/p)}\|\Box_k f\|_{p'}, \quad 2 \leqslant p \leqslant \infty. \tag{6.89}$$

Multiplying both sides of (6.89) by $\langle k\rangle^s$ and then taking ℓ^q norm, we have

$$\|S(t)f\|_{M^s_{p,q}} \lesssim (1+|t|)^{-n(1/2-1/p)}\|f\|_{M^s_{p',q}}.$$

Proposition 6.7. *Let $s \in \mathbb{R}$, $2 \leqslant p < \infty$, $1/p + 1/p' = 1$ and $0 < q < \infty$. Then we have*

$$\|S(t)f\|_{M^s_{p,q}} \lesssim (1+|t|)^{-n(1/2-1/p)}\|f\|_{M^s_{p',q}}. \tag{6.90}$$

Propositions 6.6 and 6.7 are optimal in the sense that the powers of time variable are sharp, cf. [51]. Now we consider the truncated decay estimate for the Klein-Gordon semi-group $G(t) = e^{\mathrm{i}t\omega^{1/2}}$ where $\omega = I - \Delta$.

$$\begin{aligned}\|\Box_k G(t)f\|_p &\leqslant \sum_{|\ell|_\infty \leqslant 1} \|\mathscr{F}^{-1}\sigma_{k+\ell}e^{\mathrm{i}t\langle\xi\rangle}\sigma_k\hat{f}\|_p\\ &\leqslant \sum_{|\ell|_\infty \leqslant 1} \|\mathscr{F}^{-1}(\sigma_{k+\ell}e^{\mathrm{i}t\langle\xi\rangle})\|_1\|\Box_k f\|_p.\end{aligned} \tag{6.91}$$

So, it suffices to estimate $\|\mathscr{F}^{-1}(\sigma_k e^{it\langle\xi\rangle})\|_1$. Using the Fourier multiplier estimate, we have

$$\begin{aligned}\|\mathscr{F}^{-1}(\sigma_k e^{it\langle\xi\rangle})\|_1 &= \|\mathscr{F}^{-1}(\sigma_0 e^{it\langle\xi+k\rangle})\|_1\\ &\lesssim \|\sigma_0\|_2^{1-n/2L}\sum_{|\alpha|=L}\|D^\alpha(\sigma_0 e^{it\langle\xi+k\rangle})\|_2^{n/2L}\\ &\lesssim (1+|t|^{n/2}).\end{aligned}\tag{6.92}$$

Noticing that

$$\|\Box_k G(t)f\|_2 = \|\Box_k f\|_2, \tag{6.93}$$

by a complex interpolation, we immediately have

$$\|\Box_k G(t)f\|_p \lesssim (1+|t|)^{n|1/2-1/p|}\|\Box_k f\|_p. \tag{6.94}$$

Proposition 6.8. *Let $s\in\mathbb{R}$, $1\leqslant p\leqslant\infty$ and $0<q<\infty$. Then we have*

$$\|G(t)f\|_{M^s_{p,q}} \lesssim (1+|t|)^{n|1/2-1/p|}\|f\|_{M^s_{p,q}}. \tag{6.95}$$

It is known that $G(t)$ satisfies the following $L^p-L^{p'}$ estimate

$$\|G(t)f\|_{H_p^{-2\sigma(p)}} \lesssim |t|^{-n(1/2-1/p)}\|f\|_{p'}, \tag{6.96}$$

where

$$2\leqslant p<\infty,\quad 2\sigma(p)=(n+2)\left(\frac{1}{2}-\frac{1}{p}\right). \tag{6.97}$$

From (6.96) it follows that

$$\|\Box_k G(t)f\|_{H_p^{-2\sigma(p)}} \lesssim |t|^{-n(1/2-1/p)}\|\Box_k f\|_{p'}. \tag{6.98}$$

Applying the multiplier estimate,

$$\|\Box_k (I-\triangle)^{\delta/2}g\|_p \lesssim \langle k\rangle^\delta \|g\|_p. \tag{6.99}$$

In view of (6.98) and (6.99), we have

$$\begin{aligned}\|\Box_k G(t)f\|_p &\lesssim \langle k\rangle^{2\sigma(p)}\sum_{\ell\in\Lambda}\|\Box_{k+\ell}G(t)f\|_{H_p^{-2\sigma(p)}}\\ &\lesssim \langle k\rangle^{2\sigma(p)}|t|^{-n(1/2-1/p)}\sum_{\ell\in\Lambda}\|\Box_{k+\ell}f\|_{p'}.\end{aligned}\tag{6.100}$$

On the other hand, by Hölder's and Young's inequality,

$$\begin{aligned}\|\Box_k G(t)f\|_p &\lesssim \|\sigma_k e^{it(1+|\xi|^2)^{1/2}}\widehat{f}\|_{p'}\\ &\lesssim \sum_{\ell\in\Lambda}\|\sigma_k e^{it(1+|\xi|^2)^{1/2}}\mathscr{F}\Box_{k+\ell}f\|_{p'}\end{aligned}$$

$$\lesssim \sum_{\ell\in\Lambda} \|\mathscr{F}\Box_{k+\ell}f\|_p$$
$$\lesssim \sum_{\ell\in\Lambda} \|\Box_{k+\ell}f\|_{p'}. \tag{6.101}$$

So, for any $\theta \in [0,1]$, it follows from (6.100) and (6.101) that

$$\|\Box_k G(t)f\|_p \lesssim \langle k\rangle^{2\sigma(p)\theta}|t|^{-n\theta(1/2-1/p)} \sum_{\ell\in\Lambda} \|\Box_{k+\ell}f\|_{p'}. \tag{6.102}$$

Noticing that $\sigma(p) \geqslant 0$, by (6.101) we have

$$\|\Box_k G(t)f\|_p \lesssim \langle k\rangle^{2\sigma(p)\theta} \sum_{\ell\in\Lambda} \|\Box_{k+\ell}f\|_{p'}. \tag{6.103}$$

Combining (6.102) and (6.103), we get

$$\|\Box_k G(t)f\|_p \lesssim \langle k\rangle^{2\sigma(p)\theta}(1+|t|)^{-n\theta(1/2-1/p)} \sum_{\ell\in\Lambda} \|\Box_{k+\ell}f\|_{p'}. \tag{6.104}$$

Multiplying both sides of (6.104) by $\langle k\rangle^s$ and then taking ℓ^q norm, we immediately obtain that

Proposition 6.9. *Let $s \in \mathbb{R}$, $2 \leqslant p < \infty$, $1/p + 1/p' = 1$, $0 < q < \infty$, $\theta \in [0,1]$ and $\sigma(p)$ is as in (6.97). Then we have*

$$\|G(t)f\|_{M^s_{p,q}} \lesssim (1+|t|)^{-n\theta(1/2-1/p)}\|f\|_{M^{s+2\sigma(p)\theta}_{p',q}}. \tag{6.105}$$

6.4.2 *Strichartz estimates in modulation spaces*

For convenience, we write

$$\|f\|_{\ell^{s,q}_\Box(L^\gamma(I,L^p))} = \left(\sum_{k\in\mathbb{Z}^n} \langle k\rangle^{sq}\|\Box_k f\|^q_{L^\gamma(I,L^p)}\right)^{1/q}, \tag{6.106}$$

$\ell^q_\Box(L^\gamma(I,L^p)) := \ell^{0,q}_\Box(L^\gamma(I,L^p))$, $\ell^q_\Box(L^p_{x,t\in I}) := \ell^q_\Box(L^p(I,L^p))$. Recall that the truncated decay can be generalized to the following estimate

$$\|U(t)f\|_{M^\alpha_{p,q}} \lesssim (1+|t|)^{-\delta}\|f\|_{M_{p',q}}, \tag{6.107}$$

where $2 \leqslant p < \infty$, $1 \leqslant q < \infty$, $\alpha = \alpha(p) \in \mathbb{R}$, $\delta = \delta(p) > 0$, α and δ are independent of $t \in \mathbb{R}$, $U(t)$ is a dispersive semi-group,

$$U(t) = \mathscr{F}^{-1}e^{itP(\xi)}\mathscr{F}, \tag{6.108}$$

and $P(\cdot) : \mathbb{R}^n \to \mathbb{R}$ is a real valued function. In the sequel we will assume that $U(t)$ satisfies conditions (6.107) and (6.108), from which we can get some Strichartz inequalities for $U(t)$ in modulation spaces.

Generally speaking, all of the Strichartz estimates in Sec. 3.2 hold if we replace X^* and X^α by $M_{p',2}$ and $M^\alpha_{p,2}$ in Sec. 3.2, respectively. However, do not forget that the estimate in (6.107) contains no singularity at $t = 0$, which leads that we can remove the restriction condition $\delta \leqslant 1$ in the Strichartz estimates on modulation spaces and the results are much better than those in Sec. 3.2.

Proposition 6.10 (Strichartz inequalities). *Let $U(t)$ satisfy (6.107) and (6.108). For any $\gamma \geqslant 2 \vee (2/\delta)$, we have*

$$\|U(t)f\|_{\ell_\square^{\alpha/2,q}(L^\gamma(\mathbb{R},L^p))} \lesssim \|f\|_{M_{2,q}}. \tag{6.109}$$

In addition, if $\gamma \geqslant q$, then we have

$$\|U(t)f\|_{L^\gamma(\mathbb{R},M^{\alpha/2}_{p,q})} \lesssim \|f\|_{M_{2,q}}. \tag{6.110}$$

Proof. The proof is similar to that as in Sec. 3.2. However, we need to carefully handle the indices α, p, q and γ. First, we consider the case $1 < q < \infty$ to show that

$$\int_{\mathbb{R}} (U(t)f, \psi(t))dt \lesssim \|f\|_{M_{2,q}} \|\psi\|_{\ell_\square^{-\alpha/2,q'}(L^{\gamma'}(\mathbb{R},L^{p'}))} \tag{6.111}$$

holds for all $f \in \mathscr{S}(\mathbb{R}^n)$ and $\psi \in C_0^\infty(\mathbb{R}, \mathscr{S}(\mathbb{R}^n))$. Noticing that $\mathscr{S}(\mathbb{R}^n)$ and $C_0^\infty(\mathbb{R}, \mathscr{S}(\mathbb{R}^n))$ are dense in $M_{2,q}$ and $\ell_\square^{-\alpha/2,q'}(L^{\gamma'}(\mathbb{R}, L^{p'}))$, respectively, (6.111) implies (6.109). By duality,

$$\int_{\mathbb{R}} (U(t)f, \psi(t))dt \lesssim \|f\|_{M_{2,q}} \left\| \int_{\mathbb{R}} U(-t)\psi(t)dt \right\|_{M_{2,q'}}. \tag{6.112}$$

For any $k \in \mathbb{Z}^n$,

$$\begin{aligned} &\left\| \square_k \int_{\mathbb{R}} U(-t)\psi(t)dt \right\|_2^2 \\ &\lesssim \|\square_k \psi\|_{L^{\gamma'}(\mathbb{R},L^{p'})} \left\| \square_k \int_{\mathbb{R}} U(t-s)\psi(s)ds \right\|_{L^\gamma(\mathbb{R},L^p)}. \end{aligned} \tag{6.113}$$

Recall that $\{\square_k\}_{k\in\mathbb{Z}^n}$ are almost orthogonal. By (6.107), the definition of $\|\cdot\|_{M_{p',q}}$ together with the multiplier estimate, we have

$$\begin{aligned} \|\square_k U(t)f\|_p &\lesssim \langle t\rangle^{-\delta}\langle k\rangle^{-\alpha} \sum_{|\ell|_\infty \leqslant 1} \|\square_k \square_{k+\ell} f\|_{M_{p',q}} \\ &\lesssim \langle t\rangle^{-\delta}\langle k\rangle^{-\alpha} \|\square_k f\|_{p'}. \end{aligned} \tag{6.114}$$

If $\delta \neq 1$, applying (6.114), Young's and Hardy-Littlewood-Sobolev's inequalities, we obtain that[8]

$$\left\| \Box_k \int_{\mathbb{R}} U(t-s)\psi(s)ds \right\|_{L^\gamma(\mathbb{R},L^p)} \lesssim \langle k\rangle^{-\alpha} \|\Box_k \psi\|_{L^{\gamma'}(\mathbb{R},L^{p'})}. \tag{6.115}$$

If $\delta = 1$ and $\gamma > 2$, we can use Young's inequality to get that (6.115) holds. If $\gamma = 2$ and $\delta = 1$, applying the same way as in Sec. 3.3, we can obtain (6.115). Hence, in view of (6.113) and (6.115), we have

$$\left\| \Box_k \int_{\mathbb{R}} U(-t)\psi(t)dt \right\|_2 \lesssim \langle k\rangle^{-\alpha/2} \|\Box_k \psi\|_{L^{\gamma'}(\mathbb{R},L^{p'})}. \tag{6.116}$$

Both sides in (6.116) are taken the norm in $\ell^{q'}$,

$$\left\| \int_{\mathbb{R}} U(-s)\psi(s)ds \right\|_{M_{2,q'}} \lesssim \|\psi\|_{\ell_\Box^{-\alpha/2,q'}(L^{\gamma'}(\mathbb{R},L^{p'}))}. \tag{6.117}$$

(6.112) and (6.117) imply (6.111).

If $\gamma \geqslant q$, by Minkowski's inequality, one sees that the left hand side of (6.110) can be bounded by the left hand side of (6.109).

Now we consider the case $q = 1$. It suffices to show that

$$\int_{\mathbb{R}} (U(t)f, \psi(t))dt \lesssim \|f\|_{M_{2,q}} \|\psi\|_{c_\Box^{-\alpha/2}(L^{\gamma'}(\mathbb{R},L^{p'}))} \tag{6.118}$$

holds for all $f \in \mathscr{S}(\mathbb{R}^n)$ and $\psi \in C_0^\infty(\mathbb{R}, \mathscr{S}(\mathbb{R}^n))$. Repeating the above procedure, we can prove our result. □

Denote

$$(\mathscr{U} f)(t) = \int_0^t U(t-s)f(s,\cdot)ds. \tag{6.119}$$

Proposition 6.11. *Let $U(t)$ satisfy (6.107) and (6.108). For any $\gamma \geqslant 2 \vee (2/\delta)$, we have*

$$\|\mathscr{U} f\|_{\ell_\Box^q(L^\infty(\mathbb{R},L^2))} \lesssim \|f\|_{\ell_\Box^{-\alpha/2,q}(L^{\gamma'}(\mathbb{R},L^{p'}))}. \tag{6.120}$$

In addition, if $\gamma' \leqslant q$, then

$$\|\mathscr{U} f\|_{L^\infty(\mathbb{R},M_{2,q})} \lesssim \|f\|_{L^{\gamma'}(\mathbb{R},M_{p',q}^{-\alpha/2})}. \tag{6.121}$$

Proof. We only sketch the proof. Applying the same way as in (6.113), (6.115) and (6.116), we see that

$$\|\Box_k \mathscr{U} f\|_2^2 \lesssim \langle k\rangle^{-\alpha} \|\Box_k f\|^2_{L^{\gamma'}(\mathbb{R},L^{p'})}. \tag{6.122}$$

(6.122) implies (6.120). By Minkowski' s inequality, it follows from (6.120) that (6.121) holds. □

[8]Due to $\langle t-s\rangle^{-\delta} < |t-s|^{-\delta}$, if $\delta < 1$, $\gamma = 2/\delta$, one can apply Hardy-Littlewood-Sobolev's inequality.

Proposition 6.12. *Let $U(t)$ satisfy* (6.107) *and* (6.108). *For any $\gamma \geqslant 2 \vee (2/\delta)$, we have*

$$\|\mathscr{U} f\|_{\ell_\square^{\alpha/2,q}(L^\gamma(\mathbb{R},L^p))} \lesssim \|f\|_{\ell_\square^{-\alpha/2,q}(L^{\gamma'}(\mathbb{R},L^{p'}))}. \tag{6.123}$$

In addition, we assume that $q \leqslant 2$ if $\delta = 1$ and $\gamma = 2$, then we have for any $\gamma \geqslant 2 \vee (2/\delta)$,

$$\|\mathscr{U} f\|_{L^\gamma(\mathbb{R},M_{p,q}^{\alpha/2})} \lesssim \|f\|_{L^{\gamma'}(\mathbb{R},M_{p',q}^{-\alpha/2})}. \tag{6.124}$$

Proof. We only give an outline of the proof of (6.124). By (6.107), we have

$$\|\mathscr{U} f\|_{M_{p,q}^{\alpha/2}} \lesssim \int_0^t \langle t-s\rangle^{-\delta} \|f(s)\|_{M_{p',q}^{-\alpha/2}} ds. \tag{6.125}$$

If $\delta \neq 1$, or $\delta = 1$ and $\gamma > 2$, then we can use the same way as in (6.115) to get the conclusion. If $\delta = 1$ and $\gamma = 2$, it follows from (6.115) and Minkowski's inequality that (6.124) holds. □

Proposition 6.13. *Assume that $U(t)$ satisfies* (6.107) *and* (6.108), *$\gamma \geqslant \max(2/\delta, 2)$. Then we have*

$$\|\mathscr{U} f\|_{\ell_\square^{\alpha/2,q}(L^\gamma(\mathbb{R},L^p))} \lesssim \|f\|_{\ell_\square^{q}(L^1(\mathbb{R},L^2))}. \tag{6.126}$$

In addition, if $\gamma \geqslant q$, then

$$\|\mathscr{U} f\|_{L^\gamma(\mathbb{R},M_{p,q}^{\alpha/2})} \lesssim \|f\|_{L^1(\mathbb{R},M_{2,q})}. \tag{6.127}$$

Proof. Assume that $f, \psi \in C_0^\infty(\mathbb{R}, \mathscr{S}(\mathbb{R}^n))$. By Proposition 6.11, we have

$$\begin{aligned} &\left| \int_{\mathbb{R}_+} \left(\int_0^t U(t-\tau) f(\tau) d\tau, \psi(t) \right) dt \right| \\ &\lesssim \|f\|_{L^1(\mathbb{R},M_{2,q})} \left\| \int_\cdot^\infty U(\cdot - t)\psi(t) dt \right\|_{L^\infty(\mathbb{R},M_{2,q'})} \\ &\lesssim \|f\|_{L^1(\mathbb{R},M_{2,q})} \|\psi\|_{\ell_\square^{-\alpha/2,q'}(L^{\gamma'}(\mathbb{R},L^{p'}))}. \end{aligned} \tag{6.128}$$

Since $\psi \in C_0^\infty(\mathbb{R}, \mathscr{S}(\mathbb{R}^n))$ is dense in $\ell_\square^{-\alpha/2,q'}(L^{\gamma'}(\mathbb{R}, L^{p'}))$ and $c_\square^{-\alpha/2}(L^{\gamma'}(\mathbb{R}, L^{p'}))$, by duality, we get the result, as desired. □

The Schrödinger semi-group corresponds to the cases $\alpha = 0$, $\delta = n(1/2 - 1/p)$ and $2 \leqslant p < \infty$. Taking $q = 1$ in Propositions 6.10–6.12, we immediately have

Corollary 6.3. *Let $2 \leqslant p < \infty$, $\gamma \geqslant 2 \vee \gamma(p)$, and*

$$\frac{2}{\gamma(p)} = n\Big(\frac{1}{2} - \frac{1}{p}\Big). \tag{6.129}$$

Let $S(t) = e^{it\Delta}$, $\mathscr{A} = \int_0^t S(t-s)\cdot ds$. *Then*

$$\|S(t)\varphi\|_{\ell^1_\square(L^\gamma(\mathbb{R},L^p))} \lesssim \|\varphi\|_{M_{2,1}}, \tag{6.130}$$

$$\|\mathscr{A} f\|_{\ell^1_\square(L^\gamma(\mathbb{R},L^p))\cap \ell^1_\square(L^\infty(\mathbb{R},L^2))} \lesssim \|f\|_{\ell^1_\square(L^{\gamma'}(\mathbb{R},L^{p'}))}. \tag{6.131}$$

Similar to Corollary 6.3, we have

Corollary 6.4. *Let* $2 \leqslant p < \infty$, $\theta \in (0,1]$, $1 \leqslant q < \infty$,

$$\frac{2}{\gamma_\theta(p)} = n\theta\Big(\frac{1}{2}-\frac{1}{p}\Big), \quad 2\sigma = (n+2)\theta\Big(\frac{1}{2}-\frac{1}{p}\Big). \tag{6.132}$$

Let $G(t)$ *be as in* (6.96), $\mathscr{G} = \int_0^t G(t-s)\cdot ds$. *Then for any* $\gamma \geqslant 2 \vee \gamma_\theta(p)$, *we have*

$$\|G(t)\varphi\|_{\ell^{-\sigma,q}_\square(L^\gamma(\mathbb{R},L^p))} \lesssim \|\varphi\|_{M_{2,q}}, \tag{6.133}$$

$$\|\mathscr{G} f\|_{\ell^{-\sigma,q}_\square(L^\gamma(\mathbb{R},L^p))\cap \ell^q_\square(L^\infty(\mathbb{R},L^2))} \lesssim \|f\|_{\ell^{\sigma,q}_\square(L^{\gamma'}(\mathbb{R},L^{p'}))}. \tag{6.134}$$

Related Strichartz estimates in Wiener amalgam spaces for the Schrödinger equation were obtained by Cordero and Nicola [49].

6.4.3 *Wellposedness for NLS and NLKG*

We consider the Cauchy problem for NLS,

$$iu_t + \Delta u = f(u), \quad u(0,x) = u_0(x). \tag{6.135}$$

Noticing that $B^n_{\infty,1} \subset M_{\infty,1} \subset B^0_{\infty,1} \subset L^\infty$ are sharp embeddings, up to now, we can not get the wellposedness of NLS in L^∞ or in $B^0_{\infty,1}$. However, we can obtain the local wellposedness of NLS in $M_{\infty,1}$. We have (see [11; 50])

Theorem 6.2. *Let* $n \geqslant 1$, $f(u) = \lambda|u|^\kappa u$, $\kappa \in 2\mathbb{N}$, $\lambda \in \mathbb{R}$, $u_0 \in M_{p,1}$ *and* $1 \leqslant p \leqslant \infty$. *Then there exists a* $T > 0$ *such that* (6.135) *has a unique solution* $u \in C([0,T), M_{p,1})$. *Moreover, if* $T < \infty$, *then* $\limsup_{t\nearrow T} \|u(t)\|_{M_{p,1}} = \infty$.

If the nonlinearity has an exponential growth, say $f(u) = \lambda(e^{|u|^2}-1)u$, the result in Theorem 6.2 also holds. Noticing that $B^{n/2}_{2,1} \subset M_{2,1} \subset B^0_{2,1} \cap C(\mathbb{R}^n)$ are sharp embeddings, we can get that NLS is global wellposed in $M_{2,1}$ if initial data are sufficiently small.

Theorem 6.3. *Let* $n \geqslant 1$, $f(u) = \lambda|u|^\kappa u$, $\kappa \in 2\mathbb{N}$, $\lambda \in \mathbb{R}$, $\kappa \geqslant 4/n$, $u_0 \in M_{2,1}$ *and there exists a sufficiently small* $\delta > 0$ *such that* $\|u_0\|_{M_{2,1}} \leqslant \delta$. *Then* (6.135) *has a unique solution*

$$u \in C(\mathbb{R}, M_{2,1}) \cap \ell^1_\square(L^p_{x,t\in\mathbb{R}}), \tag{6.136}$$

where $p \in [2+4/n, 2+\kappa] \cap \mathbb{N}$, $\ell^1_\square(L^p_{x,t\in\mathbb{R}})$ *is as in* (6.106).

Theorem 6.4. *Let* $n \geqslant 2$, $f(u) = \lambda(e^{\varrho|u|^2} - 1)u$, $\lambda \in \mathbb{C}$ *and* $\varrho > 0$. *Assume that* $u_0 \in M_{2,1}$ *and there exists a sufficiently small* $\delta > 0$ *such that* $\|u_0\|_{M_{2,1}} \leqslant \delta$. *Then* (6.135) *has a unique solution*

$$u \in C(\mathbb{R}, M_{2,1}) \cap \ell^1_\square(L^4_{x,t\in\mathbb{R}}). \tag{6.137}$$

The proof of Theorem 6.2 relies upon the algebra structure of $M_{p,1}$.

Lemma 6.7. *Let* $1 \leqslant p \leqslant \infty$, *then* $M_{p,1}$ *is a Banach algebra.*

The proof of Lemma 6.7 can be shown by following the same technique as that of $E^s_{2,1}$ and we omit the details.

Proof. [Proof of Theorem 6.2] Denote

$$\mathscr{A} f(t,x) = \int_0^t S(t-\tau) f(\tau,x) d\tau.$$

Consider the mapping

$$\mathscr{T} : u(t) \to S(t)u_0 - \mathrm{i}\mathscr{A} f(u). \tag{6.138}$$

We write

$$\mathcal{D} = \{u : \|u\|_{C([0,T]:M_{p,1})} \leqslant M\}, \quad d(u,v) = \|u - v\|_{C([0,T]:M_{p,1})},$$

where $M = 2C\|u_0\|_{M_{p,1}}$. If $u \in \mathcal{D}$, by Proposition 6.6 and Lemma 6.7,

$$\|\mathscr{T} u\|_{C([0,T]:M_{p,1})} \lesssim (1+T)^{n/2} \left(\|u_0\|_{M_{p,1}} + T\|u\|^{\kappa+1}_{C([0,T]:M_{p,1})} \right). \tag{6.139}$$

We can choose a sufficiently small $0 < T < 1$ such that $CTM^\kappa \leqslant 1/2$. It follows that $\mathscr{T} : (\mathcal{D}, d) \to (\mathcal{D}, d)$ is a contraction mapping. The left part of the proof is standard and we omit the details of the proof. □

Lemma 6.8. *Let* $1 \leqslant p, p_i, \gamma, \gamma_i \leqslant \infty$ *satisfy*

$$\frac{1}{p} = \frac{1}{p_1} + ... + \frac{1}{p_N}, \quad \frac{1}{\gamma} = \frac{1}{\gamma_1} + ... + \frac{1}{\gamma_N}. \tag{6.140}$$

Then we have

$$\|u_1 u_2 ... u_N\|_{\ell^1_\square(L^\gamma(\mathbb{R}, L^p))} \leqslant C^N \prod_{i=1}^N \|u_i\|_{\ell^1_\square(L^{\gamma_i}(\mathbb{R}, L^{p_i}))}. \tag{6.141}$$

Proof. It suffices to consider the case $N = 2$. We have

$$\| \Box_k(u_1 u_2)\|_p \leqslant \sum_{i,j\in\mathbb{Z}^n} \|\Box_k(\Box_i u_1 \,\Box_j u_2)\|_p. \tag{6.142}$$

Noticing that $\Box_k(\Box_i u_1 \,\Box_j u_2) = 0$ if $|k-i-j| \geqslant k_0(n)$, where $k_0(n)$ depends only on n, we have from (6.142) that

$$\| \Box_k(u_1 u_2)\|_p \leqslant \sum_{i,j\in\mathbb{Z}^n} \|\Box_k(\Box_i u_1 \,\Box_j u_2)\|_p \chi_{|k-i-j|\leqslant k_0(n)}. \tag{6.143}$$

Applying Bernstein's estimate and Hölder's inequality, by (6.143) we have

$$\| \Box_k(u_1 u_2)\|_p \leqslant \sum_{i,j\in\mathbb{Z}^n} \| \Box_i u_1\|_{p_1} \| \Box_j u_2\|_{p_2} \chi_{|k-i-j|\leqslant k_0(n)}. \tag{6.144}$$

So, by (6.144), Hölder's and Minkowski's inequalities, we get that

$$\begin{aligned} &\|u_1 u_2\|_{\ell^1_\Box(L^\gamma(\mathbb{R},L^p))} \\ &\lesssim \sum_{k\in\mathbb{Z}} \left(\int_{\mathbb{R}} \Big(\sum_{i,j\in\mathbb{Z}^n} \| \Box_i u_1(t)\|_{p_1} \| \Box_j u_2(t)\|_{p_2} \chi_{|k-i-j|\leqslant k_0(n)} \Big)^\gamma dt \right)^{1/\gamma} \\ &\lesssim \sum_{k\in\mathbb{Z}} \sum_{i,j\in\mathbb{Z}^n} \| \Box_i u_1(t)\|_{L^{\gamma_1}(\mathbb{R},L^{p_1})} \| \Box_j u_2(t)\|_{L^{\gamma_2}(\mathbb{R},L^{p_2})} \chi_{|k-i-j|\leqslant k_0(n)}. \end{aligned} \tag{6.145}$$

By Young's inequality, it follows from (6.145) that

$$\|u_1 u_2\|_{\ell^1_\Box(L^\gamma(\mathbb{R},L^p))} \lesssim \|u_1\|_{\ell^1_\Box(L^{\gamma_1}(\mathbb{R},L^{p_1}))} \|u_2\|_{\ell^1_\Box(L^{\gamma_2}(\mathbb{R},L^{p_2}))}. \tag{6.146}$$

By the induction and (6.146), we can get the result, as desired. □

Taking $p \in \mathbb{N}$, $p \in [2 + 4/n, 2 + \kappa]$ and

$$X = \ell^1_\Box(L^\infty(\mathbb{R}, L^2)) \cap \ell^1_\Box(L^p(\mathbb{R}, L^p)), \tag{6.147}$$

we easily see that

$$\frac{2}{p} \leqslant n\left(\frac{1}{2} - \frac{1}{p}\right) = \frac{n(p-2)}{2p} \tag{6.148}$$

and " = " in (6.148) holds if and only if $p = 2 + 4/n$. Using Corollary 6.3, we have

$$\|S(t)u_0\|_X \lesssim \|u_0\|_{M_{2,1}}, \tag{6.149}$$

$$\|\mathscr{A} f\|_X \lesssim \|f(u)\|_{\ell^1_\Box(L^{p'}_{x,t\in\mathbb{R}})}. \tag{6.150}$$

Proof. [Proof of Theorem 6.3] Let X be as in (6.147) and $\mathscr{T}$ be as in (6.138). By (6.149) and (6.150),

$$\|\mathscr{T}u\|_X \lesssim \|u_0\|_{M_{2,1}} + \|f(u)\|_{\ell^1_\Box(L^{p'}_{x,t\in\mathbb{R}})}. \tag{6.151}$$

Since $p \in [2+4/n, 2+\kappa]$, we have

$$\frac{1}{p'} = \frac{p-1}{p} + \frac{\kappa+2-p}{\infty}. \tag{6.152}$$

By Lemma 6.8,

$$\|\pi(u^{1+\kappa})\|_{\ell^1_\Box(L^{p'}_{x,t\in\mathbb{R}})} \lesssim \|u\|^{p-1}_{\ell^1_\Box(L^p_{x,t\in\mathbb{R}})} \|u\|^{2+\kappa-p}_{\ell^1_\Box(L^\infty_{x,t\in\mathbb{R}})}. \tag{6.153}$$

Since

$$\|\Box_i u\|_\infty \lesssim \|\Box_i u\|_2, \quad i \in \mathbb{Z}^n, \tag{6.154}$$

in view of (6.153) and (6.154), we have

$$\|\pi(u^{1+\kappa})\|_{\ell^1_\Box(L^{p'}_{x,t\in\mathbb{R}})} \lesssim \|u\|^{1+\kappa}_X. \tag{6.155}$$

It follows from (6.151) and (6.155) that

$$\|\mathscr{T}u\|_X \lesssim \|u_0\|_{M_{2,1}} + \|u\|^{1+\kappa}_X. \tag{6.156}$$

Putting

$$\mathcal{D} = \{u: \|u\|_X \leqslant M\}, \quad d(u,v) = \|u-v\|_X, \tag{6.157}$$

we easily see that, if $M>0$ is sufficiently small, $\|u_0\|_{M_{2,1}} \lesssim M/2$, then $\mathscr{T}: (\mathcal{D}, d) \to (\mathcal{D}, d)$ is a contraction mapping, which leads that (6.135) has a solution $u \in X$. The left part of the proof is standard and we omit the details. □

The proof of Theorem 6.4 is similar to that of Theorem 6.3. Let

$$Y = \ell^1_\Box(L^\infty(\mathbb{R}, L^2)) \cap \ell^1_\Box(L^4_{x,t\in\mathbb{R}}). \tag{6.158}$$

Then we have

$$\|\mathscr{T}u\|_Y \lesssim \|u_0\|_{M_{2,1}} + \sum_{k=1}^\infty \frac{\varrho^k}{k!} \||u|^{2k}u\|_{\ell^1_\Box(L^{4/3}_{x,t\in\mathbb{R}})}. \tag{6.159}$$

Using Lemma 6.8, one has that

$$\|u^{1+2k}\|_{\ell^1_\Box(L^{4/3}_{x,t\in\mathbb{R}})} \lesssim C^{2k+1}\|u\|^3_{\ell^1_\Box(L^4_{x,t\in\mathbb{R}})} \|u\|^{2k-2}_{\ell^1_\Box(L^\infty_{x,t\in\mathbb{R}})} \lesssim C^{2k+1}\|u\|^{2k+1}_Y. \tag{6.160}$$

Hence,

$$\|\mathscr{T}u\|_Y \lesssim \|u_0\|_{M_{2,1}} + \sum_{k=1}^{\infty} \frac{C^{2k+1}}{k!} \|u\|_Y^{2k+1}. \tag{6.161}$$

Applying (6.161) and combining the proof of Theorem 6.3, we can get the conclusion of Theorem 6.4.

We now consider the initial value problem for NLKG,

$$u_{tt} + (I - \Delta)u + f(u) = 0, \quad u(0) = u_0,\ u_t(0) = u_1. \tag{6.162}$$

Analogous to NLS, we have

Theorem 6.5. *Let $n \geqslant 1$, $f(u) = \lambda|u|^\kappa u$, $\kappa \in 2\mathbb{N}$, $\lambda \in \mathbb{R}$, $(u_0, u_1) \in M_{p,1} \times M_{p,1}^{-1}$ and $1 \leqslant p \leqslant \infty$. Then there exists a $T > 0$ such that (6.135) has a unique solution $(u, u_t) \in C([0,T), M_{p,1}) \times C([0,T), M_{p,1}^{-1})$. Moreover, if $T < \infty$, then $\limsup_{t \nearrow T} (\|u(t)\|_{M_{p,1}} + \|u_t(t)\|_{M_{p,1}^{-1}}) = \infty$.*

If the nonlinearity has an exponential growth, the corresponding results as in Theorem 6.5 also hold.

Theorem 6.6. *Let $n \geqslant 1$, $f(u) = \lambda u^{1+\kappa}$, $\kappa \in \mathbb{N}$ and $\kappa \geqslant 4/n$. Put*

$$\sigma = \frac{n+2}{n(2+\kappa)}. \tag{6.163}$$

Assume that $(u_0, u_1) \in M_{2,1}^\sigma \times M_{2,1}^{\sigma-1}$ and there exists a sufficiently small $\delta > 0$ such that $\|u_0\|_{M_{2,1}^\sigma} + \|u_1\|_{M_{2,1}^{\sigma-1}} \leqslant \delta$. Then (6.162) has a unique solution

$$u \in C(\mathbb{R}, M_{2,1}^\sigma) \cap \ell_\Box^1(L_{x,t\in\mathbb{R}}^{2+\kappa}). \tag{6.164}$$

Theorem 6.7. *Let $n \geqslant 2$, $f(u) = \sinh u - u$ and $\sigma = (n+2)/4n$. Assume that $(u_0, u_1) \in M_{2,1}^\sigma \times M_{2,1}^{\sigma-1}$ and there exists a sufficiently small $\delta > 0$ such that $\|u_0\|_{M_{2,1}^\sigma} + \|u_1\|_{M_{2,1}^{\sigma-1}} \leqslant \delta$. Then (6.162) has a unique solution*

$$u \in C(\mathbb{R}, M_{2,1}^\sigma) \cap \ell_\Box^1(L_{x,t\in\mathbb{R}}^4). \tag{6.165}$$

6.5 Derivative nonlinear Schrödinger equations

We study the initial value problem for the derivative nonlinear Schrödinger equation (gDNLS)

$$\mathrm{i}u_t + \Delta_\pm u = F(u, \bar{u}, \nabla u, \nabla \bar{u}), \quad u(0,x) = u_0(x), \tag{6.166}$$

where u is a complex valued function of $(t,x) \in \mathbb{R} \times \mathbb{R}^n$,

$$\Delta_{\pm} u = \sum_{i=1}^{n} \varepsilon_i \partial_{x_i}^2, \quad \varepsilon_i \in \{1, -1\}, \quad i = 1, ..., n, \tag{6.167}$$

$\nabla = (\partial_{x_1}, ..., \partial_{x_n})$, $F : \mathbb{C}^{2n+2} \to \mathbb{C}$ is a series of $z \in \mathbb{C}^{2n+2}$,

$$F(z) = F(z_1, ..., z_{2n+2}) = \sum_{m+1 \leqslant |\beta| < \infty} c_\beta z^\beta, \quad c_\beta \in \mathbb{C}, \tag{6.168}$$

$2 \leqslant m < \infty$, $m \in \mathbb{N}$, $\sup_\beta |c_\beta| < \infty$[9]. The typical nonlinear term is

$$F(u, \bar{u}, \nabla u, \nabla \bar{u}) = |u|^2 \vec{\lambda} \cdot \nabla u + u^2 \vec{\mu} \cdot \nabla \bar{u} + |u|^2 u,$$

see [48; 63; 227]. Another model is

$$F(u, \bar{u}, \nabla u, \nabla \bar{u}) = (1 + |u|^2)^{-1} (\nabla u)^2 \bar{u} = \sum_{k=0}^{\infty} (-1)^k |u|^{2k} (\nabla u)^2 \bar{u}, \quad |u| < 1,$$

which is an equivalent version of the Schrödinger flow [61; 88; 114; 261]. The non-elliptic gDNLS arises in the strongly interacting many-body systems near the criticality, where anisotropic interactions are manifested by the presence of the non-elliptic case, as well as additional residual terms which involve cross derivatives of the independent variables [48; 63; 227]. Some water wave and completely integrable system models in higher spatial dimensions are also non-elliptic, cf. [1; 152; 258; 259]. A large amount of work has been devoted to the study of gDNLS, see [100; 101; 114; 125; 126; 130; 133; 135; 143; 147; 107; 186; 209; 199].

Since the nonlinearity in gDNLS contains derivative terms and the Strichartz inequalities can not absorb any derivatives, gDNLS can not be solved if we use only the Strichartz estimate. One needs to look for some other ways to handle the derivative terms in the nonlinearity. Up to now, three kinds of methods seem to be very useful for gDNLS. One is to use the energy estimate to deal with the derivatives in the nonlinearity, the second way is to use Bourgain's space $X^{s,b}$ and the third technique is Kato's smooth effect estimates. Of course, there are some connections between these methods.

We will use the smooth effect estimates together with the frequency-uniform decomposition techniques to study gDNLS and we show that it is globally wellposed and scattering in a class of modulation spaces.

[9]In fact, we only need the condition $|c_\beta| \leqslant C^{|\beta|}$.

For convenience to the readers, we sketch the ideas. Denote

$$S(t) = e^{\mathrm{i}t\Delta_\pm} = \mathscr{F}^{-1} e^{\mathrm{i}t\sum_{j=1}^n \varepsilon_j \xi_j^2} \mathscr{F}, \quad \mathscr{A} f(t,x) = \int_0^t S(t-\tau) f(\tau, x) d\tau.$$

For the sake of brevity, we simplify the nonlinearity as $F(u, \bar{u}, \nabla u, \nabla \bar{u}) = \vec{\lambda} \cdot \nabla(|u|^2 u)$. According to the equivalent integral form of gDNLS, one needs to solve

$$u(t) = S(t)u_0 - \mathrm{i}\mathscr{A} \vec{\lambda} \cdot \nabla(|u|^2 u).$$

On the basis of the smooth effect of the Schrödinger semi-group in 1D, we easily get its smooth effect estimates in all dimensions,

$$\left\| D_{x_i}^{1/2} S(t) u_0 \right\|_{L^\infty_{x_i} L^2_{(x_j)_{j\neq i}} L^2_t(\mathbb{R}^{1+n})} \lesssim \|u_0\|_2, \tag{6.169}$$

$$\|\partial_{x_i} \mathscr{A} f\|_{L^\infty_{x_i} L^2_{(x_j)_{j\neq i}} L^2_t(\mathbb{R}^{1+n})} \lesssim \|f\|_{L^1_{x_i} L^2_{(x_j)_{j\neq i}} L^2_t}, \tag{6.170}$$

where

$$\|f\|_{L^{p_1}_{x_i} L^{p_2}_{(x_j)_{j\neq i}} L^{p_2}_t} = \left\| \|f\|_{L^{p_2}_{x_1,\ldots,x_{i-1},x_{i+1},\ldots,x_n} L^{p_2}_t(\mathbb{R}\times\mathbb{R}^{n-1})} \right\|_{L^{p_1}_{x_i}(\mathbb{R})}. \tag{6.171}$$

(6.169) and (6.170) are scaling-invariant and so, they are optimal estimates. In view of (6.170), one should choose $L^\infty_{x_i} L^2_{(x_j)_{j\neq i}} L^2_t$ as a framework to handle the partial derivative ∂_{x_i} in the nonlinearity. According to the integral equation, we have

$$\|\partial_{x_1} u\|_{L^\infty_{x_1} L^2_{(x_j)_{j\neq 1}} L^2_t} \lesssim \|D^{1/2}_{x_1} u_0\|_2 + \sum_{i=1}^n \|\partial_{x_i}(|u|^2 u)\|_{L^1_{x_1} L^2_{(x_j)_{j\neq 1}} L^2_t}. \tag{6.172}$$

So, one needs to make the following two kinds of nonlinear estimates,

$$I = \|\partial_{x_1}(|u|^2 u)\|_{L^1_{x_1} L^2_{(x_j)_{j\neq 1}} L^2_t(\mathbb{R}^{1+n})}, \quad II = \|\partial_{x_2}(|u|^2 u)\|_{L^1_{x_1} L^2_{(x_j)_{j\neq 1}} L^2_t}.$$

Let us consider the estimates of I. Using Hölder's inequality, we have

$$I \leqslant \|\partial_{x_1} u\|_{L^\infty_{x_1} L^2_{(x_j)_{j\neq 1}} L^2_t} \|u\|^2_{L^2_{x_1} L^\infty_{(x_j)_{j\neq 1}} L^\infty_t}. \tag{6.173}$$

Hence, we need to estimate $\|u\|_{L^2_{x_1} L^\infty_{(x_j)_{j\neq 1}} L^\infty_t}$. By the integral equation,

$$\|u\|_{L^2_{x_1} L^\infty_{(x_j)_{j\neq 1}} L^\infty_t} \leqslant \|S(t)u_0\|_{L^2_{x_1} L^\infty_{(x_j)_{j\neq 1}} L^\infty_t} + \|\nabla \mathscr{A}(|u|^2 u)\|_{L^2_{x_1} L^\infty_{(x_j)_{j\neq 1}} L^\infty_t}.$$

Unfortunately, we can not get a straightforward global estimates. So, we consider the frequency-localized version of $\|S(t)u_0\|_{L^2_{x_1} L^\infty_{(x_j)_{j\neq 1}} L^\infty_t}$:

$$\|\Box_k S(t) u_0\|_{L^2_{x_1} L^\infty_{(x_j)_{j\neq 1}} L^\infty_t} \lesssim \langle k_1 \rangle^{1/2} \|\Box_k u_0\|_2. \tag{6.174}$$

Since (6.174) is only a frequency-local version, one needs to localize (6.172) and introduce the following

$$\|u\|_{\ell^{1,s}_{\Box_{hi}}(L^\infty_{x_1} L^2_{(x_j)_{j\neq 1}} L^2_t)} := \sum_{k\in\mathbb{Z}^n,\, |k_1|>4} \langle k_1\rangle^s \|\Box_k u\|_{L^\infty_{x_1} L^2_{(x_j)_{j\neq 1}} L^2_t}, \quad (6.175)$$

$$\|u\|_{\ell^1_\Box(L^2_{x_1} L^\infty_{(x_j)_{j\neq 1}} L^\infty_t)} := \sum_{k\in\mathbb{Z}^n} \|\Box_k u\|_{L^2_{x_1} L^\infty_{(x_j)_{j\neq 1}} L^\infty_t}. \quad (6.176)$$

Noticing that the smooth effect estimate in the lower frequency part is worse than that of the Strichartz estimate, we have thrown away the lower frequency on the ξ_1 orientation in (6.175). If we do not consider the sharp estimate, (6.174) implies that

$$\|u\|_{\ell^1_\Box(L^2_{x_1} L^\infty_{(x_j)_{j\neq 1}} L^\infty_t)} \lesssim \|u\|_{M^{1/2}_{2,1}} + \sum_{k\in\mathbb{Z}^n} \langle k_1\rangle^{1/2} \|\Box_k \nabla(|u|^2 u)\|_{L^1_t L^2_x}.$$

Making the nonlinear estimate, we have

$$\|u\|_{\ell^1_\Box(L^2_{x_1} L^\infty_{(x_j)_{j\neq 1}} L^\infty_t)} \lesssim \|u\|_{M^{1/2}_{2,1}} + \left(\sum_{k\in\mathbb{Z}^n} \langle k\rangle^{3/2} \|\Box_k u\|_{L^3_t L^6_x}\right)^3, \quad (6.177)$$

which means that we need the following norm

$$\|u\|_{\ell^{1,3/2}_\Box(L^3_t L^6_x)} := \sum_{k\in\mathbb{Z}^n} \langle k\rangle^{3/2} \|\Box_k u\|_{L^3_t L^6_x}. \quad (6.178)$$

The local version of (6.172) is

$$\|\partial_{x_1} u\|_{\ell^{1,2}_{\Box_{hi}}(L^\infty_{x_1} L^2_{(x_j)_{j\neq 1}} L^2_t)} \lesssim \|u_0\|_{M^{5/2}_{2,1}} + \sum_{i=1}^n \langle k_1\rangle^2 \|\Box_k \partial_{x_i}(|u|^2 u)\|_{L^1_{x_1} L^2_{(x_j)_{j\neq 1}} L^2_t}. \quad (6.179)$$

After making nonlinear estimates, we find that the right hand side of (6.179) can be estimated by (6.175), (6.176) and (6.178). The estimate of II is more complicated and one needs to consider the interaction between the partial derivative ∂_{x_2} and the space $\ell^{1,2}_{\Box_{hi}}(L^\infty_{x_1} L^2_{(x_j)_{j\neq 1}} L^2_t)$, see below for details.

We now state the global wellposedness and scattering results for gDNLS. Let us recall the anisotropic Lebesgue space $L^{p_1}_{x_i} L^{p_2}_{(x_j)_{j\neq i}} L^{p_2}_t$ is defined by (6.171). For $k = (k_1, ..., k_n)$, we write

$$\|u\|_{X^s_\alpha} = \sum_{i,\ell=1}^n \sum_{k\in\mathbb{Z}^n,\, |k_i|>4} \langle k_i\rangle^{s-1/2} \left\|\partial^\alpha_{x_\ell} \Box_k u\right\|_{L^\infty_{x_i} L^2_{(x_j)_{j\neq i}} L^2_t} + \sum_{i,\ell=1}^n \sum_{k\in\mathbb{Z}^n} \left\|\partial^\alpha_{x_\ell} \Box_k u\right\|_{L^m_{x_i} L^\infty_{(x_j)_{j\neq i}} L^\infty_t}, \quad (6.180)$$

$$\|u\|_{S^s_\alpha} = \sum_{\ell=1}^{n} \sum_{k\in\mathbb{Z}^n} \langle k\rangle^{s-1} \left\|\partial^\alpha_{x_\ell} \Box_k u\right\|_{L^\infty_t L^2_x \bigcap L^3_t L^6_x}, \tag{6.181}$$

$$\|u\|_{X^s} = \sum_{\alpha=0,1} \|u\|_{X^s_\alpha}, \quad \|u\|_{S^s} = \sum_{\alpha=0,1} \|u\|_{S^s_\alpha}. \tag{6.182}$$

Theorem 6.8. *Let $n \geqslant 3$, $m = 2$, $u_0 \in M^{5/2}_{2,1}$ and there exists a suitably small $\delta > 0$ such that $\|u_0\|_{M^{5/2}_{2,1}} \leqslant \delta$. Then (6.166) has a unique solution $u \in C(\mathbb{R}, M^{5/2}_{2,1}) \cap X^{5/2} \cap S^{5/2}$, $\|u\|_{X^{5/2}\cap S^{5/2}} \lesssim \delta$. Moreover, the scattering operator S of (6.166) carries a whole zero neighborhood in $C(\mathbb{R}, M^{5/2}_{2,1})$ into $C(\mathbb{R}, M^{5/2}_{2,1})$.*

6.5.1 *Global linear estimates*

Proposition 6.14 (Smooth effect). *For any $i = 1, ..., n$, we have the following estimates*

$$\left\|D^{1/2}_{x_i} S(t) u_0\right\|_{L^\infty_{x_i} L^2_{(x_j)_{j\neq i}} L^2_t} \lesssim \|u_0\|_2, \tag{6.183}$$

$$\|\partial_{x_i} \mathscr{A} f\|_{L^\infty_{x_i} L^2_{(x_j)_{j\neq i}} L^2_t} \lesssim \|f\|_{L^1_{x_i} L^2_{(x_j)_{j\neq i}} L^2_t}, \tag{6.184}$$

$$\|\partial_{x_i} \mathscr{A} f\|_{L^\infty_t L^2_x} \lesssim \|D^{1/2}_{x_i} f\|_{L^1_{x_i} L^2_{(x_j)_{j\neq i}} L^2_t}. \tag{6.185}$$

Proof. By standard dual estimates, (6.183) implies (6.185). So, it suffices to show the first two inequalities. Denote $\bar{x} = (x_2, ..., x_n)$. By Plancherel's identity and Minkowski's inequality,

$$\|S(t)u_0\|_{L^\infty_{x_1} L^2_{\bar{x}} L^2_t} \leqslant \left\|\mathscr{F}^{-1}_{\xi_1} e^{it\varepsilon_1 \xi_1^2} \mathscr{F}_{x_1}(\mathscr{F}_{\bar{x}} u_0)\right\|_{L^2_{\bar{\xi}} L^\infty_{x_1} L^2_t}. \tag{6.186}$$

Recall the smooth effect of $S(t)$ in one spatial dimension,

$$\left\|\mathscr{F}^{-1}_{\xi} e^{it\xi^2} \mathscr{F}_x u_0\right\|_{L^\infty_x L^2_t(\mathbb{R}^{1+1})} \lesssim \|D^{-1/2}_x u_0\|_{L^2(\mathbb{R})}. \tag{6.187}$$

So, (6.186), (6.187) together with Plancherel's equality yield (6.183).

Denote

$$u = c\mathscr{F}^{-1}_{t,x} \frac{\xi_1}{|\xi|^2_\pm - \tau} \mathscr{F}_{t,x} f. \tag{6.188}$$

We can assume that $|\xi|^2_\pm = \xi_1^2 + \varepsilon_2\xi_2^2 + ... + \varepsilon_n \xi_n^n := \xi_1^2 + |\bar{\xi}|^2_\pm$. By Plancherel's identity,

$$\|u\|_{L^\infty_{x_1} L^2_{\bar{x}} L^2_t} \leqslant \left\|\mathscr{F}^{-1}_{\xi_1} \frac{\xi_1}{\xi_1^2 + |\bar{\xi}|^2_\pm - \tau} \mathscr{F}_{t,x} f\right\|_{L^2_{\bar{\xi}} L^\infty_{x_1} L^2_\tau}. \tag{6.189}$$

Making the change of the variable $\tau \to \mu + |\bar{\xi}|_{\pm}^2$, the right hand side of (6.189) becomes

$$\left\| \mathscr{F}_{\xi_1}^{-1} \frac{\xi_1}{\xi_1^2 - \mu} \mathscr{F}_{t,x_1}(e^{-\mathrm{i}t|\bar{\xi}|_{\pm}^2} \mathscr{F}_{x_2,\dots,x_n} f) \right\|_{L^2_{\bar{\xi}} L^\infty_{x_1} L^2_\mu}. \tag{6.190}$$

Recalling the smooth effect in one spatial dimension

$$\left\| \mathscr{F}_{\tau,\xi}^{-1} \frac{\xi}{\xi^2 - \tau} \mathscr{F}_{t,x} f \right\|_{L^\infty_x L^2_t(\mathbb{R}^{1+1})} \lesssim \|f\|_{L^1_x L^2_t(\mathbb{R}^{1+1})}, \tag{6.191}$$

in view of (6.189), (6.190) and (6.191) we get that

$$\|u\|_{L^\infty_{x_1} L^2_{\bar{x}} L^2_t} \lesssim \left\| e^{-\mathrm{i}t|\bar{\xi}|_{\pm}^2} \mathscr{F}_{x_2,\dots,x_n} f \right\|_{L^2_{\bar{\xi}} L^1_{x_1} L^2_t}. \tag{6.192}$$

Using Minkowski's inequality and Plancherel's identity, we immediately have

$$\|u\|_{L^\infty_{x_1} L^2_{\bar{x}} L^2_t} \lesssim \|f\|_{L^1_{x_1} L^2_{\bar{x}} L^2_t}. \tag{6.193}$$

Noticing that $\partial_{x_1} \mathscr{A} f = u - \partial_{x_1} S(t) \int_{-\infty}^{\infty} S(s)\mathrm{sgn}(s) f(s) ds$, we easily see that in (6.193), substituting u by $\partial_{x_1}\mathscr{A} f$, the result also holds. □

6.5.2 *Frequency-localized linear estimates*

We consider the frequency-localized versions of the smooth effect estimate, the maximal function estimate and their relations to the Strichartz estimate for the Schrödinger semi-group. Let $\{\eta_k\}_{k\in\mathbb{Z}} \in \Upsilon_1$ and

$$\sigma_k(\xi) := \eta_{k_1}(\xi_1)\dots\eta_{k_n}(\xi_n). \tag{6.194}$$

We have $\{\sigma_k\}_{k\in\mathbb{Z}^n} \in \Upsilon_n$. In the sequel we will always use the expression of σ_k in (6.194). Recall that $\Box_k = \mathscr{F}^{-1}\sigma_k\mathscr{F}$. For convenience, we denote $\widetilde{\Box}_k = \sum_{|\ell|_\infty \leqslant 1} \Box_{k+\ell}$.

Lemma 6.9. *Let $D^\sigma_{x_i} = (-\partial^2_{x_i})^{\sigma/2}$. For any $\sigma \in \mathbb{R}$, $k = (k_1, \dots, k_n) \in \mathbb{Z}^n$ with $|k_i| \geqslant 4$, we have*

$$\|\Box_k D^\sigma_{x_i} u\|_{L^{p_1}_{x_1} L^{p_2}_{x_2,\dots,x_n} L^{p_2}_t} \lesssim \langle k_i \rangle^\sigma \|\Box_k u\|_{L^{p_1}_{x_1} L^{p_2}_{x_2,\dots,x_n} L^{p_2}_t}.$$

If $\sigma \in \mathbb{N}$, replacing $D^\sigma_{x_i}$ by $\partial^\sigma_{x_i}$, the above estimate holds for all $k \in \mathbb{Z}^n$.

Proof. Using (6.194), we have

$$\Box_k D^\sigma_{x_i} u = \sum_{\ell=-1}^{1} \int_{\mathbb{R}} \left(\mathscr{F}^{-1}_{\xi_i} (\eta_{k_i+\ell}(\xi_i)|\xi_i|^\sigma) \right)(y_i)(\Box_k u)(x_i - y_i) dy_i.$$

By Young's inequality and

$$\|\mathscr{F}^{-1}_{\xi_i}(\eta_{k_i+\ell}(\xi_i)|\xi_i|^\sigma)\|_{L^1(\mathbb{R})} \lesssim \langle k_i \rangle^\sigma,$$

we immediately have the result, as desired. □

Proposition 6.15 (Maximal function estimate). *Let $4/n < q \leqslant \infty$ and $q \geqslant 2$. Then we have*

$$\|\Box_k S(t)u_0\|_{L^q_{x_i} L^\infty_{(x_j)_{j\neq i}} L^\infty_t} \lesssim \langle k_i\rangle^{1/q} \|\Box_k u_0\|_{L^2(\mathbb{R}^n)}. \tag{6.195}$$

Proof. For convenience, we denote $\bar{x} = (x_2, ..., x_n)$. By duality, it suffices to show that

$$\|\mathscr{F}^{-1} e^{\mathrm{i}t|\xi|^2_\pm} \eta_{k_1}(\xi_1)\eta_{\bar{k}}(\bar{\xi})\|_{L^{q/2}_{x_1} L^\infty_{\bar{x},t}(\mathbb{R}^n)} \lesssim \langle k_1\rangle^{2/q}.$$

$\Box_k S(t)$ satisfies the following decay

$$\|\mathscr{F}^{-1}_{\bar{\xi}} e^{\mathrm{i}t|\bar{\xi}|^2_\pm} \eta_{\bar{k}}(\bar{\xi})\|_{L^\infty_{\bar{x}}(\mathbb{R}^{n-1})} \lesssim (1+|t|)^{-(n-1)/2},$$
$$\|\mathscr{F}^{-1}_{\xi_1} e^{\mathrm{i}t\xi_1^2} \eta_{k_1}(\xi_1)\|_{L^\infty_{x_1}(\mathbb{R})} \lesssim (1+|t|)^{-1/2}.$$

On the other hand, integrating by parts, we can get that for $|x_1| > 4|t|\langle k_1\rangle$,

$$|\mathscr{F}^{-1}_{\xi_1} e^{\mathrm{i}t\xi_1^2} \eta_{k_1}(\xi_1)| \lesssim |x_1|^{-2}.$$

So, for any $|x_1| > 1$,

$$|\mathscr{F}^{-1} e^{\mathrm{i}t|\xi|^2_\pm} \eta_{k_1}(\xi_1)\eta_{\bar{k}}(\bar{\xi})| \lesssim (1+|x_1|)^{-2} + \langle k_1\rangle^{n/2}(\langle k_1\rangle + |x_1|)^{-n/2}.$$

This implies that

$$\begin{aligned}
&\|\mathscr{F}^{-1} e^{\mathrm{i}t|\xi|^2_\pm} \eta_{k_1}(\xi_1)\eta_{\bar{k}}(\bar{\xi})\|_{L^{q/2}_{x_1} L^\infty_{\bar{x},t}(\mathbb{R}^n)} \\
&\lesssim 1 + \langle k_1\rangle^{n/2} \|(\langle k_1\rangle + |x_1|)^{-n/2}\|_{L^{q/2}_{x_1}(\mathbb{R})} \\
&\lesssim \langle k_1\rangle^{2/q}.
\end{aligned}$$

It follows that (6.195) holds. □

By Proposition 6.14, we have

Proposition 6.16 (Frequency-localized smooth effect). *For any $k = (k_1, ..., k_n) \in \mathbb{Z}^n$, we have*

$$\|\Box_k \mathscr{A} \partial_{x_i} f\|_{L^\infty_{x_i} L^2_{(x_j)_{j\neq i}} L^2_t} \lesssim \|\Box_k f\|_{L^1_{x_i} L^2_{(x_j)_{j\neq i}} L^2_t}, \tag{6.196}$$

$$\|\Box_k \mathscr{A} \partial_{x_i} f\|_{L^\infty_t L^2_x} \lesssim \langle k_i\rangle^{1/2} \|\Box_k f\|_{L^1_{x_i} L^2_{(x_j)_{j\neq i}} L^2_t}. \tag{6.197}$$

Proof. By Proposition 6.14, we have (6.196). In view of Proposition 6.14 and Lemma 6.9, we obtain that (6.197) holds for all $|k_i| \geqslant 3$. If $|k_i| \leqslant 2$, from Proposition 6.14 it follows that

$$\begin{aligned}
\|\Box_k \mathscr{A} \partial_{x_i} f\|_{L^\infty_t L^2_x} &\lesssim \left\| D^{-1/2}_{x_i} \Box_k \mathscr{A} \partial_{x_i} f \right\|_{L^\infty_t L^2_x} \\
&\lesssim \|\Box_k f\|_{L^1_{x_i} L^2_{(x_j)_{j\neq i}} L^2_t},
\end{aligned}$$

which implies the result, as desired. □

Proposition 6.17. (Relations of Strichartz, smooth effect and maximal function estimates). *Let* $2 \leqslant r < \infty$, $2/\gamma(r) = n(1/2 - 1/r)$ *and* $\gamma > \gamma(r) \vee 2$. *We have*

$$\|\Box_k S(t)u_0\|_{L_t^\gamma L_x^r} \lesssim \|\Box_k u_0\|_{L^2(\mathbb{R}^n)}, \tag{6.198}$$

$$\|\Box_k \mathscr{A} f\|_{L_t^\infty L_x^2 \cap L_t^\gamma L_x^r} \lesssim \|\Box_k f\|_{L_t^{\gamma'} L_x^{r'}}, \tag{6.199}$$

$$\|\Box_k \mathscr{A} \partial_{x_i} f\|_{L_t^\gamma L_x^r} \lesssim \langle k_i\rangle^{1/2} \|\Box_k f\|_{L^1_{x_i} L^2_{(x_j)_{j\neq i}} L_t^2}, \tag{6.200}$$

$$\|\Box_k \mathscr{A} \partial_{x_i} f\|_{L^\infty_{x_i} L^2_{(x_j)_{j\neq i}} L_t^2} \lesssim \langle k_i\rangle^{1/2} \|\Box_k f\|_{L_t^{\gamma'} L_x^{r'}}, \tag{6.201}$$

$$\left\|\Box_k \mathscr{A} \partial^\alpha_{x_i} f\right\|_{L^2_{x_i} L^\infty_{(x_j)_{j\neq i}} L_t^\infty} \lesssim \langle k_i\rangle^{\alpha+1/2} \|\Box_k f\|_{L_t^1 L_x^2}. \tag{6.202}$$

Proof. (6.198) and (6.199) are corollaries of the Strichartz estimate. By (6.195) we can get (6.202). For convenience, we write

$$\mathcal{L}_k(f,\psi) := \left| \int_{\mathbb{R}} \left(\Box_k \int_{\mathbb{R}} S(t-\tau) f(\tau) d\tau,\ \psi(t) \right) dt \right|. \tag{6.203}$$

Now we prove (6.200). Applying the Strichartz inequality, Lemma 6.9 and Proposition 6.16, we have

$$\begin{aligned}\mathcal{L}_k(\partial_{x_1} f, \psi) &\lesssim \langle k_i\rangle^{1/2} \|\Box_k f\|_{L^1_{x_1} L^2_{x_2,\ldots,x_n} L_t^2} \|\tilde{\Box}_k \psi\|_{L_t^{\gamma'} L_x^{r'}} \\ &\lesssim \langle k_i\rangle^{1/2} \|\Box_k f\|_{L^1_{x_1} L^2_{x_2,\ldots,x_n} L_t^2} \|\psi\|_{L_t^{\gamma'} L_x^{r'}}. \end{aligned} \tag{6.204}$$

By the duality, (6.210) and Christ-Kiselev's lemma (see Appendix), we can get (6.200). Changing the role of f and ψ, we immediately get that (6.201) holds for $r > 2$. If $r = 2$, (6.201) is a straightforward consequence of the smooth effect estimate of $S(t)$. □

Corollary 6.5. *Let* $2 \leqslant q < \infty$, $q > 4/n$ *and* $4/n \leqslant p < \infty$. *We have the following results.*

$$\left\| D^{1/2}_{x_1} \Box_k S(t) u_0 \right\|_{L^\infty_{x_1} L^2_{x_2,\ldots,x_n} L_t^2} \lesssim \|\Box_k u_0\|_{L^2(\mathbb{R}^n)}, \tag{6.205}$$

$$\|\Box_k S(t) u_0\|_{L^q_{x_1} L^\infty_{x_2,\ldots,x_n} L_t^\infty} \lesssim \langle k_i\rangle^{1/q} \|\Box_k u_0\|_{L^2(\mathbb{R}^n)}, \tag{6.206}$$

$$\|\Box_k S(t) u_0\|_{L^{2+p}_{t,x} \cap L_t^\infty L_x^2} \lesssim \|\Box_k u_0\|_{L^2(\mathbb{R}^n)}, \tag{6.207}$$

$$\|\Box_k \mathscr{A} \partial_{x_1} f\|_{L^\infty_{x_1} L^2_{x_2,\ldots,x_n} L_t^2} \lesssim \|\Box_k f\|_{L^1_{x_1} L^2_{x_2,\ldots,x_n} L_t^2}, \tag{6.208}$$

$$\|\Box_k \mathscr{A} f\|_{L_t^\infty L_x^2 \cap L^{2+p}_{t,x}} \lesssim \langle k_1\rangle^{1/2} \|\Box_k f\|_{L^1_{x_1} L^2_{x_2,\ldots,x_n} L_t^2}, \tag{6.209}$$

$$\|\Box_k \mathscr{A} \partial_{x_1} f\|_{L^\infty_{x_1} L^2_{x_2,\ldots,x_n} L_t^2} \lesssim \langle k_1\rangle^{1/2} \|\Box_k f\|_{L^{(2+p)/(1+p)}_{t,x}}, \tag{6.210}$$

$$\|\Box_k \mathscr{A} \partial_{x_1} f\|_{L^q_{x_1} L^\infty_{x_2,\ldots,x_n} L_t^\infty} \lesssim \langle k_1\rangle^{1+1/q} \|\Box_k f\|_{L_t^1 L^2}, \tag{6.211}$$

$$\|\Box_k \mathscr{A} f\|_{L^\infty_t L^2_x \cap L^{2+p}_{t,x}} \lesssim \|\Box_k f\|_{L^{(2+p)/(1+p)}_{t,x}}. \tag{6.212}$$

Moreover, if $L^{2+p}_{x,t}$ is replaced by $L^3_t L^6_x$ in (6.207), (6.209) and (6.212), then corresponding results also hold.

By (6.208) in Corollary 6.5, the operator $\Box_k \partial_{x_1} \mathscr{A} : L^1_{x_1} L^2_{x_2,...,x_n} L^2_t \to L^\infty_{x_1} L^2_{x_2,...,x_n} L^2_t$ and the partial derivative ∂_{x_1} is successfully absorbed. However, $\mathscr{A}$ in the space $L^\infty_{x_1} L^2_{x_2,...,x_n} L^2_t$ can not handle the partial derivative ∂_{x_2}. So, one needs to use another way to deal with $\partial_{x_2}\mathscr{A}$ when it appears in the space $L^\infty_{x_1} L^2_{x_2,...,x_n} L^2_t$.

Proposition 6.18. *Let $i = 2, ..., n$, $2 \leqslant q \leqslant \infty$, $q > 4/n$, $2 \leqslant r < \infty$, $2/\gamma(r) = n(1/2 - 1/r)$, $\gamma \geqslant \gamma(r)$, $\gamma > 2$. Then we have*

$$\|\Box_k \partial_{x_i} \mathscr{A} f\|_{L^\infty_{x_1} L^2_{x_2,...,x_n} L^2_t} \lesssim \|\partial_{x_i} \partial^{-1}_{x_1} \Box_k f\|_{L^1_{x_1} L^2_{x_2,...,x_n} L^2_t}, \tag{6.213}$$

$$\|\Box_k \partial_{x_i} \mathscr{A} f\|_{L^\infty_{x_1} L^2_{x_2,...,x_n} L^2_t} \lesssim \|\partial_{x_i} D^{-1/2}_{x_1} \Box_k f\|_{L^{\gamma'} L^{r'}_x}, \tag{6.214}$$

$$\|\Box_k \partial_{x_i} \mathscr{A} f\|_{L^q_{x_1} L^\infty_{x_2,...,x_n} L^\infty_t} \lesssim \langle k_i \rangle \langle k_1 \rangle^{1/q} \|\Box_k f\|_{L^1_t L^2_x}. \tag{6.215}$$

Proof. (6.213) is a corollary of Proposition 6.14. We have

$$\begin{aligned} \mathcal{L}(\partial_{x_2} f, \psi) &:= \left| \int_{\mathbb{R}} \left(\int_{\mathbb{R}} S(t-\tau) \partial_{x_2} f(\tau) d\tau, \ \psi(t) \right) dt \right| \\ &\leqslant \left\| \int_{\mathbb{R}} S(-\tau) \partial_{x_2} D^{-1/2}_{x_1} f(\tau) d\tau \right\|_{L^2(\mathbb{R}^n)} \\ &\quad \times \left\| D^{1/2}_{x_1} \int_{\mathbb{R}} S(-t) \psi(t) dt \right\|_{L^2(\mathbb{R}^n)}. \end{aligned} \tag{6.216}$$

In view of the Strichartz inequality and Proposition 6.14,

$$\mathcal{L}(\partial_{x_2} f, \psi) \lesssim \|\partial_{x_2} D^{-1/2}_{x_1} f\|_{L^{\gamma'}_t L^{r'}_x} \|\psi\|_{L^1_{x_1} L^2_{x_2,...,x_n} L^2_t}. \tag{6.217}$$

If $r > 2$, by the duality, Christ-Kiselev's lemma and (6.217), we have (6.214). If $r = 2$, in view of the smooth effect estimate of $S(t)$, we see that (6.214) holds. By Proposition 6.15, we have (6.215). □

Lemma 6.10. *Let $\psi : [0, \infty) \to [0, 1]$ be a smooth cut-off function satisfying $\psi(x) = 1$ for $|x| \leqslant 1$; and $\psi(x) = 0$ for $|x| \geqslant 2$. Let $\psi_1(\xi) = \psi(\xi_2/2\xi_1)$, $\psi_2(\xi) = 1 - \psi(\xi_2/2\xi_1)$, $\xi \in \mathbb{R}^n$. For any $\sigma \geqslant 0$, we have*

$$\begin{aligned} &\sum_{k \in \mathbb{Z}^n, |k_1| > 4} \langle k_1 \rangle^\sigma \left\| \mathscr{F}^{-1}_{\xi_1,\xi_2} \psi_1 \mathscr{F}_{x_1,x_2} \Box_k \partial_{x_2} \mathscr{A} f \right\|_{L^\infty_{x_1} L^2_{x_2,...,x_n} L^2_t} \\ &\quad \lesssim \sum_{k \in \mathbb{Z}^n, |k_1| > 4} \langle k_1 \rangle^\sigma \|\Box_k f\|_{L^1_{x_1} L^2_{x_2,...,x_n} L^2_t}. \end{aligned} \tag{6.218}$$

For $\sigma \geqslant 1$,

$$\sum_{k\in\mathbb{Z}^n,\,|k_1|>4} \langle k_1\rangle^\sigma \left\| \mathscr{F}^{-1}_{\xi_1,\xi_2} \psi_2 \mathscr{F}_{x_1,x_2} \Box_k \partial_{x_2} \mathscr{A} f \right\|_{L^\infty_{x_1} L^2_{x_2,\dots,x_n} L^2_t} \lesssim \sum_{k\in\mathbb{Z}^n,\,|k_2|>4} \langle k_2\rangle^\sigma \|\Box_k f\|_{L^1_{x_1} L^2_{x_2,\dots,x_n} L^2_t}. \tag{6.219}$$

Proof. For simplicity, we denote $\bar{x} = (x_2, ..., x_n)$ and

$$I = \left\| \mathscr{F}^{-1}_{\xi_1,\xi_2} \psi_1 \mathscr{F}_{x_1,x_2} \Box_k \partial_{x_2} \mathscr{A} f \right\|_{L^\infty_{x_1} L^2_{\bar{x}} L^2_t},$$

$$II = \left\| \mathscr{F}^{-1}_{\xi_1,\xi_2} \psi_2 \mathscr{F}_{x_1,x_2} \Box_k \partial_{x_2} \mathscr{A} f \right\|_{L^\infty_{x_1} L^2_{\bar{x}} L^2_t}.$$

Let η_k be as in Lemma (6.194). For $k \in \mathbb{Z}^n$, $|k_1| > 4$, using the almost orthogonality of $\Box_k$, we have

$$I \lesssim \sum_{|\ell_1|,|\ell_2|\leqslant 1} \left\| \mathscr{F}^{-1}_{\xi_1,\xi_2} \psi\left(\frac{\xi_2}{2\xi_1}\right) \frac{\xi_2}{\xi_1} \prod_{i=1,2} \eta_{k_i+\ell_i}(\xi_i) \mathscr{F}_{x_1,x_2} \Box_k \partial_{x_1} \mathscr{A} f \right\|_{L^\infty_{x_1} L^2_{\bar{x}} L^2_t}. \tag{6.220}$$

Denote

$$(f \circledast_{12} g)(x) = \int_{\mathbb{R}^2} f(t, x_1 - y_1, x_2 - y_2, x_3, ..., x_n) g(t, y_1, y_2) dy_1 dy_2. \tag{6.221}$$

For any Banach function space X defined in $\mathbb{R}^{1+n}$, we have

$$\|f \circledast_{12} g\|_X \leqslant \|g\|_{L^1_{y_1,y_2}(\mathbb{R}^2)} \sup_{y_1,y_2} \|f(\cdot, \cdot - y_1, \cdot - y_2, \cdot, ..., \cdot)\|_X. \tag{6.222}$$

Hence, in view of (6.220) and (6.222),

$$I \lesssim \sum_{|\ell_1|,|\ell_2|\leqslant 1} \left\| \mathscr{F}^{-1}_{\xi_1,\xi_2} \psi\left(\frac{\xi_2}{2\xi_1}\right) \frac{\xi_2}{\xi_1} \prod_{i=1,2} \eta_{k_i+\ell_i}(\xi_i) \right\|_{L^1(\mathbb{R}^2)} \|\Box_k \partial_{x_1} \mathscr{A} f\|_{L^\infty_{x_1} L^2_{\bar{x}} L^2_t}. \tag{6.223}$$

Using the multiplier estimates, for any $|k_1| > 4$, we have

$$\left\| \mathscr{F}^{-1}_{\xi_1,\xi_2} \psi\left(\frac{\xi_2}{2\xi_1}\right) \frac{\xi_2}{\xi_1} \prod_{i=1,2} \eta_{k_i+\ell_i}(\xi_i) \right\|_{L^1(\mathbb{R}^2)} \lesssim \sum_{|\alpha|\leqslant 2} \left\| D^\alpha \left[\psi\left(\frac{\xi_2}{2\xi_1}\right) \frac{\xi_2}{\xi_1} \prod_{i=1,2} \eta_{k_i+\ell_i}(\xi_i) \right] \right\|_{L^2(\mathbb{R}^2)} \lesssim 1. \tag{6.224}$$

By Proposition 6.16, (6.223) and (6.224), one has that

$$I \lesssim \|\Box_k f\|_{L^1_{x_1} L^2_{\bar{x}} L^2_t}, \quad |k_1| \geqslant 4. \tag{6.225}$$

Now we consider the estimate of II. Applying Proposition 6.18, we have

$$\begin{aligned} II &\lesssim \left\| \mathscr{F}^{-1}_{\xi_1,\xi_2} (\xi_2/\xi_1)\psi_2 \mathscr{F}_{x_1,x_2} \Box_k f \right\|_{L^1_{x_1} L^2_{\bar{x}} L^2_t} \\ &\lesssim \sum_{|\ell_1|,|\ell_2|\leqslant 1} \left\| \mathscr{F}^{-1}_{\xi_1,\xi_2} \left(1-\psi\left(\frac{\xi_2}{2\xi_1}\right)\right) \frac{\xi_2}{\xi_1} \prod_{i=1,2} \eta_{k_i+\ell_i}(\xi_i) \right\|_{L^1(\mathbb{R}^2)} \\ &\quad \times \|\Box_k f\|_{L^1_{x_1} L^2_{\bar{x}} L^2_t}. \end{aligned} \tag{6.226}$$

Notice that $\mathrm{supp}\psi_2 \subset \{\xi : |\xi_2| \geqslant 2|\xi_1|\}$. If $|k_1| \geqslant 4$, then $|k_2| > 6$ and in the summation of the left hand side of (6.219), one has that $|k_2| \geqslant |k_1|$. So,

$$\sum_{k\in\mathbb{Z}^n,\ |k_1|>4} \langle k_1\rangle^\sigma II \leqslant \sum_{k\in\mathbb{Z}^n,\ |k_1|>4} \langle k_2\rangle^{\sigma-1} \langle k_1\rangle II.$$

We have

$$\begin{aligned} &\left\| \mathscr{F}^{-1}_{\xi_1,\xi_2} \left(1-\psi\left(\frac{\xi_2}{2\xi_1}\right)\right) \frac{\xi_2}{\xi_1} \prod_{i=1,2} \eta_{k_i+\ell_i}(\xi_i) \right\|_{L^1(\mathbb{R}^2)} \\ &\lesssim \sum_{|\alpha|\leqslant 2} \left\| D^\alpha \left[\mathscr{F}^{-1}_{\xi_1,\xi_2} \left(1-\psi\left(\frac{\xi_2}{2\xi_1}\right)\right) \frac{\xi_2}{\xi_1} \prod_{i=1,2} \eta_{k_i+\ell_i}(\xi_i) \right] \right\|_{L^2(\mathbb{R}^2)} \\ &\lesssim \langle k_2\rangle \langle k_1\rangle^{-1}. \end{aligned} \tag{6.227}$$

Combining (6.226) and (6.227), we can get the estimate of II. □

6.5.3 *Proof of global wellposedness for small rough data*

Denote

$$\begin{aligned} \varrho_1^{(i)}(u) &= \sum_{k\in\mathbb{Z}^n,\ |k_i|>4} \langle k_i\rangle^2 \|\Box_k u\|_{L^\infty_{x_i} L^2_{(x_j)_{j\neq i}} L^2_t}, \\ \varrho_2^{(i)}(u) &= \sum_{k\in\mathbb{Z}^n} \|\Box_k u\|_{L^m_{x_i} L^\infty_{(x_j)_{j\neq i}} L^\infty_t}, \\ \varrho_3^{(i)}(u) &= \sum_{k\in\mathbb{Z}^n} \langle k_i\rangle^{3/2} \|\Box_k u\|_{L^3_t L^6_x \cap L^\infty_t L^2_x}. \end{aligned}$$

Let

$$X := \left\{ u \in \mathscr{S}' : \ \|u\|_X := \sum_{\ell=1}^3 \sum_{\alpha=0,1} \sum_{i,j=1}^n \varrho_\ell^{(i)}(\partial^\alpha_{x_j} u) \leqslant \delta \right\}.$$

Considering the mapping

$$\mathscr{T} : u(t) \to S(t)u_0 - \mathrm{i}\mathscr{A} F(u, \bar{u}, \nabla u, \nabla \bar{u}),$$

we show that $\mathscr{T} : X \to X$ is a contraction mapping. Since $\|u\|_X = \|\bar{u}\|_X$, we can assume that

$$F(u, \bar{u}, \nabla u, \nabla \bar{u}) = F(u, \nabla u) := \sum_{m+1 \leqslant \kappa+|\nu|<\infty} c_{\kappa\nu} u^{\kappa} (\nabla u)^{\nu},$$

where $(\nabla u)^{\nu} = u_{x_1}^{\nu_1} ... u_{x_n}^{\nu_n}$. For convenience, we write

$$v_1 = ... = v_{\kappa} = u,$$
$$v_{\kappa+1} = ... = v_{\kappa+\nu_1} = u_{x_1}, ..., v_{\kappa+|\nu|-\nu_n+1} = ... = v_{\kappa+|\nu|} = u_{x_n}.$$

By (6.183), for $\alpha = 0, 1$, we have

$$\varrho_1^{(i)}(\partial_{x_j}^{\alpha} S(t)u_0) \lesssim \sum_{k\in\mathbb{Z}^n,\ |k_i|>4} \langle k_i\rangle^{1/2} \langle k_j\rangle^2 \|\Box_k u_0\|_{L^2(\mathbb{R}^n)} \leqslant \|u_0\|_{M^{5/2}_{2,1}}.$$

In view of (6.206) and (6.207), for $\alpha = 0, 1$, we have

$$\varrho_2^{(i)}(\partial_{x_j}^{\alpha} S(t)u_0) + \varrho_3^{(i)}(\partial_{x_j}^{\alpha} S(t)u_0) \lesssim \|u_0\|_{M^{5/2}_{2,1}}.$$

So,

$$\|S(t)u_0\|_X \lesssim \|u_0\|_{M^{5/2}_{2,1}}.$$

In order to estimate $\varrho_1^{(i)}(\mathscr{A}\partial_{x_j}^{\alpha}(v_1...v_{\kappa+|\nu|}))$, $i, j = 1, ..., n$, it suffices to consider the estimates of $\varrho_1^{(1)}(\mathscr{A}\partial_{x_1}^{\alpha}(v_1...v_{\kappa+|\nu|}))$ and $\varrho_1^{(1)}(\mathscr{A}\partial_{x_2}^{\alpha}(v_1...v_{\kappa+|\nu|}))$. Applying the frequency-uniform decomposition, we have

$$\begin{aligned}\Box_k(v_1...v_{\kappa+|\nu|}) = &\sum_{\mathbb{S}_1^{(i)}} \Box_k \left(\Box_{k^{(1)}} v_1 ... \Box_{k^{(\kappa+|\nu|)}} v_{\kappa+|\nu|}\right) \\ &+ \sum_{\mathbb{S}_2^{(i)}} \Box_k \left(\Box_{k^{(1)}} v_1 ... \Box_{k^{(\kappa+|\nu|)}} v_{\kappa+|\nu|}\right),\end{aligned} \tag{6.228}$$

where

$$\mathbb{S}_1^{(i)} := \{(k^{(1)}, ..., k^{(\kappa+|\nu|)}) : \ |k_i^{(1)}| \vee ... \vee |k_i^{(\kappa+|\nu|)}| > 4\},$$
$$\mathbb{S}_2^{(i)} := \{(k^{(1)}, ..., k^{(\kappa+|\nu|)}) : \ |k_i^{(1)}| \vee ... \vee |k_i^{(\kappa+|\nu|)}| \leqslant 4\}.$$

We will frequently use the almost orthogonality of $\Box_k$,

$$\begin{aligned}&\Box_k \left(\Box_{k^{(1)}} v_1 ... \Box_{k^{(\kappa+|\nu|)}} v_{\kappa+|\nu|}\right) = 0, \\ &\text{if} \quad |k - k^{(1)} - ... - k^{(\kappa+|\nu|)}|_{\infty} > \kappa + |\nu| + 1.\end{aligned} \tag{6.229}$$

By (6.196) and (6.201),

$$\begin{aligned}
&\varrho_1^{(1)}(\mathscr{A}\partial_{x_1}^{\alpha}(v_1...v_{\kappa+|\nu|}))\\
&\lesssim \sum_{k\in\mathbb{Z}^n,\ |k_1|>4} \langle k_1\rangle^2 \sum_{\mathbb{S}_1^{(1)}} \|\Box_k\left(\Box_{k^{(1)}}v_1...\Box_{k^{(\kappa+|\nu|)}}v_{\kappa+|\nu|}\right)\|_{L^1_{x_1}L^2_{\bar{x}}L^2_t}\\
&\quad+\sum_{k\in\mathbb{Z}^n,\ |k_1|>4} \langle k_1\rangle^{5/2} \sum_{\mathbb{S}_2^{(1)}} \|\Box_k\left(\Box_{k^{(1)}}v_1...\Box_{k^{(\kappa+|\nu|)}}v_{\kappa+|\nu|}\right)\|_{L_{t,x}^{\frac{\kappa+|\nu|+1}{\kappa+|\nu|}}}\\
&=: I+II. \qquad (6.230)
\end{aligned}$$

Applying the almost orthogonality of $\Box_k$ in (6.229), we have

$$\begin{aligned}
I \lesssim C^{\kappa+|\nu|} &\sum_{k^{(1)}\in\mathbb{Z}^n,\ |k_1^{(1)}|>2} \langle k_1^{(1)}\rangle^2 \|\Box_{k^{(1)}}v_1\|_{L^\infty_{x_1}L^2_{\bar{x}}L^2_t}\\
&\times \sum_{k^{(2)},...,k^{(\kappa+|\nu|)}\in\mathbb{Z}^n} \prod_{i=2}^{\kappa+|\nu|} \|\Box_{k^{(i)}}v_i\|_{L^{\kappa+|\nu|-1}_{x_1}L^\infty_{\bar{x}}L^\infty_t}. \qquad (6.231)
\end{aligned}$$

By Hölder's inequality and Lemma 6.1,

$$\begin{aligned}
&\|\Box_{k^{(i)}}v_i\|_{L^{\kappa+|\nu|-1}_{x_1}L^\infty_{\bar{x}}L^\infty_t}\\
&\leqslant \|\Box_{k^{(i)}}v_i\|^{\frac{m}{\kappa+|\nu|-1}}_{L^m_{x_1}L^\infty_{\bar{x}}L^\infty_t} \|\Box_{k^{(i)}}v_i\|^{1-\frac{m}{\kappa+|\nu|-1}}_{L^\infty_{x,t}}\\
&\lesssim \|\Box_{k^{(i)}}v_i\|^{\frac{m}{\kappa+|\nu|-1}}_{L^m_{x_1}L^\infty_{\bar{x}}L^\infty_t} \|\Box_{k^{(i)}}v_i\|^{1-\frac{m}{\kappa+|\nu|-1}}_{L^\infty_tL^2_x}. \qquad (6.232)
\end{aligned}$$

Hence, in view of $v_i = u$ or $v_i = u_{x_j}$, it follows from (6.231) and (6.232) that

$$I \lesssim (C\|u\|_X)^{\kappa+|\nu|}. \qquad (6.233)$$

By the definition of $\mathbb{S}_2^{(1)}$, one has that $|k_1| \leqslant C(\kappa+|\nu|)$ in the summation of II. Again, by Hölder's inequality and Lemma 6.1,

$$\begin{aligned}
\|\Box_{k^{(1)}}v_1...\Box_{k^{(\kappa+|\nu|)}}v_{\kappa+|\nu|}\|_{L_{x,t}^{\frac{\kappa+|\nu|+1}{\kappa+|\nu|}}} &\leqslant \prod_{i=1}^{\kappa+|\nu|} \|\Box_{k^{(i)}}v_i\|_{L^{\kappa+|\nu|+1}_{x,t}}\\
&\lesssim \prod_{i=1}^{\kappa+|\nu|} \|\Box_{k^{(i)}}v_i\|_{L^{2+m}_{x,t}\cap L^\infty_tL^2_x}, \qquad (6.234)
\end{aligned}$$

which implies that

$$II \lesssim (C\|u\|_X)^{\kappa+|\nu|}. \qquad (6.235)$$

Now we estimate $\varrho_1^{(1)}(\mathscr{A}\partial_{x_2}^\alpha(v_1...v_{\kappa+|\nu|}))$. The case $\alpha=0$ has been discussed in the above, it suffices to consider the case $\alpha=1$. Let ψ_i $(i=1,2)$ be as in Lemma 6.10 and $P_i=\mathscr{F}^{-1}\psi_i\mathscr{F}$. We have

$$\begin{aligned}
&\varrho_1^{(1)}(\mathscr{A}\partial_{x_2}(v_1...v_{\kappa+|\nu|}))\\
&\leqslant \sum_{k\in\mathbb{Z}^n,\ |k_1|>4}\langle k_1\rangle^2\|P_1\Box_k(\mathscr{A}\partial_{x_2}(v_1...v_{\kappa+|\nu|}))\|_{L^\infty_{x_1}L^2_{\bar{x}}L^2_t}\\
&\quad+\sum_{k\in\mathbb{Z}^n,\ |k_1|>4}\langle k_1\rangle^2\|P_2\Box_k(\mathscr{A}\partial_{x_2}(v_1...v_{\kappa+|\nu|}))\|_{L^\infty_{x_1}L^2_{\bar{x}}L^2_t}\\
&=: III+IV. \qquad (6.236)
\end{aligned}$$

Using the decomposition in (6.228), we have

$$\begin{aligned}
III\leqslant&\sum_{k\in\mathbb{Z}^n,\ |k_1|>4}\langle k_1\rangle^2\sum_{\mathbb{S}_1^{(1)}}\|P_1\Box_k(\mathscr{A}\partial_{x_2}(\Box_{k^{(1)}}v_1...\Box_{k^{(\kappa+|\nu|)}}v_{\kappa+|\nu|}))\|_{L^\infty_{x_1}L^2_{\bar{x}}L^2_t}\\
&+\sum_{k\in\mathbb{Z}^n,\ |k_1|>4}\langle k_1\rangle^2\sum_{\mathbb{S}_2^{(1)}}\|P_1\Box_k(\mathscr{A}\partial_{x_2}(\Box_{k^{(1)}}v_1...\Box_{k^{(\kappa+|\nu|)}}v_{\kappa+|\nu|}))\|_{L^\infty_{x_1}L^2_{\bar{x}}L^2_t}\\
=:&\ III_1+III_2. \qquad (6.237)
\end{aligned}$$

By Lemma 6.10,

$$III_1\lesssim\sum_{\mathbb{S}_1^{(1)}}\sum_{k\in\mathbb{Z}^n,\ |k_1|>4}\langle k_1\rangle^2\|\Box_k(\Box_{k^{(1)}}v_1...\Box_{k^{(\kappa+|\nu|)}}v_{\kappa+|\nu|})\|_{L^1_{x_1}L^2_{\bar{x}}L^2_t}. \qquad (6.238)$$

By the symmetry, we can assume that $|k_1^{(1)}|=\max(|k_1^{(1)}|,...,|k_1^{(\kappa+|\nu|)}|)$ in $\mathbb{S}_1^{(1)}$. So,

$$\begin{aligned}
&III_1\\
&\lesssim C^{\kappa+|\nu|}\sum_{\mathbb{S}_1^{(1)},\ |k_1^{(1)}|>4}\langle k_1^{(1)}\rangle^2\|\Box_{k^{(1)}}v_1\|_{L^\infty_{x_1}L^2_{\bar{x}}L^2_t}\prod_{i=2}^{\kappa+|\nu|}\|\Box_{k^{(i)}}v_i\|_{L^{\kappa+|\nu|-1}_{x_1}L^\infty_{\bar{x}}L^\infty_t}\\
&\lesssim C^{\kappa+|\nu|}\varrho_1^{(1)}(v_1)\prod_{i=2}^{\kappa+|\nu|}(\varrho_2^{(1)}(v_i)+\varrho_3^{(1)}(v_i))\lesssim(C\|u\|_X)^{\kappa+|\nu|}. \qquad (6.239)
\end{aligned}$$

Applying (6.214) and noticing that $|k_1|\leqslant C$ in III_2, we obtain that

$$\begin{aligned}
&III_2\\
&\lesssim\sum_{k\in\mathbb{Z}^n,\ |k_1|>4,\ |k_2|\lesssim|k_1|}\langle k_1\rangle^{5/2}\sum_{\mathbb{S}_2^{(1)}}\|\Box_k(\Box_{k^{(1)}}v_1...\Box_{k^{(\kappa+|\nu|)}}v_{\kappa+|\nu|})\|_{L^{(2+m)/(1+m)}_{t,x}}
\end{aligned}$$

$$\lesssim C^{\kappa+|\nu|} \prod_{i=1}^{\kappa+|\nu|} \varrho_3^{(1)}(v_i) \leqslant (C\|u\|_X)^{\kappa+|\nu|}. \tag{6.240}$$

We have shown that

$$III \lesssim (C\|u\|_X)^{\kappa+|\nu|}. \tag{6.241}$$

Now we estimate IV. Using (6.228), we have

$$\begin{aligned} IV \leqslant & \sum_{k\in\mathbb{Z}^n,\ |k_1|>4} \langle k_1\rangle^2 \sum_{\mathbb{S}_1^{(2)}} \|P_2\Box_k(\mathscr{A}\partial_{x_2}(\Box_{k^{(1)}}v_1...\Box_{k^{(\kappa+|\nu|)}}v_{\kappa+|\nu|}))\|_{L^\infty_{x_1}L^2_{\bar{x}}L^2_t} \\ & + \sum_{k\in\mathbb{Z}^n,\ |k_1|>4} \langle k_1\rangle^2 \sum_{\mathbb{S}_2^{(2)}} \|P_2\Box_k(\mathscr{A}\partial_{x_2}(\Box_{k^{(1)}}v_1...\Box_{k^{(\kappa+|\nu|)}}v_{\kappa+|\nu|}))\|_{L^\infty_{x_1}L^2_{\bar{x}}L^2_t} \\ =: & IV_1 + IV_2. \end{aligned} \tag{6.242}$$

By Lemma 6.10,

$$IV_1 \lesssim \sum_{\mathbb{S}_1^{(2)}} \sum_{k\in\mathbb{Z}^n,\ |k_2|>4} \langle k_2\rangle^2 \|\Box_k(\Box_{k^{(1)}}v_1...\Box_{k^{(\kappa+|\nu|)}}v_{\kappa+|\nu|})\|_{L^1_{x_1}L^2_{\bar{x}}L^2_t}. \tag{6.243}$$

We can choose some $k^{(i)}$, say $k^{(\kappa+|\nu|)}$ such that $|k_2^{(\kappa+|\nu|)}|$ does not attain or does not uniquely attain $\max_{1\leqslant i\leqslant\kappa+|\nu|}|k_2^{(i)}|$, then we can take another $k^{(i)}$, say $k^{(1)}$ such that $k_2^{(1)} = \max_{1\leqslant i\leqslant\kappa+|\nu|}|k_2^{(i)}|$. By Hölder's inequality,

$$\begin{aligned} & \|\Box_{k^{(1)}}v_1...\Box_{k^{(\kappa+|\nu|)}}v_{\kappa+|\nu|}\|_{L^1_{x_1}L^2_{\bar{x}}L^2_t} \\ & \leqslant \|\Box_{k^{(1)}}v_1...\Box_{k^{(\kappa+|\nu|-1)}}v_{\kappa+|\nu|-1}\|_{L^2_{x,t}} \|\Box_{k^{(\kappa+|\nu|)}}v_{\kappa+|\nu|}\|_{L^2_{x_1}L^\infty_{\bar{x}}L^\infty_t} \\ & \leqslant \|\Box_{k^{(1)}}v_1\|_{L^\infty_{x_1}L^2_{\bar{x}}L^2_t} \prod_{i=2}^{\kappa+|\nu|} \|\Box_{k^{(i)}}v_i\|_{(L^\infty_t L^2_x)\cap(L^2_{x_1}L^\infty_{\bar{x},t})}. \end{aligned} \tag{6.244}$$

Combining (6.243) with (6.244), we get

$$IV_1 \lesssim (C\|u\|_X)^{\kappa+|\nu|}. \tag{6.245}$$

In order to estimate IV_2, we apply (6.214) to obtain that

$$\begin{aligned} IV_2 \lesssim & \sum_{k\in\mathbb{Z}^n,\ |k_1|>4} \langle k_2\rangle^{5/2} \sum_{\mathbb{S}_2^{(2)}} \|P_2\Box_k(\Box_{k^{(1)}}v_1...\Box_{k^{(\kappa+|\nu|)}}v_{\kappa+|\nu|})\|_{L^{(2+m)/(1+m)}_{t,x}} \\ \lesssim & C^{\kappa+|\nu|} \sum_{\mathbb{S}_2^{(2)}} \|\Box_{k^{(1)}}v_1...\Box_{k^{(\kappa+|\nu|)}}v_{\kappa+|\nu|}\|_{L^{(2+m)/(1+m)}_{t,x}} \lesssim \|u\|_X^{\kappa+|\nu|}. \end{aligned} \tag{6.246}$$

Hence, in view of (6.245) and (6.246), we have

$$IV \lesssim (C\|u\|_X)^{\kappa+|\nu|}. \tag{6.247}$$

Collecting (6.233), (6.235), (6.241) and (6.247), we have shown that

$$\sum_{\alpha=0,1}\sum_{i,j=1}^{n}\varrho_1^{(i)}(\mathscr{A}\partial^{\alpha}_{x_j}(u^{\kappa}(\nabla u)^{\nu}))\lesssim (C\|u\|_X)^{\kappa+|\nu|}. \tag{6.248}$$

Now we estimate $\varrho_3^{(i)}(\mathscr{A}(u^{\kappa}(\nabla u)^{\nu}))$. By (6.212),

$$\sum_{i=1}^{n}\varrho_3^{(i)}(\mathscr{A}(u^{\kappa}(\nabla u)^{\nu}))\lesssim \sum_{k\in\mathbb{Z}^n}\langle k\rangle^{3/2}\|\Box_k(u^{\kappa}(\nabla u)^{\nu})\|_{L_{t,x}^{\frac{2+m}{1+m}}}. \tag{6.249}$$

Using Lemma 6.8, one can control the right hand side of (6.249),

$$\begin{aligned}
&\sum_{k\in\mathbb{Z}^n}\langle k\rangle^{3/2}\|\Box_k(v_1...v_{\kappa+|\nu|})\|_{L_{t,x}^{\frac{2+m}{1+m}}}\\
&\lesssim C^{\kappa+|\nu|}\prod_{i=1}^{m+1}\left(\sum_{k\in\mathbb{Z}^n}\langle k\rangle^{3/2}\|\Box_k v_i\|_{L^{2+m}_{t,x}}\right)\prod_{i=m+2}^{\kappa+|\nu|}\left(\sum_{k\in\mathbb{Z}^n}\langle k\rangle^{3/2}\|\Box_k v_i\|_{L^{\infty}_{t,x}}\right)\\
&\lesssim C^{\kappa+|\nu|}\prod_{i=1}^{m+1}\left(\sum_{k\in\mathbb{Z}^n}\langle k\rangle^{3/2}\|\Box_k v_i\|_{L^{2+m}_{t,x}}\right)\prod_{i=m+2}^{\kappa+|\nu|}\left(\sum_{k\in\mathbb{Z}^n}\langle k\rangle^{3/2}\|\Box_k v_i\|_{L^{\infty}_t L^2_x}\right)\\
&\lesssim C^{\kappa+|\nu|}\prod_{i=1}^{\kappa+|\nu|}\left(\sum_{i=1}^{n}\varrho_3^{(1)}(v_i)\right)\leqslant (C\|u\|_X)^{\kappa+|\nu|}.
\end{aligned}\tag{6.250}$$

Now we estimate $\varrho_2^{(1)}(\mathscr{A}\partial^{\alpha}_{x_1}(u^{\kappa}(\nabla u)^{\nu}))$.

$$\varrho_2^{(1)}(\mathscr{A}\partial^{\alpha}_{x_1}(v_1...v_{\kappa+|\nu|}))\lesssim\sum_{k\in\mathbb{Z}^n}\langle k\rangle^{3/2}\|\Box_k\left(v_1...v_{\kappa+|\nu|}\right)\|_{L^1_tL^2_x}. \tag{6.251}$$

Analogous to (6.250), using Lemma 6.8, we can control (6.251),

$$\begin{aligned}
\varrho_2^{(1)}(\mathscr{A}\partial^{\alpha}_{x_1}(v_1...v_{\kappa+|\nu|}))&\lesssim C^{\kappa+|\nu|}\prod_{i=1}^{\kappa+|\nu|}\left(\sum_{k\in\mathbb{Z}^n}\langle k\rangle^{3/2}\|\Box_k v_i\|_{L^{\infty}_tL^2_x\cap L^3_tL^6_x}\right)\\
&\lesssim (C\|u\|_X)^{\kappa+|\nu|}.
\end{aligned}\tag{6.252}$$

Now we estimate

$$\begin{aligned}
&\sum_{i=1}^{n}\rho_3^{(i)}(\mathscr{A}\partial_{x_1}(v_1...v_{\kappa+|\nu|}))\\
&\lesssim\sum_{|k|\leqslant 4}\|\Box_k(v_1...v_{\kappa+|\nu|})\|_{L^1_tL^2_x}\\
&\quad+\left(\sum_{k\in\mathbb{Z}^n,\ |k_1|=k_{\max}>4}+...+\sum_{k\in\mathbb{Z}^n,\ |k_n|=k_{\max}>4}\right)\langle k\rangle^{3/2}
\end{aligned}$$

$$\times \|\Box_k \mathscr{A}\partial_{x_1}(v_1...v_{\kappa+|\nu|})\|_{L_t^\infty L_x^2 \cap L_t^3 L_x^6}$$
$$=: \Upsilon_0(u) + \Upsilon_1(u) + ... + \Upsilon_n(u). \tag{6.253}$$

$\Upsilon_0(u)$ has been estimated as above. It suffices to estimate $\Upsilon_i(u)$, $i = 1, 2$. Using (6.228) and Corollary 6.5,

$$\Upsilon_1(u) \lesssim \sum_{k\in\mathbb{Z}^n,\ |k_1|>4} \langle k_1\rangle^2 \sum_{\mathbb{S}_1^{(1)}} \|\Box_k(\Box_{k^{(1)}}v_1...\Box_{k^{(\kappa+|\nu|)}}v_{\kappa+|\nu|})\|_{L^1_{x_1}L^2_{\bar{x}}L^2_t}$$
$$+ \sum_{k\in\mathbb{Z}^n,\ |k_1|>4} \langle k_1\rangle^{5/2} \sum_{\mathbb{S}_2^{(1)}} \|\Box_k(\Box_{k^{(1)}}v_1...\Box_{k^{(\kappa+|\nu|)}}v_{\kappa+|\nu|})\|_{L_{t,x}^{\frac{\kappa+|\nu|+1}{\kappa+|\nu|}}}, \tag{6.254}$$

which reduces to (6.230). Hence,

$$\Upsilon_1(u) \lesssim (C\|u\|_X)^{\kappa+|\nu|}. \tag{6.255}$$

We now estimate $\Upsilon_2(u)$. Taking notice of $|k_1| \leqslant |k_2|$ in the summation of $\Upsilon_2(u)$ and using (6.209) and (6.212), we have

$$\Upsilon_2(u) \lesssim \sum_{k\in\mathbb{Z}^n,\ |k_2|>4} \langle k_2\rangle^2 \sum_{\mathbb{S}_1^{(2)}} \|\Box_k(\Box_{k^{(1)}}v_1...\Box_{k^{(\kappa+|\nu|)}}v_{\kappa+|\nu|})\|_{L^1_{x_1}L^2_{\bar{x}}L^2_t}$$
$$+ \sum_{k\in\mathbb{Z}^n,\ |k_2|>4} \langle k_2\rangle^{5/2} \sum_{\mathbb{S}_2^{(2)}} \|\Box_k(\Box_{k^{(1)}}v_1...\Box_{k^{(\kappa+|\nu|)}}v_{\kappa+|\nu|})\|_{L_{t,x}^{\frac{\kappa+|\nu|+1}{\kappa+|\nu|}}}. \tag{6.256}$$

The first term in the right hand side has been estimated in (6.244). The second term is the same as in II of (6.230). Summarizing the above estimates, we get that[10]

$$\|\mathscr{T}u\|_X \leqslant C\|u_0\|_{M^{3/2}} + \sum_{m+1\leqslant \ell<\infty} \ell^{2n+2}C^\ell\|u\|_X^\ell. \tag{6.257}$$

We can use the contraction mapping argument to finish the proof and omit the details.

Remark 6.2.

(1) There are some recent works which have been devoted to the study of the derivative NLKG [54; 55; 76; 144; 185; 200]:

$$u_{tt} + u - \Delta u = F(u, u_t, u_{xx}, u_{xt}),$$

where the methods are quite different from those of derivative NLS. We do not know whether the frequncy-uniform decomposition techniques are applicable to the derivative NLKG.

[10] Notice that $|c_\beta| \leqslant C^{|\beta|}$.

(2) We now state some results on the Zakharov system

$$\begin{cases} \mathrm{i}\partial_t u + \Delta u = nu, \\ \partial_t^2 n - \Delta n = \Delta |u|^2. \end{cases}$$

In 2D, Bourgain and Colliander [20] used the Fourier restriction norm to prove the local well-posedness and global well-posedness for initial data in the spaces $H^1 \times L^2 \times H^{-1}$. Fang, Pecher and Zhong [71] applied the I-method to obtain the global well-posedness in $H^s \times L^2 \times H^{-1}$ with $3/4 < s < 1$. Recently, Bejenaru, Herr, Holmer, and Tataru [8] showed the local well-posedness for the Schrödinger data in L^2 and the wave data in $H^{-1/2} \times H^{-3/2}$. In higher spatial dimensions, the local well posed result can be found in Ginibre, Tsutsumi and Velo [81].

Chapter 7

Conservations, Morawetz' estimates of nonlinear Schrödinger equations

This result is too beautiful to be false; it is more important to have beauty in one's equations than to have them fit experiment. ——
Paul Dirac

On mentioning evolution partial differential equations, we will naturally think whether they satisfy some physical conservation laws, such as the mass conservation, the energy conservation and the momentum conservation, and so on. For example, as indicated in (4.2), we see that NLS (4.1) satisfies the conservations of mass and energy.

In this chapter, we will derive some conservation laws by using Nöther's theorem and Morawetz' inequalities by the virial identity. It is known that Morawetz' estimates are fundamental tools in the study of the scattering theory for NLS, see for instance, [27; 40; 43; 218].

7.1 Nöther's theorem

The energy *conservation law* means that it does not vary with time, in other word, it has a certain *symmetry* with respect to the time. This fundamental connection between conservation laws and symmetries was first discovered in 1915 (published in 1918) by Emmy Nöther[1] and was called the "Nöther's theorem". In short, Nöther's theorem states informally that *any differentiable symmetry of the action of a physical system has a corresponding conservation law.* Nöther's theorem is important, both because of the insight it gives into conservation laws, and also as a practical calculational tool. It allows us to determine the conserved quantities from

[1] Amalie Emmy Nöther (1882-1935) was a German-born mathematician known for her ground-breaking contributions to abstract algebra and theoretical physics. She revolutionized the theories of rings, fields, and algebras. In physics, Nöther's theorem explains the fundamental connection between symmetry and conservation laws.

the observed symmetries of a physical system. The latter turns out to be easier to find than the former.

Application of Nöther's theorem allows physicists to gain powerful insights into any general theories in physics, by just analyzing the various transformations that would make the form of the laws involved invariant. For example, the invariance of physical systems with respect to the spatial translation (in other words, the laws of physics do not vary with locations in space) gives the conservation law of the linear momentum; the invariance with respect to the rotation gives the conservation law of the angular momentum; the invariance with respect to the time translation gives the well-known conservation law of energy. But how to derive the associated conserved quantities of its if we know that an equation has a symmetry group? Let us begin by recalling some concepts of the calculus of variations.

Suppose the interval $I \subset \mathbb{R}$, the generalized coordinate $\mathbf{q}$ is dependent on the time t. Assume that $\mathcal{L} : \mathbb{R}^n_{\mathbf{q}} \times \mathbb{R}^n_{\dot{\mathbf{q}}} \times I \mapsto \mathbb{R}$ is smooth and convex with respect to the first slot $\dot{\mathbf{q}}$ (The dot indicates the derivative with respect to the time.), then we call the function $\mathcal{L}$ to be a *Lagrangian*[2] of the system. For some $\mathbf{q}$ satisfying the related boundary conditions, we define the *action functional*

$$\mathcal{I}[q] = \int_{t_0}^{t_1} \mathcal{L}(\mathbf{q}(t), \dot{\mathbf{q}}(t), t)dt.$$

Hamilton's principle states that the true evolution of a system described by generalized coordinates $\mathbf{q}$ between two specified states $\mathbf{q}(t_0)$ and $\mathbf{q}(t_1)$ at two specified times t_0 and t_1 is an extremum (i.e., a stationary point, a minimum, maximum or saddle point) of the action functional $\mathcal{I}[q]$, in other word, the first variation of the action functional $\mathcal{I}[q]$ is zero.

Let $\mathbf{q}(t)$ represent the true evolution of the system between two specified states $\mathbf{q}(t_0)$ and $\mathbf{q}(t_1)$ at two specified times t_0 and t_1. And let $\varepsilon(t)$ be a small perturbation that is zero at the endpoints of the trajectory

$$\varepsilon(t_0) = \varepsilon(t_1) \overset{\text{def}}{=} 0. \tag{7.1}$$

To the first order in the perturbation $\varepsilon(t)$, the change in the action functional $\delta\mathcal{I}$ would be

$$\delta\mathcal{I} = \int_{t_0}^{t_1} (\mathcal{L}(\mathbf{q}+\varepsilon, \dot{\mathbf{q}}+\dot{\varepsilon}) - \mathcal{L}(\mathbf{q}, \dot{\mathbf{q}}))\, dt = \int_{t_0}^{t_1} \left(\varepsilon \cdot \frac{\partial L}{\partial \mathbf{q}} + \dot{\varepsilon} \cdot \frac{\partial \mathcal{L}}{\partial \dot{\mathbf{q}}} \right) dt,$$

[2]It is named after Joseph Louis Lagrange (1736-1813), who was an Italian-born mathematician and astronomer, making significant contributions to all fields of analysis, to number theory, and to classical and celestial mechanics. He is one of the founders of the calculus of variations. Roughly speaking, the Lagrangian of a dynamical system is a function that summarizes the dynamics of the system. In classical mechanics, the Lagrangian is defined as the kinetic energy of the system minus its potential energy.

where we have expanded the Lagrangian $\mathcal{L}$ to the first order in the perturbation $\varepsilon(t)$. Applying the integration by parts to the last term results in

$$\delta\mathcal{I} = \left(\varepsilon \cdot \frac{\partial \mathcal{L}}{\partial \dot{\mathbf{q}}}\right)_{t_0}^{t_1} + \int_{t_0}^{t_1} \left(\varepsilon \cdot \frac{\partial \mathcal{L}}{\partial \mathbf{q}} - \varepsilon \cdot \frac{d}{dt}\frac{\partial \mathcal{L}}{\partial \dot{\mathbf{q}}}\right) dt.$$

By the boundary condition (7.1), the first term vanishes, and then

$$\delta\mathcal{I} = \int_{t_0}^{t_1} \varepsilon \cdot \left(\frac{\partial \mathcal{L}}{\partial \mathbf{q}} - \frac{d}{dt}\frac{\partial \mathcal{L}}{\partial \dot{\mathbf{q}}}\right) dt = 0.$$

Noticing that we do not make any assumptions to the generalized coordinates $\mathbf{q}$ except that all of the generalized coordinates are independent of each other. Thus, we can apply the fundamental theorem of variational calculus to obtain an Euler-Lagrange equation

$$\frac{d}{dt}\frac{\partial \mathcal{L}}{\partial \dot{\mathbf{q}}} - \frac{\partial \mathcal{L}}{\partial \mathbf{q}} = \mathbf{0}, \tag{7.2}$$

which is a system involving n second-order equations with respect to $\mathbf{q}(t)$. The *Legendre transformation*[3] of $\mathcal{L}$ is

$$\mathcal{H}(\mathbf{q}(t), \mathbf{p}(t), t) = \max_{\dot{\mathbf{q}}}[\mathbf{p}(t) \cdot \dot{\mathbf{q}} - \mathcal{L}(\mathbf{q}(t), \dot{\mathbf{q}}(t), t)].$$

Let $M(\mathbf{p}, \dot{\mathbf{q}}) = \mathbf{p}(t) \cdot \dot{\mathbf{q}} - \mathcal{L}(\mathbf{q}(t), \dot{\mathbf{q}}(t), t)$. In order to reach the maximum of M with respect to $\dot{\mathbf{q}}$, we have to require that the first-order partial derivative of $M(\mathbf{p}, \dot{\mathbf{q}})$ with respect to $\dot{\mathbf{q}}$ is zero and that its second partial derivative is less than zero (this can be guaranteed by the convexity of $\mathcal{L}$, i.e. $\frac{\partial^2 \mathcal{L}}{\partial \dot{\mathbf{q}}^2} > 0$). Thus, if $\mathbf{p}(t) = \frac{\partial \mathcal{L}}{\partial \dot{\mathbf{q}}}(\mathbf{q}(t), \dot{\mathbf{q}}(t), t)$, then $M(\mathbf{p}, \dot{\mathbf{q}})$ reaches the maximum, i.e.

$$\begin{cases} \mathbf{p}(t) = \dfrac{\partial \mathcal{L}}{\partial \dot{\mathbf{q}}}(\mathbf{q}(t), \dot{\mathbf{q}}(t), t), \\ \mathcal{H}(\mathbf{q}(t), \mathbf{p}(t), t) = \mathbf{p}(t)\dot{\mathbf{q}} - \mathcal{L}(\mathbf{q}(t), \dot{\mathbf{q}}(t), t), \end{cases}$$

where $\mathcal{H}$ is called the *Hamiltonian.* The second equation yields

$$\frac{\partial \mathcal{H}}{\partial \mathbf{q}} = -\frac{\partial \mathcal{L}}{\partial \mathbf{q}}, \quad \frac{\partial \mathcal{H}}{\partial \mathbf{p}} = \dot{\mathbf{q}}.$$

Thus, the Euler-Lagrange equation (7.2) can be rewritten as

$$\begin{cases} \dot{\mathbf{p}} = -\dfrac{\partial \mathcal{H}}{\partial \mathbf{q}}, \\ \dot{\mathbf{q}} = \dfrac{\partial \mathcal{H}}{\partial \mathbf{p}}, \end{cases} \tag{7.3}$$

[3] Legendre transformation is an operation that transforms one real-valued function of a real variable into another. Specifically, the Legendre transformation of a function f is the function $f^\star$ defined by $f^\star(p) = \max_x(px - f(x))$.

which are called *Hamilton's equation.* This is a system of $2n$ first-order equations in $(\mathbf{p}(t), \mathbf{q}(t))$. Denote $\mathbf{u}(t) = (\mathbf{p}(t), \mathbf{q}(t)) \in \mathbb{R}^{2n}$, $\mathcal{H}_{\mathbf{u}} = (\mathcal{H}_{\mathbf{p}}, \mathcal{H}_{\mathbf{q}})$ and define the matrix

$$\mathbb{J} = \begin{pmatrix} \mathbf{0} & -\mathbb{I} \\ \mathbb{I} & \mathbf{0} \end{pmatrix},$$

where $\mathbb{I}$ denotes the unit matrix, then Hamilton's equation (7.3) can be written as[4]

$$\dot{\mathbf{u}} = \mathbb{J}\, \mathcal{H}_{\mathbf{u}}. \tag{7.4}$$

Notice that $\mathbb{J}^2 = -\mathbb{I}$ and Hamilton's equations are defined on an even dimensional space. It suggests that there may be a nice connection with complex numbers.

For $(x, y) \in \mathbb{R}^2$, define $z = x + iy$, $\bar{z} = x - iy$, $i^2 = -1$. Define $\partial_z = \frac{1}{2}(\partial_x - i\partial_y)$ and $\partial_{\bar{z}} = \frac{1}{2}(\partial_x + i\partial_y)$. It is clear that $\partial_z z = \partial_{\bar{z}} \bar{z} = 1$ and $\partial_z \bar{z} = \partial_{\bar{z}} z = 0$, that is, $\bar{z}$ and z are independent of each other. Let $\mathbf{z} = \mathbf{q} + i\mathbf{p}$, then (7.3) yields

$$\dot{\mathbf{z}} = \dot{\mathbf{q}} + i\dot{\mathbf{p}} = -i(\mathcal{H}_{\mathbf{q}} + i\mathcal{H}_{\mathbf{p}}) = -2i\mathcal{H}_{\bar{\mathbf{z}}}.$$

We give an example. Let I be a time interval, $u : \mathbb{R}^n \times I \mapsto \mathbb{C}$, u and its derivatives are smooth and vanish at the infinity. Assume that the Hamiltonian is defined by

$$\mathcal{H} = \mathcal{H}[u, \bar{u}] = \frac{1}{2} \int_{\mathbb{R}} |\nabla u|^2 dx = \frac{1}{2} \int_{\mathbb{R}} \nabla u \nabla \bar{u} dx.$$

The usual calculus of variations argument shows

$$\lim_{\tau \to 0} \frac{\mathcal{H}[u, \bar{u} + \tau\bar{v}] - \mathcal{H}[u, \bar{u}]}{\tau} = \frac{1}{2} \int_{\mathbb{R}} (-\Delta u)\bar{v} dx = \frac{1}{2} \langle -\Delta u, v \rangle.$$

Therefore, $\mathcal{H}_{\bar{u}}[u, \bar{u}] = -\frac{1}{2}\Delta u$ and the associated Hamilton's equation is

$$\dot{u} = -2i\mathcal{H}_{\bar{u}} = i\Delta u,$$

which is precisely the linear Schrödinger equation.

Now, we consider the n-dimensional nonlinear Schrödinger equation

$$iu_t + \Delta u = F'(|u|^2)u. \tag{7.5}$$

Assume that u and its derivatives are smooth and vanish as $|x| \to +\infty$. The nonlinearity $F'(\cdot)$ is a smooth real-valued function of its argument and define

$$F(\lambda) = \int_0^\lambda F'(s) ds.$$

[4] It is convenient to write $\mathbf{u}$ and $\mathcal{H}_{\mathbf{u}}$ as column vectors in the computation.

Define the Lagrangian

$$\mathcal{L} = \mathcal{L}(u, \bar{u}, u_t, \bar{u}_t, \nabla u, \nabla \bar{u}) = \frac{i}{2}(\bar{u}u_t - u\bar{u}_t) - [|\nabla u|^2 + F(|u|^2)], \tag{7.6}$$

and the associated action functional

$$\mathcal{I}[u] = \int_{t_0}^{t_1} \int_{\mathbb{R}^n} \mathcal{L}\, dxdt, \quad \forall u \in \mathcal{A}, \tag{7.7}$$

where $\mathcal{A}$ is some appropriate class of admissible functions. The usual calculus of variations argument shows that if u is a smooth critical point of $\mathcal{I}(\cdot)$, u satisfies the associated Euler-Lagrange equation

$$\frac{\partial \mathcal{L}}{\partial u} - \frac{\partial}{\partial t}\frac{\partial \mathcal{L}}{\partial(\partial_t u)} - \sum_{j=1}^{n} \frac{\partial}{\partial x_j}\frac{\partial \mathcal{L}}{\partial(\partial_{x_j} u)} = 0. \tag{7.8}$$

Substituting (7.6) into the above equation and taking the complex conjugate, we can obtain the nonlinear Schrödinger equation (7.5).

Now, we give another statement for Nöther's theorem:

Theorem 7.1 (Nöther's theorem). *If the action functional is invariant under a family of transformations, then the solutions of the associated Euler-Lagrange equation satisfy a conservation law.*

We mainly consider the case of the one parameter family of transformations and apply Nöther's theorem to the nonlinear Schrödinger equation (7.5). For convenience, denote $\xi = (t, x) = (\xi_0, \xi_1, \cdots, \xi_n)$, $\partial_0 = \partial_t$, $\partial = (\partial_t, \nabla_x) = (\partial_0, \partial_1, \cdots, \partial_n)$, and $\mathbf{u} = (u_1, u_2) = (u, \bar{u})$, and denote the integral domain $\mathcal{D} := [t_0, t_1] \times \mathbb{R}^n$. A one parameter group of transformations T^ε is defined by

$$\xi \mapsto \tilde{\xi}(\xi, \mathbf{u}, \varepsilon), \quad \mathbf{u} \mapsto \tilde{\mathbf{u}}(\xi, \mathbf{u}, \varepsilon), \tag{7.9}$$

where we assume $\tilde{\xi}$ and $\tilde{\mathbf{u}}$ are differentiable with respect to ε, and it degenerates into an identical transformation for the case $\varepsilon = 0$. For infinitesimal ε, denote

$$\tilde{\xi} = \xi + \delta\xi, \quad \tilde{\mathbf{u}} = \mathbf{u} + \delta\mathbf{u}, \tag{7.10}$$

where $\delta\xi$ and $\delta\mathbf{u}$ are functions of $(\xi, \mathbf{u}, \varepsilon)$, both $\delta\xi = O(\varepsilon)$ and $\delta\mathbf{u} = O(\varepsilon)$ tend to zero as $\varepsilon \to 0$, and Jacobi's determinant $\frac{\partial(\tilde{\xi}_0, \cdots, \tilde{\xi}_n)}{\partial(\xi_0, \cdots, \xi_n)} = 1 + \sum_{j=0}^{n} \frac{\partial(\delta\xi)_j}{\partial \xi_j} + o(\varepsilon)$. Applied T^ε, $\mathbf{u}(\xi)$ changes into $\tilde{\mathbf{u}}(\tilde{\xi})$, the domain $\mathcal{D}$ becomes into $\tilde{\mathcal{D}}$, and the action functional

$$\mathcal{I}[u] = \int_{\mathcal{D}} \mathcal{L}(\mathbf{u}, \partial\mathbf{u}) d\xi \tag{7.11}$$

turns into

$$\tilde{\mathcal{I}}[\tilde{u}] = \int_{\tilde{\mathcal{D}}} \mathcal{L}(\tilde{\mathbf{u}}, \tilde{\partial}\tilde{\mathbf{u}}) d\tilde{\xi} = \int_{\mathcal{D}} \mathcal{L}(\tilde{\mathbf{u}}, \tilde{\partial}\tilde{\mathbf{u}}) \left(1 + \sum_{j=0}^{n} \frac{\partial(\delta\xi)_j}{\partial\xi_j} + o(\varepsilon)\right) d\xi$$

$$= \int_{\mathcal{D}} \mathcal{L}(\tilde{\mathbf{u}}, \tilde{\partial}\tilde{\mathbf{u}}) d\xi + \int_{\mathcal{D}} \mathcal{L}(\mathbf{u}, \partial\mathbf{u}) \sum_{j=0}^{n} \frac{\partial(\delta\xi)_j}{\partial\xi_j} d\xi + o(\varepsilon), \tag{7.12}$$

where ∂ and $\tilde{\partial}$ denote the differential with respect to ξ and $\tilde{\xi}$, respectively. The last two terms can be obtained by expanding with the help of (7.6) and subsuming the higher order terms of ε in $o(\varepsilon)$ in the expansion. We only consider the transformations under which $\mathcal{I}$ is invariant. Denote

$$\begin{aligned} \delta\mathcal{I} :=& \tilde{\mathcal{I}}[\tilde{u}] - \mathcal{I}[u] \\ =& \int_{\mathcal{D}} \Big(\mathcal{L}(\tilde{\mathbf{u}}, \tilde{\partial}\tilde{\mathbf{u}}) - \mathcal{L}(\mathbf{u}, \partial\mathbf{u})\Big) d\xi + \int_{\mathcal{D}} \mathcal{L}(\mathbf{u}, \partial\mathbf{u}) \sum_{j=0}^{n} \frac{\partial(\delta\xi)_j}{\partial\xi_j} d\xi + o(\varepsilon), \end{aligned} \tag{7.13}$$

where the first integrand can be written as

$$\begin{aligned} &\mathcal{L}(\tilde{\mathbf{u}}, \tilde{\partial}\tilde{\mathbf{u}}) - \mathcal{L}(\mathbf{u}, \partial\mathbf{u}) \\ =& \sum_{k=1}^{2} \left(\frac{\partial\mathcal{L}}{\partial u_k} (\tilde{u}_k(\tilde{\xi}) - u_k(\xi)) + \sum_{j=0}^{n} \frac{\partial\mathcal{L}}{\partial(\partial_j u_k)} \Big(\tilde{\partial}_j \tilde{u}_k(\tilde{\xi}) - \partial_j u_k(\xi)\Big) \right). \end{aligned} \tag{7.14}$$

Denote

$$\delta\tilde{u}_k \equiv \tilde{u}_k(\tilde{\xi}) - u_k(\xi) = \sum_{j=0}^{n} \partial_j u_k (\delta\xi)_j + \delta u_k(\xi). \tag{7.15}$$

Since

$$\begin{aligned} \partial_j \tilde{u}_k(\tilde{\xi}) &= \sum_{l=0}^{n} \tilde{\partial}_l \tilde{u}_k(\tilde{\xi}) \frac{\partial\tilde{\xi}_l}{\partial\xi_j} = \sum_{l=0}^{n} (\delta_{jl} + \frac{\partial(\delta\xi)_l}{\partial\xi_j}) \tilde{\partial}_l \tilde{u}_k(\tilde{\xi}) \\ &= \left(\tilde{\partial}_j + \sum_{l=0}^{n} \frac{\partial(\delta\xi)_l}{\partial\xi_j} \tilde{\partial}_l \right) \tilde{u}_k(\tilde{\xi}), \end{aligned} \tag{7.16}$$

we have

$$\begin{aligned} \tilde{\partial}_j \tilde{u}_k(\tilde{\xi}) - \partial_j u_k(\xi) =& (\tilde{\partial}_j - \partial_j) \tilde{u}_k(\tilde{\xi}) + \partial_j (\tilde{u}_k(\tilde{\xi}) - u_k(\xi)) \\ =& - \sum_{l=0}^{n} \frac{\partial(\delta\xi)_l}{\partial\xi_j} \tilde{\partial}_l \tilde{u}_k(\tilde{\xi}) + \partial_j \left(\sum_{l=0}^{n} \partial_l u_k (\delta\xi)_l + \delta u_k \right). \end{aligned} \tag{7.17}$$

Hence,

$$\mathcal{L}(\tilde{\mathbf{u}}, \tilde{\partial}\tilde{\mathbf{u}}) - \mathcal{L}(\mathbf{u}, \partial\mathbf{u}) = \sum_{k=1}^{2} \frac{\partial \mathcal{L}}{\partial u_k} \left(\sum_{j=0}^{n} \partial_j u_k (\delta\xi)_j + \delta u_k \right)$$
$$+ \sum_{k=1}^{2} \sum_{j=0}^{n} \frac{\partial \mathcal{L}}{\partial(\partial_j u_k)} \left(- \sum_{l=0}^{n} \frac{\partial(\delta\xi)_l}{\partial \xi_j} \tilde{\partial}_l \tilde{u}_k(\tilde{\xi}) + \partial_j \left(\sum_{l=0}^{n} \partial_l u_k (\delta\xi)_l + \delta u_k \right) \right). \tag{7.18}$$

Noticed that

$$\frac{\partial}{\partial \xi_j}(\mathcal{L}(\delta\xi)_j) = \mathcal{L}\frac{\partial(\delta\xi)_j}{\partial \xi_j} + \sum_{k=1}^{2} \left(\frac{\partial \mathcal{L}}{\partial u_k} \partial_j u_k + \sum_{l=0}^{n} \frac{\partial \mathcal{L}}{\partial(\partial_l u_k)} \partial^2_{lj} u_k \right) (\delta\xi)_j, \tag{7.19}$$

and

$$\frac{\partial \mathcal{L}}{\partial(\partial_j u_k)} \partial_j \delta u_k = \frac{\partial}{\partial \xi_j} \left(\frac{\partial \mathcal{L}}{\partial(\partial_j u_k)} \delta u_k \right) - \frac{\partial}{\partial \xi_j} \left(\frac{\partial \mathcal{L}}{\partial(\partial_j u_k)} \right) \delta u_k, \tag{7.20}$$

it is easily to obtain

$$\begin{aligned} &\mathcal{L}(\tilde{\mathbf{u}}, \tilde{\partial}\tilde{\mathbf{u}}) - \mathcal{L}(\mathbf{u}, \partial\mathbf{u}) \\ &= \sum_{j=0}^{n} \frac{\partial}{\partial \xi_j}(\mathcal{L}(\delta\xi)_j) - \sum_{j=0}^{n} \mathcal{L}\frac{\partial(\delta\xi)_j}{\partial \xi_j} + \sum_{k=1}^{2} \frac{\partial \mathcal{L}}{\partial u_k} \delta u_k \\ &+ \sum_{k=1}^{2} \sum_{j=0}^{n} \left(\frac{\partial}{\partial \xi_j} \left(\frac{\partial \mathcal{L}}{\partial(\partial_j u_k)} \delta u_k \right) - \frac{\partial}{\partial \xi_j} \left(\frac{\partial \mathcal{L}}{\partial(\partial_j u_k)} \right) \delta u_k \right) \\ &- \sum_{k=1}^{2} \sum_{j=0}^{n} \frac{\partial \mathcal{L}}{\partial(\partial_j u_k)} \sum_{l=0}^{n} \frac{\partial(\delta\xi)_l}{\partial \xi_j} (\tilde{\partial}_l \tilde{u}_k(\tilde{\xi}) - \partial_l u_k). \end{aligned} \tag{7.21}$$

Therefore, it yields

$$\delta\mathcal{I} = \int_{\mathcal{D}} \sum_{k=1}^{2} \left(\frac{\partial \mathcal{L}}{\partial u_k} - \sum_{j=0}^{n} \frac{\partial}{\partial \xi_j} \left(\frac{\partial \mathcal{L}}{\partial(\partial_j u_k)} \right) \right) \delta u_k d\xi$$
$$+ \int_{\mathcal{D}} \sum_{j=0}^{n} \frac{\partial}{\partial \xi_j} \left(\mathcal{L}(\delta\xi)_j + \sum_{k=1}^{2} \frac{\partial \mathcal{L}}{\partial(\partial_j u_k)} \delta u_k \right) d\xi + o(\varepsilon). \tag{7.22}$$

(7.8) implies

$$\frac{\partial \mathcal{L}}{\partial u_k} - \sum_{j=0}^{n} \frac{\partial}{\partial \xi_j} \left(\frac{\partial \mathcal{L}}{\partial(\partial_j u_k)} \right) = 0, \quad k = 1, 2. \tag{7.23}$$

Since $\mathcal{D}$ is arbitrary, we have the following result in order to guarantee that $\mathcal{I}$ remains invariant under the infinitesimal transformation T^ε.

Theorem 7.2. *Let* $\xi = (t, x_1, \cdots, x_n)$, $\mathbf{u} = (u_1, u_2)$. *If the action functional* (7.11) *is invariant under the infinitesimal transformation* T^ε: $\xi \mapsto \tilde{\xi}(\xi, \mathbf{u}, \varepsilon)$, $\mathbf{u} \mapsto \tilde{\mathbf{u}}(\xi, \mathbf{u}, \varepsilon)$, *then the following conservation law holds:*

$$\sum_{j=0}^{n} \frac{\partial}{\partial \xi_j} \left(\mathcal{L}(\delta\xi)_j + \sum_{k=1}^{2} \frac{\partial \mathcal{L}}{\partial(\partial_j u_k)} \delta u_k \right) = 0, \tag{7.24}$$

where $\delta u_k(\xi) = \tilde{u}_k(\tilde{\xi}) - u_k(\xi) - \nabla_\xi u_k \cdot \delta\xi$.

Integrating with respect to the spatial variables yields the following result.

Theorem 7.3. *If the action functional* (7.11) *is invariant under the infinitesimal transformation*

$$t \mapsto \tilde{t} = t + \delta t(t, x, u), \tag{7.25}$$

$$x \mapsto \tilde{x} = x + \delta x(t, x, u), \tag{7.26}$$

$$u(t, x) \mapsto \tilde{u}(\tilde{t}, \tilde{x}) = u(t, x) + \delta u(t, x), \tag{7.27}$$

then

$$\int_{\mathbb{R}^n} \left(\frac{\partial \mathcal{L}}{\partial u_t}(u_t \delta t + \nabla u \cdot \delta x - \delta u) + \frac{\partial \mathcal{L}}{\partial \bar{u}_t}(\bar{u}_t \delta t + \nabla \bar{u} \cdot \delta x - \delta \bar{u}) - \mathcal{L}\delta t \right) dx$$

is a conserved quantity. In particular, in the case of nonlinear Schrödinger equations, since $\frac{\partial \mathcal{L}}{\partial u_t} = \frac{i}{2}\bar{u}$, $\frac{\partial \mathcal{L}}{\partial \bar{u}_t} = -\frac{i}{2}u$, *it implies that*

$$\int_{\mathbb{R}^n} \left(\frac{i}{2}\bar{u}(u_t \delta t + \nabla u \cdot \delta x - \delta u) - \frac{i}{2}u(\bar{u}_t \delta t + \nabla \bar{u} \cdot \delta x - \delta \bar{u}) - \mathcal{L}\delta t \right) dx = const. \tag{7.28}$$

7.2 Invariance and conservation law

In this subsection, we apply the formalism (7.28) to identify some invariant quantities of the action functional for nonlinear Schrödinger equations and thereby infer some conservation laws for nonlinear Schrödinger equations.

(i) *Invariance by the phase shift*: $\tilde{u} = e^{i\varepsilon}u$. For the infinitesimal quantity ε, let $\delta u = i\varepsilon u$, $\delta t = 0$, and $\delta x = 0$. From (7.28), we get *the mass conservation law* (or the charge conservation law, or the L^2-norm conservation law)

$$N(t) := \int_{\mathbb{R}^n} |u(t, x)|^2 dx = const. \tag{7.29}$$

(ii) *Invariance by the time translation*: $t \mapsto t+\delta t$, $\delta x = 0$ and $\delta u = \delta \bar{u} = 0$, where δt is an infinitesimal quantity independent of $(t, x, u, \bar{u})$. (7.28) and (7.6) yield *the energy conservation law*

$$H(t) := \int_{\mathbb{R}^n} |\nabla u(t,x)|^2 + F(|u(t,x)|^2)dx = const. \tag{7.30}$$

(iii) *Invariance by the spatial translation*: $x \mapsto x + \delta x$ and $\delta t = \delta u = \delta \bar{u} = 0$, where δx is an infinitesimal quantity independent of $(t, x, u, \bar{u})$. (7.28) implies *the momentum conservation law*

$$\vec{P}(t) := i\int_{\mathbb{R}^n} (u(t,x)\nabla \bar{u}(t,x) - \bar{u}(t,x)\nabla u(t,x))dx = constant\ vector. \tag{7.31}$$

(iv) *Invariance by the spatial rotation*: $\delta x = \delta\theta(\vec{a} \times x)$ and $\delta t = \delta u = \delta \bar{u} = 0$, which mean the spatial variables rotate anticlockwise for a angle of $\delta\theta$ along the axis $\vec{a}$ where $\delta\theta$ is an infinitesimal quantity independent of $(t, x, u, \bar{u})$. From (7.28), we have *the angular momentum conservation law*

$$i\int_{\mathbb{R}^n} x \times (\bar{u}\nabla u - u\nabla \bar{u})dx = constant\ vector. \tag{7.32}$$

(v) *Invariance by the Galilean transformation*:

$$\begin{cases} x \mapsto \tilde{x} = x - \vec{c}t, \\ t \mapsto \tilde{t} = t, \\ u \mapsto \tilde{u}(\tilde{t}, \tilde{x}) = e^{-i[\frac{1}{2}\vec{c}\cdot\tilde{x} + \frac{1}{4}|\vec{c}|^2\tilde{t}]}u(\tilde{t}, \tilde{x} + \vec{c}t), \end{cases} \tag{7.33}$$

in other words, for infinitesimal velocity $\vec{c}$, let

$$\delta t = 0, \quad \delta x = -\vec{c}t, \quad \delta u = -\frac{i}{2}\vec{c}\cdot xu(t,x).$$

From (7.28), it is easily to gain *the renormalized centroid conservation law*

$$\int_{\mathbb{R}^n} x|u(t,x)|^2 dx - t\vec{P}(t) = constant\ vector. \tag{7.34}$$

(vi) *Invariance by the pseudo-conformal transformation*:

$$\begin{cases} x \mapsto \tilde{x} = \dfrac{x}{\ell(t)}, \\ t \mapsto \tilde{t} = \displaystyle\int_0^t \frac{1}{\ell^2(\tau)}d\tau, \\ u \mapsto \tilde{u}(\tilde{t}, \tilde{x}) = \ell^{n/2}u(t,x)\exp(-i\dfrac{\ell_t}{\ell}\dfrac{|x|^2}{4}). \end{cases} \tag{7.35}$$

Assume $\ell_{tt} = 0$ and $F'(|\tilde{u}|^2) = \ell^2 F'(|u|^2)$, then the equation is invariant under this pseudo-conformal transformation. Let $\varepsilon \in (0, 1/|t|)$ be an infinitesimal quantity and $\ell(t) = 1 - \varepsilon t$, then $\delta x = \varepsilon tx$, $\delta t = \varepsilon t^2$ and $\delta u = \varepsilon(-\frac{n}{2}t + \frac{i}{4}|x|^2)u$. From (7.28), we have *the pseudo-conformal conservation law*

$$\int_{\mathbb{R}^n} \left(|xu + 2it\nabla u|^2 + 4t^2 F(|u|^2)\right) dx = const. \tag{7.36}$$

Remark 7.1. 1) In fact, $x + 2it\nabla$ is a Galilean operator, which is commutable with $i\partial_t + \Delta$. Let $J(t) = x + 2it\nabla$ and $M(t) = e^{\frac{|x|^2}{4it}}$, then we have the following relations between $J(t)$ and the Schrödinger semigroup $S(t) = e^{it\Delta}$ or $M(t)$:

$$J(t) = S(-t)xS(t), \quad J(t)u = 2itM(-t)\nabla(M(t)u).$$

2) For the general nonlinearity $F'(|u|^2)u$, we can differentiate (7.36) with respect to the time t,

$$\begin{aligned}
&\frac{d}{dt}\int_{\mathbb{R}^n} \left(|xu + 2it\nabla u|^2 + 4t^2 F(|u|^2)\right) dx \\
=&4t \int_{\mathbb{R}^n} \left((n+2)F(|u|^2) - nF'(|u|^2)|u|^2\right) dx,
\end{aligned} \tag{7.37}$$

to obtain, for any time t in the time interval where the solution exists, the following pseudo-conformal conservation law

$$\begin{aligned}
&\int_{\mathbb{R}^n} \left(|xu + 2it\nabla u|^2 + 4t^2 F(|u|^2)\right) dx \\
=&\int_{\mathbb{R}^n} |xu(0,x)|^2 dx + 4\int_0^t \tau \int_{\mathbb{R}^n} \left((n+2)F(|u|^2) - nF'(|u|^2)|u|^2\right) dx d\tau.
\end{aligned} \tag{7.38}$$

7.3 Virial identity and Morawetz inequality

In the previous subsection, it involves $|u|^2$, $x|u|^2$, $|x|^2|u|^2$, $\cdots$, in the conservation laws. Have they some kind of rules to follow? What will happen if we replace the coefficient in front of $|u|^2$ by a general function? Can we obtain some other conservation laws or estimates?

Now, let $a(t,x)$ be an arbitrary real-valued function on $\mathbb{R}^{1+n}$, and define the *virial potential* $V_a(t)$ and the *Morawetz*[5] *action* $M_a(t)$ associated to a as

$$V_a(t) := \int_{\mathbb{R}^n} a(t,x)|u(t,x)|^2 dx, \quad M_a(t) := \partial_t V_a(t).$$

For convenience, denote $G(|u|^2) = F'(|u|^2)|u|^2 - F(|u|^2)$. By (7.5) and the integration by parts, we have

$$M_a(t) = \partial_t V_a(t) = \int_{\mathbb{R}^n} (a_t|u|^2 + 2\mathrm{Im}(\bar{u}\nabla a \cdot \nabla u))\, dx, \tag{7.39}$$

$$\begin{aligned} \partial_t M_a(t) = \partial_{tt} V_a(t) = & \int_{\mathbb{R}^n} (a_{tt} - \Delta^2 a)|u|^2 dx \\ & + 4\int_{\mathbb{R}^n} \mathrm{Im}(\bar{u}\nabla a_t \cdot \nabla u) dx \\ & + 4\int_{\mathbb{R}^n} \sum_{j,\,k=1}^{n} a_{jk}\mathrm{Re}(u_k \bar{u}_j) dx \\ & + 2\int_{\mathbb{R}^n} \Delta a G(|u|^2) dx. \end{aligned} \tag{7.40}$$

In the time independent case $a(t,x) = a(x)$, we have

$$\begin{aligned} \partial_{tt} V_a(t) = & -\int_{\mathbb{R}^n} \Delta^2 a|u|^2 dx + 4\int_{\mathbb{R}^n} \sum_{j,\,k=1}^{n} a_{jk}\mathrm{Re}(u_k \bar{u}_j) dx \\ & + 2\int_{\mathbb{R}^n} \Delta a G(|u|^2) dx. \end{aligned} \tag{7.41}$$

On the other hand, in the time dependent case $a(t,x) \neq a(x)$, we get, for any positive function $Q = Q(t,x)$, that

$$4\mathrm{Im}(\bar{u}\nabla a_t \cdot \nabla u) = |Q^{-1}(\nabla a_t)u - 2iQ\nabla u|^2 - Q^{-2}|\nabla a_t|^2|u|^2 - 4Q^2|\nabla u|^2.$$

Thus,

$$\begin{aligned} \partial_{tt} V_a(t) = & \int_{\mathbb{R}^n} (a_{tt} - \Delta^2 a - Q^{-2}|\nabla a_t|^2)|u|^2 dx \\ & + \int_{\mathbb{R}^n} |Q^{-1}(\nabla a_t)u - 2iQ\nabla u|^2 dx \end{aligned}$$

[5]Cathleen Synge Morawetz (1923-) is a mathematician born in Toronto, Canada. Morawetz' research was mainly in the study of the partial differential equations governing fluid flow, particularly those of mixed type occurring in transonic flow. She is a professor emerita at the Courant Institute of Mathematical Sciences at the New York University, where she had also served as the director from 1984 to 1988.

$$+4\int_{\mathbb{R}^n}\sum_{j,k=1}^{n}(a_{jk}-Q^2\delta_{jk})\mathrm{Re}(u_k\bar{u}_j)dx$$
$$+2\int_{\mathbb{R}^n}\Delta a G(|u|^2)dx. \tag{7.42}$$

The problem is then to choose Q and a such that as many as possible of these terms are non-negative up to some manageable errors. Next, we show some examples.

Example 7.1 (Virial identity). Let $a(t,x)=|x|^2$, then $a_{jk}=2\delta_{jk}$ and $\Delta^2 a=0$, thus from (7.41), it yields the *virial identity*

$$\partial_{tt}\int_{\mathbb{R}^n}|x|^2|u|^2dx=8n\int_{\mathbb{R}^n}|\nabla u|^2dx+4n\int_{\mathbb{R}^n}G(|u|^2)dx.$$

For the defocusing case $F(|u|^2)=\frac{2}{p+1}|u|^{p+1}$ where $p>1$, i.e. $G(|u|^2)=\frac{p-1}{p+1}|u|^{p+1}$, it implies that the convexity of $\int_{\mathbb{R}^n}|x|^2|u|^2dx$ can be controlled by the energy. For the focusing case $F(|u|^2)=-\frac{2}{p+1}|u|^{p+1}$ with $p\geqslant 1+\frac{4}{n}$, i.e. $G(|u|^2)=-\frac{p-1}{p+1}|u|^{p+1}$, the quantity $\partial_{tt}\int_{\mathbb{R}^n}|x|^2|u|^2dx$ can be bounded from above by a multiple of the energy, thus if the energy is negative and $\int_{\mathbb{R}^n}|x|^2|u|^2dx$ is finite, the solution of the equation necessarily has a blowup in finite time.

Example 7.2 (Morawetz' inequality). *Let* $a(t,x)=|x|$, $n>1$, *then* $\nabla a=\frac{x}{|x|}$, $a_{jk}=\frac{\delta_{jk}}{|x|}-\frac{x_jx_k}{|x|^3}$ *and* $\Delta a=\frac{n-1}{|x|}$. *From* (7.39) *and* (7.41), *we have*

$$\begin{aligned}M_a(t)=&2\int_{\mathbb{R}^n}\mathrm{Im}(\bar{u}\frac{x}{|x|}\cdot\nabla u)dx,\\ \partial_t M_a(t)=&-(n-1)\int_{\mathbb{R}^n}(\Delta\frac{1}{|x|})|u|^2dx+4\int_{\mathbb{R}^n}\frac{|\nabla\!\!\!/_0 u|^2}{|x|}dx\\ &+2(n-1)\int_{\mathbb{R}^n}\frac{G(|u|^2)}{|x|}dx,\end{aligned}$$

where

$$|\nabla\!\!\!/_y u|^2:=|\nabla u|^2-|\frac{x-y}{|x-y|}\cdot\nabla u|^2.$$

In the three-dimensional case $n=3$, *we have* $\Delta(\frac{1}{|x|})=-4\pi\delta$, *and so*

$$\partial_t M_a(t)=8\pi|u(t,0)|^2+4\int_{\mathbb{R}^3}\frac{|\nabla\!\!\!/_0 u|^2+G(|u|^2)}{|x|}dx.$$

Integrating over $[T_*, T^*]$ with respect to t, we get the Morawetz identity

$$2\int_{T_*}^{T^*}\int_{\mathbb{R}^3}\frac{|\nabla_0 u|^2+G(|u|^2)}{|x|}dxdt+4\pi\int_{T_*}^{T^*}|u(t,0)|^2dt$$
$$=\int_{\mathbb{R}^3}\mathrm{Im}(\bar{u}(T^*,x)\frac{x}{|x|}\cdot\nabla u(T^*,x))dx-\int_{\mathbb{R}^3}\mathrm{Im}(\bar{u}(T_*,x)\frac{x}{|x|}\cdot\nabla u(T_*,x))dx.$$

In the three-dimensional defocusing case, it implies the following Morawetz inequality

$$\int_{T_*}^{T^*}\int_{\mathbb{R}^3}\frac{|\nabla_0 u|^2+G(|u|^2)}{|x|}dxdt+2\pi\int_{T_*}^{T^*}|u(t,0)|^2dt$$
$$\lesssim \sup_{T_*\leqslant t\leqslant T^*}\|u(t)\|^2_{\dot{H}^{1/2}}. \tag{7.43}$$

This inequality can be used to demonstrate scattering results for defocusing Schrödinger equations.

In higher dimensional defocusing case ($n\geqslant 4$), we have $-\Delta(\frac{1}{|x|})=\frac{(n-3)}{|x|^3}>0$ and the Morawetz inequality

$$(n-1)(n-3)\int_{T_*}^{T^*}\int_{\mathbb{R}^n}\frac{|u|^2}{|x|^3}dxdt+4\int_{T_*}^{T^*}\int_{\mathbb{R}^n}\frac{|\nabla_0 u|^2}{|x|}dxdt$$
$$+2(n-1)\int_{T_*}^{T^*}\int_{\mathbb{R}^n}\frac{G(|u|^2)}{|x|}dxdt$$
$$\lesssim \sup_{T_*\leqslant t\leqslant T^*}\|u(t)\|^2_{\dot{H}^{1/2}}. \tag{7.44}$$

But in lower dimensions $n=1,2$, the distribution $-\Delta(\frac{1}{|x|})$ is quite nasty so that we can not obtain a simple result as in higher dimensional cases.

Example 7.3 (Nakanishi-Morawetz inequality). Let $a(t,x)=\lambda:=|(t,x)|=\sqrt{t^2+|x|^2}$, then we can compute

$$a_t=\frac{t}{\lambda},\quad a_j=\frac{x_j}{\lambda},\quad a_{tj}=-\frac{tx_j}{\lambda^3},\quad a_{tt}=\frac{|x|^2}{\lambda^3},$$
$$a_{jk}=\frac{\delta_{jk}t^2}{\lambda^3}+\frac{\delta_{jk}|x|^2-x_jx_k}{\lambda^3},\quad \Delta a=\frac{n}{\lambda}-\frac{|x|^2}{\lambda^3}=\frac{n-1}{\lambda}+\frac{t^2}{\lambda^3}.$$

Substituting these into (7.39) and (7.42), we obtain

$$M_a(t)=\int_{\mathbb{R}^n}\left(\frac{t}{\lambda}|u|^2+2\mathrm{Im}(\bar{u}\frac{x}{\lambda}\cdot\nabla u)\right)dx,$$
$$\partial_t M_a(t)=\int_{\mathbb{R}^n}(\frac{|x|^2}{\lambda^3}-\Delta^2 a-\frac{t^2|x|^2}{\lambda^6 Q^2})|u|^2dx$$

$$
\begin{aligned}
&+\int_{\mathbb{R}^n} |-\frac{tx}{Q\lambda^3}u - 2iQ\nabla u|^2 dx \\
&+4\int_{\mathbb{R}^n} (\frac{t^2}{\lambda^3} - Q^2)|\nabla u|^2 + \frac{|\nabla_0 u|^2|x|^2}{\lambda^3} dx \\
&+2\int_{\mathbb{R}^n} \Delta a G(|u|^2) dx.
\end{aligned}
$$

Taking $Q^2 := \frac{t^2}{\lambda^3}$, we can eliminate some terms to reduce the above identity as

$$
\begin{aligned}
\partial_t M_a(t) = &-\int_{\mathbb{R}^n} \Delta^2 a|u|^2 dx + \int_{\mathbb{R}^n} \frac{|xu+2it\nabla u|^2}{\lambda^3} dx + 4\int_{\mathbb{R}^n} \frac{|\nabla_0 u|^2|x|^2}{\lambda^3} dx \\
&+2\int_{\mathbb{R}^n} (\frac{n-1}{\lambda} + \frac{t^2}{\lambda^3}) G(|u|^2) dx.
\end{aligned}
$$

Noticing $\Delta^2 a = O(\lambda^{-3})$, we have, for $G(|u|^2) \geqslant 0$, that

$$
\begin{aligned}
&\int_{\mathbb{R}^n} \frac{|xu+2it\nabla u|^2}{\lambda^3} dx + 4\int_{\mathbb{R}^n} \frac{|\nabla_0 u|^2|x|^2}{\lambda^3} dx \\
&+2\int_{\mathbb{R}^n} \frac{t^2 G(|u|^2)}{\lambda^3} dx \\
&\lesssim \partial_t M_a(t) + \int_{\mathbb{R}^n} O(\lambda^{-3})|u|^2 dx.
\end{aligned}
$$

Integrating over $(1,T]$ with respect to t (we can deal with $[-T,-1)$ in a similar way), we obtain

$$
\begin{aligned}
&\int_1^T \int_{\mathbb{R}^n} \frac{|xu+2it\nabla u|^2}{\lambda^3} dxdt + 4\int_1^T \int_{\mathbb{R}^n} \frac{|\nabla_0 u|^2|x|^2}{\lambda^3} dxdt \\
&+2\int_1^T \int_{\mathbb{R}^n} \frac{t^2 G(|u|^2)}{\lambda^3} dxdt \\
&\lesssim M_a(T) - M_a(1) + \int_1^T \int_{\mathbb{R}^n} O(\lambda^{-3})|u|^2 dxdt \\
&\lesssim C(E(0), N(0)) + \int_1^T \int_{\mathbb{R}^n} O(\lambda^{-3})|u|^2 dxdt,
\end{aligned}
$$

where $H(t)$ and $N(t)$ are the energy and the L^2-norm defined before, respectively. For $|t| > 1$, we have

$$
\int_{|t|>1} \int_{\mathbb{R}^n} \frac{|u|^2}{\lambda^3} dxdt \lesssim \int_{|t|>1} \int_{\mathbb{R}^n} \frac{|u|^2}{|t|^3} dxdt \lesssim N(0).
$$

Taking $T \to \infty$, we get

$$
\int_{|t|>1} \int_{\mathbb{R}^n} \frac{|xu+2it\nabla u|^2}{\lambda^3} dxdt + 4\int_{|t|>1} \int_{\mathbb{R}^n} \frac{|\nabla_0 u|^2|x|^2}{\lambda^3} dxdt
$$

$$+2\int_{|t|>1}\int_{\mathbb{R}^n}\frac{t^2G(|u|^2)}{\lambda^3}dxdt$$
$$\lesssim C(E(0),N(0)).$$

In particular, let $F(|u|^2)=\frac{2}{p+1}|u|^{p+1}$. Assume that $p\in(1,\infty)$ if $n=1,2$, and $p\in(1,1+\frac{4}{n-2})$ if $n\geqslant 3$, then $F'(|u|^2)|u|^2-F(|u|^2)=\frac{p-1}{p+1}|u|^{p+1}$. From the previous estimates, it yields

$$\int_{|t|>1}\int_{\mathbb{R}^n}\frac{t^2|u|^{p+1}}{|(t,x)|^3}dxdt\lesssim C(E(0),N(0)). \tag{7.45}$$

For the case $|t|\leqslant 1$, we split it into two cases. In the case $|x|>1$, we have

$$\iint_{\substack{|t|\leqslant 1\\|x|>1}}\frac{t^2|u|^{p+1}}{|(t,x)|^3}dxdt\lesssim\iint_{\substack{|t|\leqslant 1\\|x|>1}}|u|^{p+1}dxdt\lesssim C(E(0),N(0)). \tag{7.46}$$

In the case $|x|\leqslant 1$, we get

$$\begin{aligned}\iint_{\substack{|t|\leqslant 1\\|x|\leqslant 1}}\frac{t^2|u|^{p+1}}{|(t,x)|^3}dxdt&\lesssim\iint_{\substack{|t|\leqslant 1\\|x|\leqslant 1}}\frac{|u|^{p+1}}{|(t,x)|}dxdt\\&\lesssim\iint_{\substack{|t|\leqslant 1\\|x|\leqslant 1}}\frac{|u|^{p+1}}{|t|^{1-(p+1)\varepsilon}|x|^{(p+1)\varepsilon}}dxdt\\&\lesssim\int_{|t|\leqslant 1}\frac{\||x|^{-\varepsilon}u\|^{p+1}_{L^{p+1}(|x|\leqslant 1)}}{|t|^{1-(p+1)\varepsilon}}dt\\&\lesssim\int_{-1}^{1}\frac{1}{|t|^{1-(p+1)\varepsilon}}\|u\|^{p+1}_{H^1(|x|\leqslant 1)}dt\\&\lesssim C(E(0),N(0)),\end{aligned} \tag{7.47}$$

where we have used the following Hardy's[6] inequality

$$\||x|^{-\varepsilon}u\|_{L^q(|x|\leqslant 1)}\leqslant C\|u\|_{H^1},\quad\forall 2\leqslant q<2^*. \tag{7.48}$$

Here $2^*:=\begin{cases}2n/(n-2), & n\geqslant 3,\\ \infty, & n\leqslant 2,\end{cases}$ $0<\varepsilon<\begin{cases}n/q-n/2^*, & n\neq 2,\\ n/2q, & n=2.\end{cases}$ In fact, by Hölder's[7] inequality, we have, for $n\neq 2$, that

$$\||x|^{-\varepsilon}u\|_{L^q(|x|\leqslant 1)}\leqslant\||x|^{-\varepsilon}\|_{L^{q2^*/(2^*-q)}(|x|\leqslant 1)}\|u\|_{L^{2^*}(|x|\leqslant 1))}$$

[6] Godfrey Harold Hardy (1877-1947) was a prominent English mathematician, known for his achievements in number theory and mathematical analysis.

[7] Otto Ludwig Hölder (1859-1937) was a German mathematician born in Stuttgart. He is famous for many things including: Hölder's inequality, the Jordan-Hölder theorem, Hölder's theorem and the Hölder condition which is used e.g. in the theory of partial differential equations and function spaces.

$$\leqslant C\left(\int_0^1 r^{-\varepsilon\frac{q2^*}{2^*-q}+n-1}dr\right)^{\frac{2^*-q}{q2^*}}\|u\|_{L^{2^*}(|x|\leqslant 1))}$$
$$\leqslant C\|u\|_{H^1}.$$

For the case $n=2$, from Hölder's inequality, we get

$$\||x|^{-\varepsilon}u\|_{L^q(|x|\leqslant 1)} \leqslant \||x|^{-\varepsilon}\|_{L^{2q}(|x|\leqslant 1)}\|u\|_{L^{2q}(|x|\leqslant 1))}$$
$$\leqslant C\left(\int_0^1 r^{-2q\varepsilon+n-1}dr\right)^{1/2q}\|u\|_{L^{2q}(|x|\leqslant 1))}$$
$$\leqslant C\|u\|_{H^1}.$$

From (7.45), (7.45) and (7.45), we obtain the Nakanishi-Morawetz inequality (cf. [176])

$$\int_{\mathbb{R}^{1+n}} \frac{t^2|u|^{p+1}}{|(t,x)|^3}dxdt \lesssim C(E(0),N(0)), \quad \forall n\in\mathbb{N}. \tag{7.49}$$

7.4 Morawetz' interaction inequality

Let $a(x)$ be an arbitrary real-valued function on $\mathbb{R}^n$, and define the virial interaction potential $V^a(t)$ and Morawetz' interaction potential $M^a(t)$ associated to a as

$$V^a(t) := \int_{\mathbb{R}^n}\int_{\mathbb{R}^n} a(x-y)|u(t,x)|^2|u(t,y)|^2dxdy, \quad M^a(t) := \partial_t V^a(t).$$

From (7.5) and the integration by parts, we have

$$M_a(t) = 2\int_{\mathbb{R}^n}\int_{\mathbb{R}^n} \nabla a(x-y)\cdot\Big[\mathrm{Im}(\nabla u(t,x)\bar{u}(t,x))|u(t,y)|^2 - \mathrm{Im}(\nabla u(t,y)\bar{u}(t,y))|u(t,x)|^2\Big]dxdy, \tag{7.50}$$

and

$$\begin{aligned}
&\partial_t M_a(t)\\
=&-2\int_{\mathbb{R}^n}\int_{\mathbb{R}^n} \Delta^2 a(x-y)|u(t,x)|^2|u(t,y)|^2dxdy\\
&+2\int_{\mathbb{R}^n}\int_{\mathbb{R}^n} \Delta a(x-y)[G(|u(x)|^2)|u(y)|^2+G(|u(y)|^2)|u(x)|^2]dxdy\\
&+4\int_{\mathbb{R}^n}\int_{\mathbb{R}^n}\sum_{j,k=1}^n a_{jk}(x-y)\Big[\mathrm{Re}(u_k(t,x)\bar{u}_j(t,x))|u(t,y)|^2\\
&\qquad+\mathrm{Re}(u_k(t,y)\bar{u}_j(t,y))|u(t,x)|^2
\end{aligned}$$

$$- \operatorname{Im}(u_k(t,y)\bar{u}(t,y))\operatorname{Im}(u_j(t,x)\bar{u}(t,x))$$
$$- \operatorname{Im}(u_k(t,x)\bar{u}(t,x))\operatorname{Im}(u_j(t,y)\bar{u}(t,y))\Big] dxdy. \quad (7.51)$$

Denote

$$A_j(t,x,y) := u(t,x)\bar{u}_j(t,y) + u_j(t,x)\bar{u}(t,y),$$
$$B_j(t,x,y) := u(t,x)u_j(t,y) + u_j(t,x)u(t,y),$$

then, the last integral can be reduced to

$$2\int_{\mathbb{R}^n}\int_{\mathbb{R}^n}\sum_{j,\,k=1}^{n} a_{jk}(x-y)[A_j(t,x,y)\overline{A_k(t,x,y)} + B_j(t,x,y)\overline{B_k(t,x,y)}]dxdy,$$

where we can exchange k and j since $a_{jk} = a_{kj}$.

For convenience, we denote the above quantity by I. If a is a radial function, i.e. $a(x) = a(|x|)$, then $a_{jk}(x) = \frac{a'(|x|)}{|x|}\delta_{jk} + (a''(|x|) - \frac{a'(|x|)}{|x|})\frac{x_k}{|x|}\frac{x_j}{|x|}$. From the Cauchy-Schwartz inequality, we obtain

$$\begin{aligned}
I =& 2\int_{\mathbb{R}^n}\int_{\mathbb{R}^n}\frac{a'(|x-y|)}{|x-y|}\sum_{j=1}^{n}[|A_j(t,x,y)|^2 + |B_j(t,x,y)|^2]dxdy \\
&+ 2\int_{\mathbb{R}^n}\int_{\mathbb{R}^n}(a''(|x-y|) - \frac{a'(|x-y|)}{|x-y|})\Big[\Big|\sum_{j=1}^{n}\frac{x_j-y_j}{|x-y|}A_j(t,x,y)\Big|^2 \\
&\qquad + \Big|\sum_{j=1}^{n}\frac{x_j-y_j}{|x-y|}B_j(t,x,y)\Big|^2\Big]dxdy \\
\geqslant& 2\int_{\mathbb{R}^n}\int_{\mathbb{R}^n} a''(|x-y|)\Big[\Big|\sum_{j=1}^{n}\frac{x_j-y_j}{|x-y|}A_j(t,x,y)\Big|^2 \\
&\qquad + \Big|\sum_{j=1}^{n}\frac{x_j-y_j}{|x-y|}B_j(t,x,y)\Big|^2\Big]dxdy.
\end{aligned}$$

It is clear that $I \geqslant 0$ if $a''(\lambda) \geqslant 0$ for $\lambda \geqslant 0$. Hence, (7.51) and (7.50) imply

Theorem 7.4 (Morawetz' interaction inequality). *Let $a(x)$ be a real-valued radial convex function, and $u(t,x)$ be a solution of (7.5), then it holds*

$$\begin{aligned}
&\int_{\mathbb{R}^n}\int_{\mathbb{R}^n}(-\Delta^2 a(x-y))|u(t,x)|^2|u(t,y)|^2 dxdy \\
&+ \int_{\mathbb{R}^n}\int_{\mathbb{R}^n}\Delta a(x-y)[G(|u(t,x)|^2)|u(t,y)|^2 + G(|u(t,y)|^2)|u(t,x)|^2]dxdy \\
&\leqslant \frac{1}{2}\partial_t M_a(t), \qquad (7.52)
\end{aligned}$$

where $M_a(t)$ is defined in (7.50).

Next, we consider an example of three dimensional nonlinear Schrödinger equations.

Example 7.4 (Morawetz' interaction inequality of 3D NLS). *Let $n = 3$, $a(x) = |x|$, then $\nabla a = \frac{x}{|x|}$, $\Delta a = \frac{2}{|x|} > 0$, and $-\Delta^2 a = 8\pi\delta(x)$. From (7.52), we get*

$$\begin{aligned}
&8\pi \int_{T_*}^{T^*} \int_{\mathbb{R}^3} |u(t,x)|^4 dxdt \\
&+2\int_{T_*}^{T^*} \int_{\mathbb{R}^3} \int_{\mathbb{R}^3} \frac{G(|u(t,x)|^2)|u(t,y)|^2 + G(|u(t,y)|^2)|u(t,x)|^2}{|x-y|} dxdydt \\
&\leqslant \frac{1}{2}(M_a(T^*) - M_a(T_*)) \\
&= \int_{\mathbb{R}^n} \int_{\mathbb{R}^n} \Big[\mathrm{Im}(\frac{x-y}{|x-y|} \cdot \nabla u(T^*,x)\bar{u}(T^*,x))|u(T^*,y)|^2 \\
&\quad - \mathrm{Im}(\frac{x-y}{|x-y|} \cdot \nabla u(T^*,y)\bar{u}(T^*,y))|u(T^*,x)|^2 \Big] dxdy \\
&\quad - \int_{\mathbb{R}^n} \int_{\mathbb{R}^n} \Big[\mathrm{Im}(\frac{x-y}{|x-y|} \cdot \nabla u(T_*,x)\bar{u}(T_*,x))|u(T_*,y)|^2 \\
&\quad - \mathrm{Im}(\frac{x-y}{|x-y|} \cdot \nabla u(T_*,y)\bar{u}(T_*,y))|u(T_*,x)|^2 \Big] dxdy \\
&\lesssim N^2(0) \sup_{T_* \leqslant t \leqslant T^*} \|u(t)\|^2_{\dot{H}^{\frac{1}{2}}}.
\end{aligned} \tag{7.53}$$

7.5 Scattering results for NLS

The Morawetz estimates are time-decaying estimates, which are very useful for us to study the existence of the scattering operators. In this section, we only state some results on the existence of the scattering operators for NLS

$$iu_t + \Delta u = |u|^p u, \quad u(0,x) = u_0(x). \tag{7.54}$$

The existence of the scattering operators in the H^1-subcritical but L^2-supercritical cases was shown by Nakanishi [176] in 1 and 2 spatial dimensions and by Ginibre and Velo [79] in higher spatial dimensions $n \geqslant 3$:

Theorem 7.5. *Let $p \in (4/n, 4/(n-2))$ for $n \geqslant 3$, and $p \in (4/n, \infty)$ for $n = 1, 2$. For any initial datum $u_0 \in H^1$ with $E(u_0) < \infty$, (7.54) has a unique global solution $u \in C_t^0(\dot{H}_x^1) \cap L_{t,x}^{p(n+2)/2}$ satisfying*

$$\int_{-\infty}^{\infty} \int_{\mathbb{R}^n} |u(t,x)|^{p(n+2)/2} dxdt \leqslant C(E(u_0)),$$

where

$$E(u(t)) = \int_{\mathbb{R}^n} \frac{1}{2}|\nabla u(t,x)|^2 + \frac{2}{2+p}|u(t,x)|^{2+p}dx$$

is the energy conserved quantity. Moreover, there exist finite-energy solutions $u_\pm$ *of the linear Schrödinger equation* $(i\partial_t + \Delta)u_\pm = 0$ *such that*

$$\|u_\pm(t) - u(t)\|_{\dot{H}^1_x} \to 0, \text{ as } t \to \pm\infty,$$

and the mapping $u_0 \mapsto u_\pm(0)$ *is a homeomorphism from* $\dot{H}^1(\mathbb{R}^n)$ *to itself.*

Next, we introduce briefly the related results for the scattering theory of H^1-critical defocusing NLS

$$iu_t + \Delta u = |u|^{\frac{4}{n-2}}u, \quad u(0,x) = u_0(x). \tag{7.55}$$

In 1999, J. Bourgain [19] proved the global wellposedness of the solution with radial initial data in 3 and 4 dimensions. Then M. Grillakis [84] gave a different proof for Bourgain's results in 2000. In 2005, these results were generalized to any spherically symmetric initial data in any dimensions $n \geqslant 3$ by T. Tao [217]. At the same time (the paper was published in 2008), J. Colliander, M. Keel, G. Staffilani, H. Takaoka, and T. Tao [44] obtained the global well-posedness and the scattering theory for any general initial data in three dimensions based on the induction arguments on the energy of Bourgain's. They applied the induction argument not only on the frequency space but also on the physical space, and replaced the original Morawetz' inequality (7.43) by Morawetz' interaction inequality (7.53) in order to deal with the non-radial cases. In addition, the Morawetz' interaction inequality, together with the almost conserved quantity of the momentum which controlled the mass in the frequency space, described the possibilities of the energy concentration. For the higher dimensional case, it is generalized by E. Ryckman and M. Visan [196] for $n = 4$ and by M. Visan [232] for $n \geqslant 5$. Thus, the scattering in the energy class is now known for (7.55) with any general initial data in all dimensions $n \geqslant 3$, namely, as stated in the following results [44; 196; 232].

Theorem 7.6. *Let* $n \geqslant 3$. *For any initial datum* $u_0 \in H^1$ *with* $E(u_0) < \infty$, *then* (7.55) *yields a unique global solution* $u \in C_t^0(\dot{H}^1_x) \cap L_{t,x}^{\frac{2(n+2)}{n-2}}$ *such that*

$$\int_{-\infty}^{\infty}\int_{\mathbb{R}^n} |u(t,x)|^{\frac{2(n+2)}{n-2}}dxdt \leqslant C(E(u_0)),$$

where

$$E(u(t)) = \int_{\mathbb{R}^n} \frac{1}{2}|\nabla u(t,x)|^2 + \frac{n-2}{2n}|u(t,x)|^{\frac{2n}{n-2}} dx$$

is the energy conserved quantity.

The above theorem implies the scattering results [44; 232; 196]:

Corollary 7.1. *Let* $n \geqslant 3$. *Assume that* u_0 *has finite energies,* u *is the unique global solution of* (7.55) *on* $u \in C_t^0(\dot{H}_x^1) \cap L_{t,x}^{\frac{2(n+2)}{n-2}}$. *Then, there exist finite-energy solutions* $u_\pm$ *of the linear Schrödinger equation* $(i\partial_t + \Delta)u_\pm = 0$ *such that*

$$\|u_\pm(t) - u(t)\|_{\dot{H}_x^1} \to 0, \quad t \to \pm\infty,$$

and the mapping $u_0 \mapsto u_\pm(0)$ *is a homeomorphism from* $\dot{H}^1(\mathbb{R}^n)$ *to itself.*

Along with the perfect solution of the scattering theory of the energy-critical defocusing NLS, most researchers transfer their interests into another important problem – the global well-posedness and scattering theories of the mass-critical defocusing NLS

$$iu_t + \Delta u = |u|^{\frac{4}{n}} u, \quad u(0,x) = u_0(x). \tag{7.56}$$

At present, there are some results in this aspect:

- $n \geqslant 1$, for sufficiently small initial datum $u_0 \in L^2(\mathbb{R}^n)$, the solution of (7.56) is global well-posed and scatters.
- $n \geqslant 2$, for radial initial data $u_0 \in L^2(\mathbb{R}^n)$, the solution of (7.56) is global well-posed and scatters, cf. [138; 140].
- $n \geqslant 1$, for any initial data $u_0 \in L^2(\mathbb{R}^n)$, the solution of (7.56) is global well-posed and scatters, cf. [64; 65; 66].

For other nonlinearities, there are also some interesting results. K. Nakanishi established the scattering theory for the mixed power nonlinearity $|u|^{p_1}u + |u|^{p_2}u$ for $4/n < p_1 \leqslant p_2 < 4/(n-2)$ and $4/n \leqslant p_1 < p_2 \leqslant 2^* - 2$ in the energy space H^1, cf. [176; 179; 220]. For the nonlinearity $(V * |u|^2)u$, i.e. the Hartree equation, he also proved the existence of the scattering operator in H^1 for $V \in L^{p_1} + L^{p_2}$, $p_1, p_2 \geqslant 1$ and $n/4 < p_2 \leqslant p_1 < n/2$, cf. [175]. For the nonlinear Schrödinger equation with the following nonlinearity

$$f(u) = \mu \sum_{k \geqslant 1+2/n} \frac{\lambda^{2k}}{(2k)!}|u|^{2k}u, \quad \mu, \lambda > 0, \quad n = 1, 2, \tag{7.57}$$

the scattering operator also exists under some assumptions, cf. [244].

In addition, there is a great process for focusing nonlinear Schrödinger equations

$$\begin{cases} iu_t + \Delta u = -|u|^p u, \\ u(0,x) = u_0(x). \end{cases} \tag{7.58}$$

For $\dot{H}^1$-critical nonlinearities in $n = 3,4,5$ dimensions, i.e. $p = 4/(n-2)$, the solution of (7.58) is global well-posed and scatters for any radial initial datum $u_0 \in \dot{H}^1(\mathbb{R}^n)$ whose energy and kinetic energy are less than those of the ground state, cf. [124]; For the $n \geqslant 5$ dimensional case, the condition "radial" can be eliminated and it holds that the solution of (7.58) is global well-posed and scatters for any initial datum $u_0 \in \dot{H}^1(\mathbb{R}^n)$ whose energy and kinetic energy are less than those of the ground state, cf. [139]. Otherwise, in the $n \geqslant 4$ dimensions, for a global solution u to the focusing mass-critical nonlinear Schrödinger equation (7.58) (where $p = 4/n$) with spherically symmetric H^1 initial data and mass equal to that of the ground state Q, u does not scatter then, up to the phase rotation and the scaling, u is the solitary wave e^{itQ}. And the only spherically symmetric minimal-mass non-scattering solutions are, up to the phase rotation and the scaling, the pseudo-conformal ground state and the ground state solitary wave, cf. [137]. Recently, scattering for the focusing energy-subcritical NLS has been studied by Fang, Xie and Cazenave [72].

Chapter 8

Boltzmann equation without angular cutoff

The kinetic theory of the gas is a theory devoted to the study of evolutionary behaviors of the gas in the one-particle phase space of position and velocity. Recently, the kinetic theory is getting more and more recognized to be significant both in mathematics and practical applications as a key theory connecting the microscopic and macroscopic theory of gases and fluids. In this sense, the kinetic theory is in-between or mesoscopic.

In the macroscopic scales where the gas and fluid are regarded as a continuum, their motion is described by the macroscopic quantities such as macroscopic mass density, bulk velocity, temperature, pressure, stresses, heat flux and so on. The Euler and Navier-Stokes equations, compressible or incompressible, are the most famous equations among governing equations proposed so far in fluid dynamics.

The extreme contrary is the microscopic scale where the gas, fluid, and hence any matter, are looked at as a many-body system of microscopic particles (atom/molecule). Thus, the motion of the system is governed by the coupled Newton equations, within the framework of the classical mechanics. The number of the involved equations is $6N$ if the total number of the microscopic particles is N.

Although the Newton equation is the first principle of the classical mechanics, it is not of practical use because the number of the equations is so enormous ($N \sim$ the Avogadro number 6×10^{23}) that it is hopeless to specify all the initial data, and we must appeal to statistics. On the other hand, the macroscopic (fluid dynamical) quantities mentioned above are related to statistical average of quantities depending on the microscopic state. Thus, the kinetic theory that gives the mesoscopic descriptions of the gas and fluid is noticed to be a key theory that links the microscopic and macroscopic scales. The Boltzmann equation, which is the subject of these notes, is the most classical but fundamental equation in the mesoscopic kinetic theory.

The goal of this chapter is to introduce, and make a preliminary discussion of the Boltzmann equation. We mainly use harmonic analysis to consider the Boltzmann equation without angular cutoff. We will show that the Boltzmann collision operator without angular cutoff behaves like a singular integral operator or pseudo-differential operator whose leading term is characterized by the operator $(-\Delta)^{\nu/2}$.

This was first pointed out by Pao [188], see also [228], and was formulated explicitly by Lions [160]. The optimal Sobolev exponent $\nu/2$ is due to Villani [231]. In the middle of 1990s, Desvillettes managed to prove the regularity of solutions to some simplified models for the spatially homogeneous problems, cf. [57; 58]. Around 2000s, the regularity induced by the grazing collision was analyzed in terms of the entropy production integral, cf. the work [2] and others. In particular, [2] establishes several elegant formulations associated with the collision operator which have been essentially used to the study of the spatially homogeneous problem.

8.1 Models for collisions in kinetic theory

In this section, we will first give the derivation of the Boltzmann equation, in details, one can refer to [28; 59; 229].

8.1.1 *Transport model*

The object of kinetic theory is the modelling of a gas (or plasma, or any system made up of a large number of particles) by a distribution function in the particle phase space. This phase space includes macroscopic variables, i.e. the position in physical space, but also microscopic variables, which describe the *state* of the particles. In the present survey, we shall restrict ourselves, most of the time, to systems made of a single species of particles (no mixtures), and which obey the laws of classical mechanics (non-relativistic, non-quantum).

Assume that the gas is contained in a (bounded or unbounded) domain $\Omega \subset \mathbb{R}^n$ (n = 3 in applications) and observed on a time interval $[0;T]$, or $[0;+\infty)$. Then, under the above simplifying assumptions, the corresponding kinetic model is a nonnegative function $f(t;x;v)$, defined on $[0;T]\times\Omega\times\mathbb{R}^n$. Here the space $\mathbb{R}^n = \mathbb{R}^n_v$ is the space of possible velocities, and should be thought of as the tangent space to Ω. $f(t;x;v)$ is the phase space density of particles which at time t and point x move with velocity v, where $x = (x_1, x_2, ...x_n)$ and $v = (v_1, v_2, ...v_n)$.

When there is no interaction between the particles and their surrounding environment (including themselves), they will move at a constant velocity and along straight lines. In other words, for all times t and s, point x and velocity v, a particle which at time t sits at point x and move with velocity v will sit at time $t+s$ at point $x+vs$ and will keep its velocity v. This entails that

$$f(t+s, x+vs, v) = f(t, x, v), \quad \text{for any } s, \tag{8.1}$$

or, after differentiation with respect to s

$$\partial_t f + v \cdot \nabla_x f = 0. \tag{8.2}$$

(8.2) is free transport model. $v\cdot\nabla_x$ is the transport operator, where $v\cdot\nabla_x = \sum_{k=1}^n v_k \frac{\partial}{\partial x_k}$. (8.2) implies that the change rate of density function is zero if there is no collisions of the particles.

When a given force $F(t,x)$ acts on the particles (such a force can also depend on v in specific situations, for example when the particles are charged and feel the action of a magnetic field, or when the force is the drag force due to a surrounding gas), the particles will follow the trajectories of the following system of differential equations:

$$\dot{x}(t) = v(t) \tag{8.3}$$

$$\dot{v}(t) = F(t, x(t)) \tag{8.4}$$

and the corresponding partial differential equation satisfied by f (that is, the PDE whose characteristic curves are exactly the solutions of (8.3) and (8.4)), is the Vlasov equation

$$\partial_t f + v \cdot \nabla_x f + F(t,x) \cdot \nabla_v f = 0, \tag{8.5}$$

where $F(t,x)\cdot\nabla_v = \sum_{k=1}^n F_k \frac{\partial}{\partial v_k}$.

8.1.2 *Boltzmann model*

When the forces acting on the particles are mainly due to the collisions of the particles between themselves, one is led to write down the Boltzmann equation. (Very often, particles will be assumed to interact via a given interaction potential $\phi(r)$, r distance between particles). We first shall make several postulates:

(1). We assume that particles interact via binary collisions : this is a vague term describing the process in which two particles happen to come very close to each other, so that their respective trajectories are strongly deviated in a very short time. Underlying this hypothesis is an implicit assumption that the gas is dilute enough that the effect of interactions involving more than two particles can be neglected. Typically, if we deal with a three-dimensional gas of n hard spheres of radius r, this would mean

$$Nr^3 \ll 1, \quad Nr^2 \simeq 1. \tag{8.6}$$

(2). Moreover, we assume that these collisions are localized both in space and time, i.e. they are brief events which occur at a given position x and a given time t. This means that the typical duration of a collision is very small compared to the typical time scale of the description, and also quantities such as the impact parameter (see below) are negligible in front of the typical space scale (say, a space scale on which variations due to the transport operator are of order unity).

(3). Next, we further assume these collisions to be elastic : momentum and kinetic energy are preserved in a collision process.

(4). We also assume collisions to be microreversible. This word can be understood in a purely deterministic way : microscopic dynamics are time-reversible; or in a probabilistic way : if let (v, v_*) stand for the velocities before collision, and (v', v'_*) stand for the velocities after collision; the probability that velocities $(v, v_*)$are changed into (v', v'_*) in a collision process, is the same as the probability that (v', v'_*) are changed into (v, v_*).

(5). Finally, we make the Boltzmann chaos assumption : the velocities of two particles which are about to collide are uncorrelated. Roughly speaking, this means that if we randomly pick up two particles at position x, which have not collided yet, then the joint distribution of their velocities will be given by a tensor product (in velocity space) of f with itself. Note that this assumption implies an asymmetry between past and future: indeed, in general if the pre-collisional velocities are uncorrelated, then post-collisional velocities have to be correlated!

Thus we only take into account interactions between particles of rarefied gas, equation (8.2) can be written by a generalized form

$$\partial_t f + v \cdot \nabla_x f = Q(f)(v). \tag{8.7}$$

(8.7) means that change rate of density function of particles depends on itself, is a function of f, and is not zero when there exist collision between particles. Next we will derive the form of $Q(f)$.

We denote by $f_2(v_1, v_2)$ the joint density of two particles with respective velocities v_1 and v_2. It follows (due to the assumption that the collisions are binary) that we must take into account only two distinct phenomena which modify the number density of particles with velocity v.

First, because of a possible collision with a particle of velocity v_*, a particle which had v for velocity will end up with a velocity v' (its partner in the collision will end up with velocity v'_*).

Secondly, some particle with a velocity w will encounter a particle with velocity w_* and will end up with a velocity v after the collision (its partner in the collision will end up with velocity w'_*).

We now denote by $p(v_1, v_2 \to v_3, v_4)$ the (density of) probability that for two particles sitting at the same point x at a given time t, a collision occurs and transforms the ingoing velocities v_1 and v_2 in the outgoing velocities v_3 and v_4(we shall see that in the so-called non cutoff case, this quantity is in fact far from being a probability density, since it is not integrable).

We see that $Q(f)$ is the sum of two terms $-Q^-(f)$ and $Q^+(f)$ which respectively correspond to the two phenomena described above. According to their definition $Q^-(f)$ and $Q^+(f)$ write

$$Q^-(f)(v) = \int_{v_*}\int_{v'}\int_{v'_*} f_2(v, v_*)p(v, v_* \to v', v'_*)dv'_*dv'dv_*, \tag{8.8}$$

$$Q^+(f)(v) = \int_{w}\int_{w_*}\int_{w'_*} f_2(w, w_*)p(w, w_* \to v, w'_*)dw'_*dw_*dw. \tag{8.9}$$

From the assumption (4), it follows that

$$\forall\ v_1, v_2, v_3, v_4,\ p(v_1, v_2 \to v_3, v_4) = p(v_3, v_4 \to v_1, v_2).$$

Moreover, using the assumption (5), since v and v_* are irrelative variables, w and w_* are irrelative variables; we have that $f_2(v, v_*) = f(v)f(v_*)$ and $f_2(w, w_*) = f(w)f(w_*)$. Let $Q(f) = Q(f, f) = Q^+(f, f) - Q^-(f, f)$, where

$$Q^-(f, f)(v) = \int_{v_*}\int_{v'}\int_{v'_*} f(v)f(v_*)p(v, v_* \to v', v'_*)dv'_*dv'dv_*, \tag{8.10}$$

$$Q^+(f, f)(v) = \int_{v_*}\int_{v'}\int_{v'_*} f(v')f(v'_*)p(v', v'_* \to v, v_*)dv'_*dv'dv_*. \tag{8.11}$$

We have

$$Q(f, f)(v) = \int_{v_*}\int_{v'}\int_{v'_*} \Big(f(v')f(v'_*) - f(v)f(v_*)\Big)p(v, v_* \to v', v'_*)dv'_*dv'dv_*. \tag{8.12}$$

Let us explain (8.12) a little bit. This operator can formally be split, in a self-evident way, into a gain and a loss term, $Q(f;f) = Q^+(f;f) - Q^-(f;f)$: The loss term *counts* all collisions in which a given particle of velocity v will encounter another particle, of velocity v_*. As a result of such a collision, this particle will in general change its velocity, and this will make less particles with velocity v. On the other hand, each time particles collide with respective velocities v' and v'_*, then the v' particle may acquire v as new velocity after the collision, and this will make more particles with velocity v : this is the meaning of the gain term.

Next, we only need to give the expression of $p(v, v_* \to v', v'_*)$. Using the conservation of momentum and kinetic energy in a collision, we have

$$\begin{cases} v + v_* = v' + v'_*, \\ |v|^2 + |v_*|^2 = |v'|^2 + |v'_*|^2. \end{cases} \tag{8.13}$$

Since this is a system of $n+1$ scalar equations for $2n$ scalar unknowns, it is natural to expect that its solutions can be defined in terms of $n-1$ parameters. Here is a convenient representation of all these solutions, which we shall sometimes call the σ-representation:

$$v' = \frac{v + v_*}{2} + \frac{|v - v_*|}{2}\sigma, \quad v'_* = \frac{v + v_*}{2} - \frac{|v - v_*|}{2}\sigma. \tag{8.14}$$

Here the parameter $\sigma \in S^{n-1}$ varies in the $n-1$ unit sphere. One can see the following figure

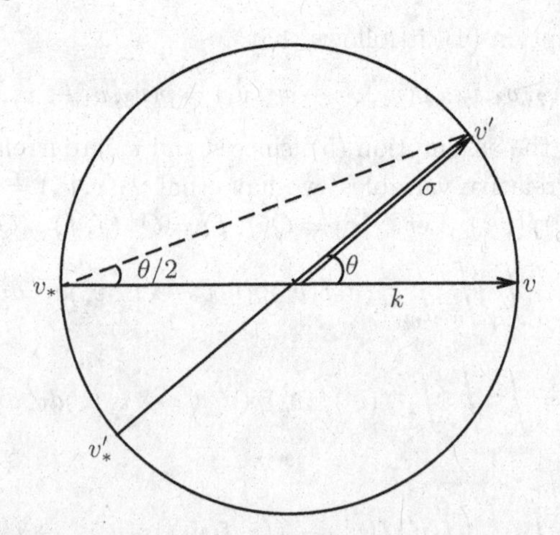

Fig. 8.1 The geometry of collisions

The Galilean invariance which holds in the context of binary collisions entails that the measure $p(v, v_* \to v', v'_*)$ can only depend on $|v - v_*|$ and $\frac{v-v_*}{|v-v_*|} \cdot \sigma$, $p(v, v_* \to v', v'_*)$ can be written by $B(\cdot, \cdot)$. We now can write down the *final* form of Boltzmann's collision operator :

$$Q(f,f) = \int_{\mathbb{R}^n} \int_{S^{n-1}} B(|v - v_*|, \frac{v - v_*}{|v - v_*|} \cdot \sigma)\{f(v'_*)f(v') - f(v_*)f(v)\}d\sigma dv_*, \tag{8.15}$$

where function B is called the Boltzmann collision cross section (or the collision kernel). For convenience, we shall also use the bilinear form $Q(g, f)$ related to the quadratic form $Q(f, f)$ and defined by

$$Q(g,f) = \int_{\mathbb{R}^n} \int_{S^{n-1}} B(|v - v_*|, \sigma)\{g(v'_*)f(v') - g(v_*)f(v)\}d\sigma dv_*. \tag{8.16}$$

For simplicity, we sometimes use standard abbreviations: $g(v'_*)f(v') = g'_*f'$ and $g(v_*)f(v) = g_*f$. Finally, we write down the standard form of the Boltzmann equation

$$\partial_t f + v \cdot \nabla_x f = Q(f, f)(v), \tag{8.17}$$

where $Q(f, f)$ is given by (8.15). The spatially homogeneous Boltzmann equation describes the behavior of a dilute gas, in which the velocity distribution of particles is assumed to be independent of the position; it reads

$$\partial_t f = Q(f, f)(v), \tag{8.18}$$

where the unknown $f = f(t, v)$ is assumed to be nonnegative, and stands for the density of particles at time t with velocity v. These equations are called homogeneous since f is assumed to be independent of the position. This model is simple, in this chapter we only consider (8.18).

8.1.3 *Cross section*

In most cases, the collision kernel B cannot be expressed explicitly. However, to capture its main properties, it is usually assumed to have the form:

$$B(|v - v_*|, \sigma) = \Phi(|v - v_*|)b(\cos\theta), \quad \cos\theta = \langle \frac{v - v_*}{|v - v_*|}, \sigma \rangle, \ 0 \leqslant \theta \leqslant \frac{\pi}{2}, \tag{8.19}$$

where the deviation angle θ is the angle between pre- and post-collisional velocities. Using n-dimensional spherical coordinates, we have

$$\int_{S^{n-1}} B(|v - v_*|, \sigma)d\sigma = |S^{n-2}| \int_0^{\frac{\pi}{2}} B(|v - v_*|, \cos\theta) \sin^{n-2}\theta d\theta,$$

where $|S^{n-2}|$ denotes the surface area of unit sphere in n dimension.

The collision kernel should be computed in terms of the interaction potential ϕ. Here, we consider the important potential: inverse-power law potentials. It is possible to (almost) explicitly compute the cross section if potential $\phi(r)$ is inverse-power law potentials, that is, the inter particle force is proportional to r^{-s} (with r denoting the interparticle distance and $s > 2$). In such a case (and in dimension 3) B writes

$$B(|v - v_*|, \sigma) = |v - v_*|^{\frac{s-5}{s-1}} b(\cos\theta) \tag{8.20}$$

with $b > 0$ a smooth function except at point 1 and satisfying:

$$\int_0^{\frac{\pi}{2}} \sin^n \theta b(\cos\theta) d\theta < \infty \text{ and } \sin^{n-2}\theta b(\cos\theta) \approx \frac{K}{\theta^{\frac{s+1}{s-1}}} \tag{8.21}$$

when $\theta \to 0$, $K > 0$. Since $\frac{s+1}{s-1} > 1$, b is not integrable at $\theta = 0$. This means

$$\int_{S^{n-1}} b(k\cdot\sigma) d\sigma = |S^{n-2}| \int_0^{\frac{\pi}{2}} \sin^{n-2}\theta b(\cos\theta) d\theta = \infty.$$

Because of the difficulties entailed by this singularity, Grad has proposed to introduce an angular cutoff near $\theta = 0$. It means that we replace B in (8.20) and (8.21) by a new cross section cross

$$\tilde{B}(|v - v_*|, \sigma) = |v - v_*|^{\frac{s-5}{s-1}} \tilde{b}(\cos\theta), \tag{8.22}$$

where $\tilde{b}$ is smooth, or at least such that $\sin^{n-2}\theta\tilde{b}(\cos\theta)$ is integrable near $\theta = 0$. In our model case, this means

$$\int_{S^{n-1}} \tilde{b}(k\cdot\sigma) d\sigma = |S^{n-2}| \int_0^{\frac{\pi}{2}} \sin^{n-2}\theta\tilde{b}(\cos\theta) d\theta < \infty.$$

In the sequel,we shall speak of cutoff cross sections (cutoff potentials or Grad's angular cut-off) when B is locally integrable, and of non cutoff cross sections (or non cutoff potentials) when B has a singularity like in (8.21).

In this chapter, we mainly consider the Boltzmann equation without cutoff cross section. Note that the separation $Q(f,f) = Q^+(f,f) - Q^-(f,f)$ make sense if B is cutoff cross section, where

$$Q^+(f,f) = \int_{\mathbb{R}^n}\int_{S^{n-1}} B(|v - v_*|, \frac{v - v_*}{|v - v_*|}\cdot\sigma) f(v'_*) f(v') d\sigma dv_*, \tag{8.23}$$

$$Q^-(f,f) = \int_{\mathbb{R}^n}\int_{S^{n-1}} B(|v - v_*|, \frac{v - v_*}{|v - v_*|}\cdot\sigma) f(v_*) f(v) d\sigma dv_*. \tag{8.24}$$

If B is non cutoff cross section, for convenience, we sometimes need to calculate the gain and loss terms $Q^\pm$ in the collision operator separately. This separation is not allowed for the non-cutoff cross section but is legitimate as an intermediate step in the calculation.

In this chapter, we mainly consider the inverse-power law potentials, the cross section B have the form:

$$B(|v-v_*|,\sigma)=|v-v_*|^\gamma b(\cos\theta). \tag{8.25}$$

If $\gamma>0$, it is called hard potentials; in general, we consider the case of $0<\gamma\leqslant 2$.

If $\gamma=0$, it is called Maxwellian potentials.

If $-n<\gamma<0$, it is called soft potentials; for $-2<\gamma<0$, it is called moderately soft potentials, for $\gamma\leqslant -2$ it is called very soft potentials.

Note that if $s=2$ then $\phi(r)$ becomes Coulomb potential, which leads to Fokker-Planck-Landau equation, we do not consider this model in this book.

8.2 Basic surgery tools for the Boltzmann operator

In this section, we give some basic tools which one often needs for a fine study of the Boltzmann operator. In details, one can refer to [59; 229].

1.Symmetrization of the collision kernel. In view of formulas (8.14), the quantity $f'_*f'-f_*f$ is invariant under the change of variables $\sigma\to -\sigma$. From the physical point of view this reflects the undiscernability of particles. Thus one can replace (from the very beginning, if necessary) B by its symmetrized version

$$\bar{B}(z,\sigma)=\{B(z,\sigma)+B(z,-\sigma)\}I_{z\cdot\sigma>0}.$$

This is the reason why we assume that angle θ varies from 0 to $\frac{\pi}{2}$ in (8.19). In fact, we get rid of collisions with deflexion angle larger than $\frac{\pi}{2}$ by this symmetrization trick. B in (8.19) is indeed $\bar{B}$ in the above equality.

2. Pre-postcollisional change of variables. A universal tool in the Boltzmann theory is the involutive change of variables with unit Jacobian $(v,v_*,\sigma)\to(v',v'_*,k)$, where $k=\frac{v-v_*}{|v-v_*|}$ is the unit vector along $v-v_*$. Indeed, denote the mapping $\Phi:(v,v_*,\sigma)\to(v',v'_*,k)$, then it's easy to verify that $\Phi^2=I$. Thus the Jacobian of Φ is 1. Then for any function f we have

$$\begin{aligned}&\int_{\mathbb{R}^n}\int_{\mathbb{R}^n}\int_{S^{n-1}}f(v,v_*,v',v'_*,\sigma)d\sigma dv_*dv\\&=\int_{\mathbb{R}^n}\int_{\mathbb{R}^n}\int_{S^{n-1}}f(v',v'_*,v,v_*,\frac{v-v_*}{|v-v_*|})d\sigma dv_*dv.\end{aligned} \tag{8.26}$$

3. The change of variables $(v, v_*, \sigma) \to (v_*, v, \sigma)$ is clearly involutive and has unit Jacobian. For function $f = f(v, v_*, v', v'_*, \sigma)$, one has

$$\begin{aligned}&\int_{\mathbb{R}^n}\int_{\mathbb{R}^n}\int_{S^{n-1}} f(v, v_*, v', v'_*, \sigma) d\sigma dv_* dv \\ &= \int_{\mathbb{R}^n}\int_{\mathbb{R}^n}\int_{S^{n-1}} f(v_*, v, v', v'_*, \sigma) d\sigma dv_* dv.\end{aligned} \tag{8.27}$$

Next, we give some applications of the two change of variables above. We first give the conservation laws of the Boltzmann equation. As a consequence, if $\varphi(v)$ is an arbitrary continuous function of the velocity v, then we get the following various weak formulations for Boltzmann's kernel Q

$$\begin{aligned}&\int_{\mathbb{R}^n} Q(g, f)\varphi(v) dv \\ &= \int_{\mathbb{R}^{2n}}\int_{S^{n-1}} B(|v - v_*|, \frac{v - v_*}{|v - v_*|}\cdot\sigma)\{g(v'_*)f(v') - g(v_*)f(v)\}\varphi(v) d\sigma dv_* dv \\ &= \int_{\mathbb{R}^{2n}}\int_{S^{n-1}} B(|v - v_*|, \frac{v - v_*}{|v - v_*|}\cdot\sigma) g(v_*)f(v)(\varphi(v') - \varphi(v)) d\sigma dv_* dv.\end{aligned} \tag{8.28}$$

In the special case $g = f$, we have that

$$\begin{aligned}&\int_{\mathbb{R}^n} Q(f, f)\varphi(v) dv \\ &= \frac{1}{2}\int_{\mathbb{R}^n\times\mathbb{R}^n}\int_{S^{n-1}} Bf(v_*)f(v)(\varphi(v'_*) + \varphi(v') - \varphi(v_*) - \varphi(v)) d\sigma dv_* dv.\end{aligned}$$

From the mathematical point of view, it is interesting because expressions like (8.28) may be well-defined in situations where $Q(f, f)$ is not. From the physical point of view, it expresses the change in the integral $\int_{\mathbb{R}^n} Q(f, f)(v) dv$ which is due to the action of collisions. Let f be a solution of the Boltzmann equation (8.17), set in the whole space $\mathbb{R}^n$ to simplify. By the conservative properties of the transport operator $v \cdot \nabla_x$,

$$\frac{d}{dt}\int_{\mathbb{R}^n} f(t, x, v)\varphi(v) dx dv = \int_{\mathbb{R}^n} Q(f, f)\varphi(v) dx dv \tag{8.29}$$

and the right-hand side is just the x-integral of any one of the expressions in formulas (8.28). As an immediate consequence, whenever φ satisfies the functional equation

$$\forall (v, v_*, \sigma) \in \mathbb{R}^n \times \mathbb{R}^n \times S^{n-1},\ \varphi(v') + \varphi(v'_*) = \varphi(v) + \varphi(v_*), \tag{8.30}$$

then, at least formally,

$$\frac{d}{dt}\int_{\mathbb{R}^n} f(t, x, v)\varphi(v) dx dv = 0 \tag{8.31}$$

along solutions of the Boltzmann equation. The words at least formally of course mean that the preceding equations must be rigorously justified with the help of some integrability estimates on the solutions to the Boltzmann equation.

Pluging $\varphi(v) = 1, v_1, ..., v_n, |v|^2/2$, in formula (8.28) and (8.31), by (8.30), we get the conservation of mass, momentum and energy at the level of the Boltzmann operator:

$$\int_{\mathbb{R}^n} Q(f,f)(v)dv = 0, \int_{\mathbb{R}^n} Q(f,f)(v)v_j dv = 0, \int_{\mathbb{R}^n} Q(f,f)(v)\frac{|v|^2}{2}dv = 0, \tag{8.32}$$

or,

$$\frac{d}{dt}\int_{\mathbb{R}^n} f(t,x,v)dxdv = \frac{d}{dt}\int_{\mathbb{R}^n} f(t,x,v)v_j dv = \frac{d}{dt}\int_{\mathbb{R}^n} f(t,x,v))\frac{|v|^2}{2}dv = 0. \tag{8.33}$$

Next, we give the Boltzmann's H theorem. Without caring about integrability issues, we plug $\varphi(v) = \log f(v)$ into (8.28),

$$\int_{\mathbb{R}^n} Q(f,f)(v)\log f(v)dv = -D(f),$$

where D is the entropy dissipation functional. Using the fact that $(F - G)(\log F - \log G) \geqslant 0$, we have

$$D(f) = \frac{1}{4}\int_{\mathbb{R}^n\times\mathbb{R}^n}\int_{S^{n-1}} B(|v-v_*|, \frac{v-v_*}{|v-v_*|}\cdot\sigma)\Big(f(v'_*)f(v') - f(v_*)f(v)\Big)$$
$$\times \log\frac{(f(v'_*)f(v')}{f(v_*)f(v)} d\sigma dv_* dv \geqslant 0. \tag{8.34}$$

Now, we introduce Boltzmann's H functional,

$$H(f) = \int_{\mathbb{R}^n_x\times\mathbb{R}^n_v} f\log f dxdv.$$

Of course, the transport operator $-v\cdot\nabla_x$ does not contribute in any change of the H functional in time. As a consequence, if $f = f(t,x,v)$ is a solution of the Boltzmann equation, then $H(f)$ will evolve in time because of the effects of the collision operator:

$$\frac{d}{dt}H(f(t,\cdot,\cdot)) = -\int_{\mathbb{R}^n_x} D(f(t,\cdot,\cdot))dx \leqslant 0. \tag{8.35}$$

This is the first part of Boltzmann's H-theorem. Next we assume that $B(|v-v_*|,\sigma) > 0$ for almost all (v, v_*, σ), which is always the case in applications of interest. Then equality in Boltzmann's H theorem occurs if and only if for almost all x, v, v_*, σ

$$f(v'_*)f(v') = f(v_*)f(v). \tag{8.36}$$

Then, it is possible to prove (under suitable, but rather weak assumptions on B and f) that

$$D(f) = 0 \Longleftrightarrow Q(f, f) = 0 \Longleftrightarrow f = M(v), \tag{8.37}$$

$$M(v) = M_{\rho,u,T}(v) = \frac{\rho}{(2\pi T)^{n/2}} e^{-\frac{|v-u|^2}{2T}}, \tag{8.38}$$

where M is called the Maxwellian and is known to describe the velocity distribution of a gas in an equilibrium state with the mass density $\rho(x)$, bulk velocity $u(x)$ and temperature T. Here, (ρ, u, T) are taken to be parameters. If (ρ, u, T) are constants,then M is called a global (absolute) Maxwellian ; if they are functions of (x, t), then it is called a local Maxwellian. Evidently, the global Maxwellian is a stationary solution Boltzmann equation (8.17) without force F. This is the second part of Boltzmann's H-theorem.

4. Bobylev's identities . We now turn to more intricate tools introduced by Bobylev, who first made Fourier transform an extremely powerful tool in the study of the Boltzmann operator with Maxwellian collision kernel. Now we give the Fourier transform of $Q(g, f)$ in (8.19). We first perform the calculation of the Fourier transform of the gain term in a general Boltzmann collision operator $Q^+(g, f)$.

For any test function $\varphi(v)$, using the pre-postcollisional change of variables in (8.26), we have

$$\begin{aligned} &\int_{\mathbb{R}^n} Q^+(g, f)\varphi(v)dv \\ &= \int_{\mathbb{R}^n\times\mathbb{R}^n}\int_{S^{n-1}} B(|v - v_*|, \frac{v - v_*}{|v - v_*|}\cdot\sigma)g(v_*)f(v)\varphi(v')d\sigma dv_* dv. \end{aligned} \tag{8.39}$$

Plugging $\varphi(v) = e^{-iv\xi}$ in (8.39), we have

$$\begin{aligned} &\mathcal{F}(Q^+(g, f))(\xi) \\ &= \int_{\mathbb{R}^{2n}}\int_{S^{n-1}} B(|v - v_*|, \frac{v - v_*}{|v - v_*|}\cdot\sigma)g(v_*)f(v)e^{-i\frac{v+v_*}{2}\cdot\xi}e^{-i\frac{|v-v_*|}{2}\sigma\cdot\xi}d\sigma dv_* dv. \end{aligned} \tag{8.40}$$

Using the following general equality:

$$\int_{S^{n-1}} F(k\cdot\sigma, l\cdot\sigma)d\sigma = \int_{S^{n-1}} F(l\cdot\sigma, k\cdot\sigma)d\sigma, \ |l| = |k| = 1, \tag{8.41}$$

(due to the existence of an isometry on S^{n-1} exchanging l and k) we have

$$\begin{aligned}&\int_{S^{n-1}} B(|v-v_*|, \frac{v-v_*}{|v-v_*|}\cdot\sigma)e^{-i\frac{|v-v_*|}{2}\sigma\cdot\xi}d\sigma \\ &= \int_{S^{n-1}} B(|v-v_*|, \frac{\xi}{|\xi|}\cdot\sigma)e^{-i\frac{|\xi|}{2}\sigma\cdot(v-v_*)}d\sigma.\end{aligned} \tag{8.42}$$

Then, using (8.42) and the Fourier inversion formula, we have

$$\begin{aligned}&\mathcal{F}(Q^+(g,f))(\xi) \\ &= \int_{\mathbb{R}^{2n}}\int_{S^{n-1}} B(|v-v_*|, \frac{\xi}{|\xi|}\cdot\sigma)g(v_*)f(v)e^{-i\frac{v+v_*}{2}\cdot\xi}e^{-i\frac{|\xi|}{2}\sigma\cdot(v-v_*)}d\sigma dv_* dv \\ &= \int_{\mathbb{R}^{2n}}\int_{S^{n-1}} B(|v-v_*|, \frac{\xi}{|\xi|}\cdot\sigma)g(v_*)f(v)e^{-iv\xi^+}e^{-iv_*\xi^-}d\sigma dv_* dv \\ &= \int\Big(\int_{\mathbb{R}^{2n}} B(|v-v_*|, \frac{\xi}{|\xi|}\cdot\sigma) \\ &\qquad \times \hat{g}(\eta_*)\hat{f}(\eta)e^{iv\eta}e^{iv_*\eta_*}e^{-iv\xi^+}e^{-iv_*\xi^-}d\eta_* d\eta\Big)d\sigma dv_* dv \\ &= \int_{\mathbb{R}^{2n}\times S^{n-1}} \hat{g}(\eta_*)\hat{f}(\eta) \\ &\qquad \times\Big(\int B(|v-v_*|, \frac{\xi}{|\xi|}\cdot\sigma)e^{iv(\eta-\xi^+)}e^{iv_*(\eta_*-\xi^-)}dv_* dv\Big)d\sigma d\eta_* d\eta,\end{aligned} \tag{8.43}$$

where

$$\xi^+ = \frac{\xi+|\xi|\sigma}{2}, \quad \xi^- = \frac{\xi-|\xi|\sigma}{2}. \tag{8.44}$$

Let δ be the Dirac measure, $\hat{B}(|\xi|,\cos\theta) = \int_{\mathbb{R}^n} B(|q|,\cos\theta)e^{-iq\cdot\xi}$ denotes the Fourier transform of B in the relative velocity variable. By the change of variables $q = v - v_*$,

$$\begin{aligned}&\int_{\mathbb{R}^{2n}} B(|v-v_*|, \frac{\xi}{|\xi|}\cdot\sigma)e^{iv(\eta-\xi^+)}e^{iv_*(\eta_*-\xi^-)}dv_* dv \\ &= \int_{\mathbb{R}^{2n}} B(|q|, \frac{\xi}{|\xi|}\cdot\sigma)e^{iv(\eta_*+\eta-\xi)}e^{-iq(\eta_*-\xi^-)}dq dv \\ &= \hat{B}(|\eta_*-\xi^-|, \frac{\xi}{|\xi|}\cdot\sigma)\delta(\eta=\xi-\eta_*).\end{aligned} \tag{8.45}$$

By (8.45), we give the Fourier transform of $Q^+(g,f)$

$$\mathcal{F}(Q^+(g,f))(\xi) = \int_{\mathbb{R}^n\times S^{n-1}} \hat{g}(\eta_*)\hat{f}(\xi-\eta_*)\hat{B}(|\eta_*-\xi^-|, \frac{\xi}{|\xi|}\cdot\sigma))d\sigma d\eta_*. \tag{8.46}$$

Let $\xi_* = \eta_* - \xi^-$, we find

$$\mathcal{F}(Q^+(g,f))(\xi) = \int_{\mathbb{R}^n \times S^{n-1}} \hat{g}(\xi^- + \xi_*)\hat{f}(\xi^+ - \xi_*)\hat{B}(|\xi_*|, \frac{\xi}{|\xi|} \cdot \sigma))d\sigma d\xi_*. \tag{8.47}$$

In Maxwellian potentials (Maxwellian molecules) case, that is, $B(|z|, \cos\theta) = b(\cos\theta)$, we have

$$\hat{B}(|\xi_*|, \cos\theta) = \delta(\xi_* = 0)b(\cos\theta). \tag{8.48}$$

and as a consequence

$$\mathcal{F}(Q^+(g,f))(\xi) = \int_{S^{n-1}} \hat{g}(\xi^-)\hat{f}(\xi^+)b(\frac{\xi}{|\xi|} \cdot \sigma)d\sigma. \tag{8.49}$$

For the Fourier transform of $Q^-(g,f)$, if $B(|z|, \cos\theta) = b(\cos\theta)$, then we easily have

$$\mathcal{F}(Q^-(g,f))(\xi) = \int_{S^{n-1}} \hat{g}(0)\hat{f}(\xi)b(\frac{\xi}{|\xi|} \cdot \sigma)d\sigma, \tag{8.50}$$

which follows from the fact that $\int_{S^{n-1}} b(k \cdot \sigma)$ does not depend on k.

8.3 Properties of Boltzmann collision operator without cutoff

In this section, we will consider the properties of the Boltzmann collision operator $Q(g,f)$ without cutoff. We prove that for a given distribution function $g \in L^1$, the Boltzmann operator $Q(g,f)$ behaves essentially as a fractional power of the Laplacian:

$$Q(g,f) = -C_g(-\Delta)^{\nu/2} f + \text{more regular terms}.$$

In details, one can refer to [2]. In the following discussion, we also adopt the notations for the weighted function spaces,

$$\|f\|_{L^p_r} = \|f(v)\langle v\rangle^r\|_{L^p}, \qquad 1 \leqslant p \leqslant \infty, \quad r \in \mathbb{R},$$

$$\|f\|_{L\log L} = \int_{\mathbb{R}^n} f\log(1+f)dv.$$

We first give the weak formulation of $Q(g,f)$. For $f \in H^{\frac{n}{2}+1}(\mathbb{R}^n)$, we have

$$\begin{aligned}
&\Big(Q(g,f), f\Big)_{L^2} \\
&= \int_{\mathbb{R}^{2n} \times S^{n-1}} B(|v - v_*|, \frac{v - v_*}{|v - v_*|} \cdot \sigma)g(v_*)f(v)(f(v') - f(v))d\sigma dv_* dv \\
&= \frac{1}{2}\int_{\mathbb{R}^{2n} \times S^{n-1}} B(|v - v_*|, \frac{v - v_*}{|v - v_*|} \cdot \sigma)g(v_*)(f^2(v') - f^2(v))d\sigma dv_* dv \\
&\quad - \frac{1}{2}\int_{\mathbb{R}^{2n} \times S^{n-1}} B(|v - v_*|, \frac{v - v_*}{|v - v_*|} \cdot \sigma)g(v_*)(f(v') - f(v))^2 d\sigma dv_* dv.
\end{aligned} \tag{8.51}$$

For the first term in (8.51), we have the following lemma.

Lemma 8.1. (Cancellation) *For a.e.* $v_* \in \mathbb{R}^n$,

$$\int_{\mathbb{R}^n \times S^{n-1}} B(|v-v_*|, k\cdot\sigma)(f^2(v') - f^2(v))d\sigma dv = (f^2 * A)(v_*), \quad (8.52)$$

where

$$k = \frac{v-v_*}{|v-v_*|}, \quad (8.53)$$

$$A(z) = S^{n-2}\int_0^{\frac{\pi}{2}} \sin^{n-2}\theta\Big(\frac{1}{\cos^n\frac{\theta}{2}}B(\frac{|z|}{\cos\frac{\theta}{2}}, \cos\theta) - B(|z|, \cos\theta)\Big)d\theta. \quad (8.54)$$

Proof. We do the calculation as if B were integrable and apply a limiting procedure to conclude in the general case, that is, we choose a sequence integrable function $\tilde{B}_j$ in S^{n-1}, such that $\lim_{j\to\infty}\tilde{B}_j = B$. In fact the right-hand side of (8.52) should be taken as a definition of the left-hand side. Recall that

$$v' = \frac{v+v_*}{2} + \frac{|v-v_*|}{2}\sigma = v_* + \frac{|v-v_*|}{2}(k + |k|\sigma).$$

For each σ and with v_* still fixed, we perform the change of variables $v \to v'$. This change of variables is well-defined on the set $\cos\theta > 0$, and it follows either by a direct calculation or by using the cylindrical symmetry of this transformation that its Jacobian determinant is

$$\left|\frac{dv'}{dv}\right| = \left|\frac{1}{2}I + \frac{1}{2}k\otimes\sigma\right| = \frac{1}{2^n}(1 + k\cdot\sigma) = \frac{(k'\cdot\sigma)}{2^{n-1}}, \quad (8.55)$$

where $k' = \frac{v'-v_*}{|v'-v_*|}$. Then

$$k'\cdot\sigma = \cos\frac{\theta}{2} \geqslant \frac{1}{\sqrt{2}}.$$

Define the inverse transformation $\psi_\sigma : v' \to \psi_\sigma(v') = v$, from the figure 8.1, it follows that $|v_* - \psi_\sigma(v')| = \frac{|v'-v_*|}{k'\cdot\sigma}$. Then changing the name v' for v,

$$|v_* - \psi_\sigma(v)| = \frac{|v-v_*|}{k\cdot\sigma}. \quad (8.56)$$

Applying this change of variable to the first part in the left-hand side of (8.52), then changing the name v' for v,

$$\int_{\mathbb{R}^n\times S^{n-1}} B(|v-v_*|, k\cdot\sigma)g(v_*)f^2(v')d\sigma dv$$

$$= \int_{\mathbb{R}^n \times S^{n-1}} B(|v - v_*|, 2(k' \cdot \sigma)^2 - 1) g(v_*) f^2(v') |\frac{dv}{dv'}| d\sigma dv'$$

$$= \int_{k' \cdot \sigma \geqslant \frac{1}{\sqrt{2}}} B(|\psi_\sigma(v') - v_*|, 2(k' \cdot \sigma)^2 - 1) g(v_*) f^2(v') \frac{2^{n-1}}{(k' \cdot \sigma)^2} d\sigma dv'$$

$$= \int_{k \cdot \sigma \geqslant \frac{1}{\sqrt{2}}} B(|\psi_\sigma(v) - v_*|, 2(k \cdot \sigma)^2 - 1) g(v_*) f^2(v) \frac{2^{n-1}}{(k \cdot \sigma)^2} d\sigma dv. \quad (8.57)$$

Using the fact that $2^{n-2} \sin^{n-2} \frac{\theta}{2} \cos^{n-2} \frac{\theta}{2} = \sin^{n-2} \theta$ and n-dimensional spherical coordinates, we have

$$\int_{k \cdot \sigma \geqslant \frac{1}{\sqrt{2}}} B(|\psi_\sigma(v) - v_*|, 2(k \cdot \sigma)^2 - 1) \frac{2^{n-1}}{(k \cdot \sigma)^2} d\sigma$$

$$= \int_{k \cdot \sigma \geqslant \frac{1}{\sqrt{2}}} B(\frac{|v - v_*|}{k \cdot \sigma}, 2(k \cdot \sigma)^2 - 1) \frac{2^{n-1}}{(k \cdot \sigma)^2} d\sigma$$

$$= |S^{n-2}| \int_0^{\frac{\pi}{4}} \sin^{n-2} \theta B(\frac{|v - v_*|}{\cos \theta}, \cos 2\theta) \frac{2^{n-1}}{\cos^2 \theta} d\theta$$

$$= |S^{n-2}| \int_0^{\frac{\pi}{2}} \sin^{n-2} \frac{\theta}{2} B(\frac{|v - v_*|}{\cos \frac{\theta}{2}}, \cos \theta) \frac{2^{n-2}}{\cos^2 \frac{\theta}{2}} d\theta$$

$$= |S^{n-2}| \int_0^{\frac{\pi}{2}} \frac{\sin^{n-2} \theta}{\cos^n \frac{\theta}{2}} B(\frac{|v - v_*|}{\cos \frac{\theta}{2}}, \cos \theta) d\theta. \quad (8.58)$$

Define

$$A(v - v_*)$$
$$= S^{n-2} \int_0^{\frac{\pi}{2}} \sin^{n-2} \theta \Big(\frac{1}{\cos^n \frac{\theta}{2}} B(\frac{|v - v_*|}{\cos \frac{\theta}{2}}, \cos \theta) - B(|v - v_*|, \cos \theta) \Big) d\theta. \quad (8.59)$$

This completes the proof. □

Remark 8.1. If $B(|v - v_*|, k \cdot \sigma) = |v - v_*|^\gamma b(k \cdot \sigma)$, then

$$A(v - v_*)$$
$$= S^{n-2} \int_0^{\frac{\pi}{2}} \sin^{n-2} \theta \Big(\frac{1}{\cos^n \frac{\theta}{2}} (\frac{|v - v_*|}{\cos \frac{\theta}{2}})^\gamma b(\cos \theta) - |v - v_*|^\gamma b(\cos \theta) \Big) d\theta$$

$$= S^{n-2} \int_0^{\frac{\pi}{2}} \sin^{n-2} \theta |v - v_*|^\gamma b(\cos \theta) (\frac{1}{\cos^{\gamma+n} \frac{\theta}{2}} - 1) d\theta. \quad (8.60)$$

Note that if $\gamma > -n$, then (8.60) is nonnegative; if $\gamma = -n$, it is Coulomb potentials, they require a more careful analysis. Using Lemma 8.1, we find that if $0 \leqslant \gamma \leqslant 2$

$$\int_{\mathbb{R}^{2n}\times S^{n-1}} B(|v-v_*|, k\cdot\sigma)g(v_*)(f^2(v')-f^2(v))d\sigma dv dv_* \\ \leqslant C\|g\|_{L^1_2}\|f\|_{L^2_1}, \tag{8.61}$$

which follows from the fact that $1-\cos^n\frac{\theta}{2} \leqslant n(1-\cos\frac{\theta}{2}) = 2n\sin\frac{\theta}{4}$.

For the second term in (8.51), for simplicity, we only consider Maxwellian molecules case, that is, cross section $B(|v-v_*|, k\cdot\sigma) = b(k\cdot\sigma)$.

Lemma 8.2. *Assume $g \in L^1(\mathbb{R}^n)$ and $f \in L^2(\mathbb{R}^n)$, the following Plancherel-type identity holds*

$$\begin{aligned}&\int_{\mathbb{R}^{2n}\times S^{n-1}} b(\frac{v-v_*}{|v-v_*|}\cdot\sigma)g(v_*)(f(v')-f(v))^2 d\sigma dv_* dv \\ &= \int_{\mathbb{R}^n\times S^{n-1}} b(\frac{\xi}{|\xi|}\cdot\sigma)\Big(\hat{g}(0)|\hat{f}(\xi)|^2 + \hat{g}(0)|\hat{f}(\xi^+)|^2 \\ &\qquad - \hat{g}(\xi^-)\hat{f}(\xi^+)\overline{\hat{f}(\xi)} - \overline{\hat{g}(\xi^-)\hat{f}(\xi^+)}\hat{f}(\xi)\Big)d\xi d\sigma.\end{aligned} \tag{8.62}$$

Proof. We shall do the proof only in the case when b is integrable and the result will follow by monotonicity. First, we have

$$(f(v')-f(v))^2 = f^2(v') - 2f(v')f(v) + f^2(v). \tag{8.63}$$

We begin with the middle term $2f(v')f(v)$ in (8.63). By the pre-postcollisional change of variables and Parseval's identity and Bobylev's identity, we have

$$\begin{aligned}&\int b(k\cdot\sigma)g(v_*)f(v')f(v)d\sigma dv_* dv = \int Q^+(g,f)f dv \\ &= \frac{1}{2}(Q^+(g,f),f)_{L^2} + \frac{1}{2}(f,Q^+(g,f))_{L^2} \\ &= \frac{1}{2}(\mathcal{F}Q^+(g,f),\mathcal{F}f)_{L^2} + \frac{1}{2}(\mathcal{F}f,\mathcal{F}Q^+(g,f))_{L^2} \\ &= \frac{1}{2}\int b(\frac{\xi}{|\xi|}\cdot\sigma)\Big(\hat{g}(\xi^-)\hat{f}(\xi^+)\overline{\hat{f}(\xi)} + \overline{\hat{g}(\xi^-)\hat{f}(\xi^+)}\hat{f}(\xi)\Big)d\xi d\sigma.\end{aligned} \tag{8.64}$$

For the third term $f^2(v)$ in (8.63), using the fact that $\int_{S^{n-1}} b(k\cdot\sigma)d\sigma$ does not depend on the unit vector k, we have

$$\begin{aligned}&\int b(k\cdot\sigma)g(v_*)f^2(v)d\sigma dv_* dv = \int_{S^{n-1}} b(k\cdot\sigma)d\sigma\int_{\mathbb{R}^n} g(v_*)dv_*\int_{\mathbb{R}^n} f^2(v)dv \\ &= \int b(\frac{\xi}{|\xi|}\cdot\sigma)\hat{g}(0)|\hat{f}(\xi)|^2 d\xi d\sigma.\end{aligned} \tag{8.65}$$

For the first term $f^2(v')$ in (8.63), we first make the change of variables $(v - v_*, v_*) \to (v_1, v_*)$, change the name v_1 for v, and then use the change of variables $v \to v'$ as in Lemma 8.1 to obtain

$$\begin{aligned}
&\int b(\frac{v-v_*}{|v-v_*|}\cdot\sigma)g(v_*)f^2(\frac{v+v_*}{2}+\frac{|v-v_*|}{2}\sigma)d\sigma dv_* dv \\
&= \int b(\frac{v_1}{|v_1|}\cdot\sigma)g(v_*)|\tau_{-v_*}f(\frac{v_1+|v_1|\sigma}{2})|^2 d\sigma dv_* dv_1 \\
&= \int b(\frac{v}{|v|}\cdot\sigma)g(v_*)|\tau_{-v_*}f(\frac{v+|v|\sigma}{2})|^2 d\sigma dv_* dv \\
&= \int b(\psi(v',\sigma)g(v_*)\frac{2^{n-1}}{(\frac{v'}{|v'|}\cdot\sigma)^2}|\tau_{-v_*}f(v')|^2 d\sigma dv_* dv',
\end{aligned} \tag{8.66}$$

where

$$\psi(v',\sigma) = 2(\frac{v'}{|v'|}\cdot\sigma)^2 - 1, \quad \tau_{-v_*}f = f(v_*+).$$

Using the fact that $\int_{S^{n-1}} b(k\cdot\sigma)$ does not depend on k and $|\mathcal{F}(\tau_h f)| = |\mathcal{F}(f)|$, and reversing the change of variables $\xi \to \xi^+$ as in Lemma 8.1, we have

$$\begin{aligned}
(8.66) &= \int g(v_*)\Big(\int b(\psi(\xi,\sigma)\frac{2^{n-1}}{(\frac{\xi}{|\xi|}\cdot\sigma)^2}|f(\xi)|^2 d\sigma d\xi\Big)dv_* \\
&= \hat{g}(0)\int b(\frac{\xi}{|\xi|}\cdot\sigma)|f(\xi^+)|^2 d\sigma d\xi.
\end{aligned} \tag{8.67}$$

Collecting (8.63), (8.64), (8.65) and (8.67), we can obtain (8.62). □

Corollary 8.1. *Assume $g \in L^1(\mathbb{R}^n)$, $f \in L^2(\mathbb{R}^n)$ and $f \geqslant 0$. Then*

$$\begin{aligned}
&\int_{\mathbb{R}^{2n}\times S^{n-1}} b(\frac{v-v_*}{|v-v_*|}\cdot\sigma)g(v_*)(f(v')-f(v))^2 d\sigma dv_* dv \\
&\geqslant \int_{\mathbb{R}^n}|\hat{f}(\xi)|^2\int_{S^{n-1}} b(\frac{\xi}{|\xi|}\cdot\sigma)(\hat{g}(0)-|\hat{g}(\xi^-)|)d\sigma d\xi.
\end{aligned} \tag{8.68}$$

Proof. Using Lemma 8.2 and the following inequality,

$$|\hat{f}(\xi^+)|^2 + |\hat{f}(\xi)|^2 \geqslant |\hat{f}(\xi)|^2,$$

we can obtain Corollary 8.1. □

Lemma 8.3. *Assume that b satisfies (8.21), Then there exists a positive constant C_g depending on n, $\|g\|_{L^1_1}$, $\|g\|_{L\log L}$ and b, such that for $|\xi| \geqslant 1$*

$$\int_{S^{n-1}} b(\frac{\xi}{|\xi|}\cdot\sigma)(\hat{g}(0)-|\hat{g}(\xi^-)|)d\sigma \geqslant C_g|\xi|^\nu. \tag{8.69}$$

Lemma 8.3 is a consequence of the two lemmas below, next we will use C_g to denote a positive constant depending on n, $\|g\|_{L^1_1}$, $\|g\|_{L\log L}$ and b.

Lemma 8.4. *There exists a positive constant $\tilde{C}_g$ depending on n, $\|g\|_{L^1_1}$ and $\|g\|_{L\log L}$, such that for all $\xi \in \mathbb{R}$*

$$\hat{g}(0) - \hat{g}(\xi) \geqslant \tilde{C}_g(|\xi|^2 \wedge 1). \tag{8.70}$$

Proof. For some $\theta \in \mathbb{R}$, we have

$$\begin{aligned}
&\hat{g}(0) - \hat{g}(\xi) \\
&= \int_{\mathbb{R}^n} g(v)(1 - \cos(v \cdot \xi + \theta))dv \\
&= 2\int_{\mathbb{R}^n} g(v)\sin^2(\frac{v\cdot\xi+\theta}{2})dv \\
&\geqslant 2\sin^2\varepsilon \int_{|v|\leqslant r, \forall p\in\mathbb{Z}, |v\cdot\xi+\theta-2p\pi|\geqslant 2\varepsilon} g(v)dv \qquad (8.71)\\
&\geqslant 2\sin^2\varepsilon\Big\{\|g\|_{L^1} - \frac{\|g\|_{L^1_1}}{r} - \int_{|v|\leqslant r, \forall p\in\mathbb{Z}, |v\cdot\xi+\theta-2p\pi|\leqslant 2\varepsilon} g(v)dv\Big\} \\
&\geqslant 2\sin^2\varepsilon\Big\{\|g\|_{L^1} - \frac{\|g\|_{L^1_1}}{r} - \sup_{|A|\leqslant \frac{4\varepsilon}{|\xi|}(2r)^{n-1}(1+\frac{r|\xi|}{\pi})} \int_A g(v)dv\Big\}.
\end{aligned}$$

If $|\xi| \geqslant 1$, Lemma 8.4 holds with the following constant

$$\tilde{C}_g = 2\sin^2\varepsilon\Big\{\|g\|_{L^1} - \frac{\|g\|_{L^1_1}}{r} - \sup_{|A|\leqslant 4\varepsilon(2r)^{n-1}+\frac{2\varepsilon}{\pi}(2r)^n} \int_A g(v)dv\Big\}, \tag{8.72}$$

$\varepsilon > 0$ and $r > 0$ being chosen in such a way that this quantity is positive.

If $|\xi| \leqslant 1$, let $\delta = \frac{\varepsilon}{|\xi|}$ in (8.71), Lemma 8.4 holds with the following constant

$$C_g = 2\delta^2 \inf_{|\xi|\leqslant 1} \frac{\sin^2(\delta|\xi|)}{\delta^2|\xi|^2}\Big\{\|g\|_{L^1} - \frac{\|g\|_{L^1_1}}{r} - \sup_{|A|\leqslant 4\varepsilon(2r)^{n-1}(1+\frac{r}{\pi})} \int_A g(v)dv\Big\}, \tag{8.73}$$

where $\delta > 0$ and $r > 0$ are chosen in such a way that this quantity is positive. □

Lemma 8.5. *If b satisfies*

$$\sin^{n-2}\theta b(\cos\theta) \approx \frac{K(\nu)}{\theta^{1+\nu}} \quad as \ \theta \to 0, \ K(\nu) > 0; \tag{8.74}$$

then for $|\xi| \geqslant 1$,

$$\int_{S^{n-1}} b(\frac{\xi}{|\xi|}\cdot\sigma)(|\xi^-|^2 \wedge 1)d\sigma \geqslant K(\nu)|\xi|^\nu. \tag{8.75}$$

Proof. Recall that

$$|\xi^-|^2 = \frac{|\xi|^2}{2}(1 - \frac{\xi}{|\xi|} \cdot \sigma).$$

Passing to n-dimensional spherical coordinates, we find for some $\theta_0 > 0$,

$$\begin{aligned}
&\int_{S^{n-1}} b(\frac{\xi}{|\xi|} \cdot \sigma)(|\xi^-|^2 \wedge 1)d\sigma \\
&= S^{n-2} \int_0^{\frac{\pi}{2}} \sin^{n-2}\theta b(\cos\theta)\Big(\frac{|\xi|^2}{2}(1-\cos\theta) \wedge 1\Big)d\theta \qquad (8.76)\\
&\geqslant \frac{K}{2}|S^{n-2}| \int_0^{\theta_0} (\frac{|\xi|^2\theta^2}{2} \wedge 1)\frac{d\theta}{\theta^{1+\nu}}.
\end{aligned}$$

By the change of variables $\theta \to |\xi|\theta$, the integral in (8.76) is also

$$|\xi|^\nu \int_0^{\theta_0} (\frac{\theta^2}{2} \wedge 1)\frac{d\theta}{\theta^{1+\nu}} \tag{8.77}$$

so that when $|\xi| \geqslant 1$, Lemma 8.5 holds with

$$K(v) = \frac{K}{2}|S^{n-2}| \int_0^{\theta_0} (\frac{\theta^2}{2} \wedge 1)\frac{d\theta}{\theta^{1+\nu}}.$$

□

By Lemma 8.2, Corollary8.1 and Lemma 8.3, for the second term in (8.51), we have

$$\int_{\mathbb{R}^{2n}\times S^{n-1}} b(\frac{v-v_*}{|v-v_*|} \cdot \sigma)g(v_*)(f(v') - f(v))^2 d\sigma dv_* dv \geqslant C_g\|f\|^2_{H^{\nu/2}}. \tag{8.78}$$

8.4 Regularity of solutions for spatially homogeneous case

In this section, we will give the application of the results of Section 3 in this chapter; and prove the regularity of solutions for the spatially homogeneous Boltzmann equation without angular cutoff

$$\begin{cases} \partial_t f(t,v) = Q(f,f)(t,v), \quad t \geqslant 0, v \in \mathbb{R}^n, \\ \qquad f(0,v) = f_0(v), \end{cases} \tag{8.79}$$

where cross section B

$$B(|v-v_*|, \frac{v-v_*}{|v-v_*|} \cdot \sigma) = |v-v_*|^\gamma b(\frac{v-v_*}{|v-v_*|} \cdot \sigma),$$

$$\sin^{n-2}\theta b(\cos\theta) \approx \frac{K(\nu)}{\theta^{1+\nu}} \quad \text{when} \quad \theta \to 0, \ K(\nu) > 0.$$

From (8.33), it follows that the solutions $f(t,v)$ of the Cauchy problem (8.79) have the following conservation of mass, momentum and energy,

$$\int_{\mathbb{R}^n} f(t,v)dv = \int_{\mathbb{R}^n} f_0(v)dv, \tag{8.80}$$

$$\int_{\mathbb{R}^n} f(t,v)v_j dv = \int_{\mathbb{R}^n} f_0(v)v_j dv, \ j = 1, 2...n, \tag{8.81}$$

$$\int_{\mathbb{R}^n} f(t,v)\frac{|v|^2}{2}dv = \int_{\mathbb{R}^n} f_0(v)\frac{|v|^2}{2}dv. \tag{8.82}$$

For the existence of weak solutions for Cauchy problem (8.79), in 1998 Villani [230] showed that if the initial data have the finite mass, energy and entropy,

$$\int_{\mathbb{R}^n} f_0(v)[1 + |v|^2 + \log(1 + f_0(v))]dv < +\infty, \tag{8.83}$$

then he constructed a weak solution $f(t,v)$ of the Cauchy problem (8.79), which satisfies the following:

$$f(t,v) \geqslant 0, \ f(t,v) \in C(\mathbb{R}^+, \mathcal{S}'); \ t \geqslant 0, \ f(t,v) \in L^1_2 \cap L\log L, \tag{8.84}$$

$$f(t,v) \in L^1([0,T]; L^1_{2+\gamma}), \tag{8.85}$$

$$f(0,v) = f_0(v), \tag{8.86}$$

$$\int_{\mathbb{R}^n} f(t,v)\psi(v)dv = \int_{\mathbb{R}^n} f_0(v)\psi(v)dv, \ \psi(v) = 1, v_1, ..., v_n, \frac{|v|^2}{2}, \tag{8.87}$$

$$\int_{\mathbb{R}^n} f(t,v)\log f(t,v)dv \leqslant \int_{\mathbb{R}^n} f_0(v)\log f_0(v)dv, \tag{8.88}$$

$$\begin{aligned} &\int_{\mathbb{R}^n} f(t,v)\varphi(t,v)dv - \int_{\mathbb{R}^n} f_0(v)\varphi(0,v)dv - \int_0^t d\tau \int_{\mathbb{R}^n} f(\tau,v)\partial_\tau\varphi(\tau,v)dv \\ &= \int_0^t d\tau \int_{\mathbb{R}^n} Q(f,f)(\tau,v)\varphi(\tau,v)dv, \ \ \forall \varphi(t,v) \in C^1(\mathbb{R}^+; C_0^\infty(\mathbb{R}^n)), \end{aligned} \tag{8.89}$$

where the last integral in the right-hand side being defined by the following formulae

$$\begin{aligned} &\int_{\mathbb{R}^n} Q(f,f)(\tau,v)\varphi(\tau,v)dv \\ &= \frac{1}{2}\int_{\mathbb{R}^{2n}}\int_{S^{n-1}} Bf(v_*)f(v)(\varphi(v_*') + \varphi(v') - \varphi(v_*) - \varphi(v))dvdv_*d\sigma. \end{aligned}$$

These make sense provided that f satisfies (8.84) and test function $\varphi \in L^\infty([0,T]; W^{2,\infty})$.

In this section, we assume that a weak solution to the Boltzmann equation (8.79) has already been constructed and that it satisfies the usual entropic estimates, to say,

$$\int_{\mathbb{R}^n} f(t,v)[1+|v|^2+\log(1+f(t,v))]dv < +\infty, \tag{8.90}$$

we are interested in regularity issues associated to such solutions. That is, is this weak solution more regular than the initial datum, and if so, can we have estimates on this regularity?

In this section, we will show that the weak solutions above constructed $f(t,v)$ of the Cauchy problem (8.79) with Maxwellian molecules case belongs to $H_v^{+\infty}$. For hard potentials case, there exist some results only in *modified* hard potentials cases now, in details, one can refer to [4; 60; 111]. For soft potentials cases, there seems no any results now.

In this section, we will introduce two methods of multiplier to consider it. That is, we first choose a team of suitable multiplier, then mainly give the sharp estimates of the commutators of the collision operator $Q(f,f)$ and pseudo-differential operators composed by multipliers.

First, we give the method of the Littlewood-Paley decomposition, in details, one can refer to [3]. The Littlewood-Paley decomposition is defined as in Section 3 in Chapter 1: Assume that $\operatorname{supp}\varphi_k \subset \{\xi : 2^{k-1} \leqslant |\xi| \leqslant 2^{k+1}\}$, $k \in \mathbb{N}$, $\operatorname{supp}\varphi_0 \subset \{\xi : |\xi| \leqslant 2\}$, $\sum_{k=0}^\infty \varphi_k = 1$; we can assume φ_k is a radial function. Recall that for every multi-index α, there exists a positive number C_α such that

$$2^{k|\alpha|}|D^\alpha \varphi_k(\xi)| \leqslant C_\alpha, \quad k = 0,1,2...; \ \xi \in \mathbb{R}^n. \tag{8.91}$$

Then the Littlewood-Paley projection operator is defined by

$$\widehat{\triangle_k f}(\xi) = \varphi_k(\xi)\hat{f}(\xi), \quad k = 0,1,2...; \ \xi \in \mathbb{R}^n.$$

Lemma 8.6. *Assume that the initial data $f_0(v)$ satisfies* (8.83), *cross section $B(|v-v_*|, k\cdot\sigma) = b(k\cdot\sigma)$ with b satisfying the following*

$$\sin^{n-2}\theta b(\cos\theta) \approx \frac{K}{\theta^{1+\nu}} \quad \textit{when} \ \ \theta \to 0, \ K > 0. \tag{8.92}$$

Let $f(t,v)$ be any weak non negative solution of the Cauchy problem (8.79), *satisfying* (8.80)-(8.82) *and* (8.90). *Then*

$$(\triangle_k Q(f,f), \triangle_k f)_{L^2} - (Q(f,\triangle_k f), \triangle_k f)_{L^2} \leqslant C\|f_0\|_{L^1}\Big(\sum_{j=k-2}^{k+1} \|\triangle_k f\|_{L^2}\Big), \tag{8.93}$$

$$(\triangle_k f, \triangle_k Q(f,f))_{L^2} - (\triangle_k f, Q(f, \triangle_k f))_{L^2} \leqslant C\|f_0\|_{L^1}(\sum_{j=k-2}^{k+1} \|\triangle_k f\|_{L^2}). \tag{8.94}$$

Remark 8.2. Let $f(t,v)$ be a weak solution of the Boltzmann equation. If we choose $\triangle_k f$ as test function ($\triangle_k f \in L^\infty([0,T]; W^{2,\infty})$), then by the definition of weak solutions, the inner products in the left side of (8.93) and (8.94) make sense. Unless $f \in H^\infty(\mathbb{R}^n)$, $f(t,v)$ cannot be chosen as a test function in the definition of weak solutions. This is also one of the reasons why we use the method of multiplier.

Proof. We only prove (8.93), the proof of (8.94) is similar with one of (8.93). Applying Fourier transform on Boltzmann equation (8.79), using (8.49) and (8.50) (Bobylev identity), we have

$$\partial_t \hat{f}(\xi) = \widehat{Q(f,f)}(\xi) = \int_{S^{n-1}} b(\frac{\xi}{|\xi|}\cdot\sigma)\Big(\hat{f}(\xi^-)\hat{f}(\xi^+) - \hat{f}(0)\hat{f}(\xi)\Big)d\sigma. \tag{8.95}$$

Multiplying (8.95) by $\varphi_k(\xi)$, we have

$$\begin{aligned}\partial_t \widehat{\triangle_k f}(\xi) &= \triangle_k \widehat{Q(f,f)}(\xi)\\ &= \int_{S^{n-1}} b(\frac{\xi}{|\xi|}\cdot\sigma)\Big(\hat{f}(\xi^-)\hat{f}(\xi^+)\varphi_k(\xi) - \hat{f}(0)\hat{f}(\xi)\varphi_k(\xi)\Big)d\sigma.\end{aligned} \tag{8.96}$$

Then,

$$\begin{aligned}&\partial_t(\widehat{\triangle_k f}, \widehat{\triangle_k f})_{L^2} = \int_{\mathbb{R}^n} (\partial_t \widehat{\triangle_k f})\overline{\widehat{\triangle_k f}} d\xi + \int_{\mathbb{R}^n} \widehat{\triangle_k f}\partial_t \overline{\widehat{\triangle_k f}} d\xi\\ &= (\triangle_k \widehat{Q(f,f)}, \widehat{\triangle_k f})_{L^2} + (\widehat{\triangle_k f}, \triangle_k \widehat{Q(f,f)})_{L^2}\\ &= \int_{\mathbb{R}^n}\int_{S^{n-1}} b(\frac{\xi}{|\xi|}\cdot\sigma)\Big(\hat{f}(\xi^-)\hat{f}(\xi^+)\varphi_k^2(\xi)\overline{\hat{f}}(\xi) - \hat{f}(0)\hat{f}(\xi)\varphi_k^2(\xi)\overline{\hat{f}}(\xi)\Big)d\sigma d\xi\\ &+ \int_{\mathbb{R}^n}\int_{S^{n-1}} b(\frac{\xi}{|\xi|}\cdot\sigma)\Big(\overline{\hat{f}}(\xi^-)\overline{\hat{f}}(\xi^+)\varphi_k^2(\xi)\hat{f}(\xi) - \overline{\hat{f}}(0)\overline{\hat{f}}(\xi)\varphi_k^2(\xi)\hat{f}(\xi)\Big)d\sigma d\xi.\end{aligned} \tag{8.97}$$

Moreover, we have

$$\widehat{Q(f, \triangle_k f)}(\xi) = \int_{S^{n-1}} b(\frac{\xi}{|\xi|}\cdot\sigma)\Big(\hat{f}(\xi^-)\varphi_k(\xi^+)\hat{f}(\xi^+) - \hat{f}(0)\varphi_k(\xi)\hat{f}(\xi)\Big)d\sigma, \tag{8.98}$$

$$
\begin{aligned}
&(\widehat{Q(f,\triangle_k f)},\widehat{\triangle_k f})_{L^2}+(\widehat{\triangle_k f},\widehat{Q(f,\triangle_k f)})_{L^2}\\
&=\iint b(\frac{\xi}{|\xi|}\cdot\sigma)\Big(\hat f(\xi^-)\varphi_k(\xi^+)\hat f(\xi^+)\varphi_k(\xi)\overline{\hat f}(\xi)-\hat f(0)\hat f(\xi)\varphi_k^2(\xi)\overline{\hat f}(\xi)\Big)d\sigma d\xi\\
&+\iint b(\frac{\xi}{|\xi|}\cdot\sigma)\Big(\overline{\hat f}(\xi^-)\varphi_k(\xi^+)\overline{\hat f}(\xi^+)\varphi_k(\xi)\hat f(\xi)-\overline{\hat f}(0)\overline{\hat f}(\xi)\varphi_k^2(\xi)\hat f(\xi)\Big)d\sigma d\xi.
\end{aligned}
\tag{8.99}
$$

Thus, from (8.97) and (8.99) it follows that

$$
\begin{aligned}
&(\widehat{\triangle_k Q(f,f)},\widehat{\triangle_k f})_{L^2}-(\widehat{Q(f,\triangle_k f)},\widehat{\triangle_k f})_{L^2}\\
&=\int_{\mathbb{R}^n}\int_{S^{n-1}}b(\frac{\xi}{|\xi|}\cdot\sigma)\hat f(\xi^-)\hat f(\xi^+)(\varphi_k(\xi)-\varphi_k(\xi^+))\varphi_k(\xi)\overline{\hat f}(\xi)d\sigma d\xi.
\end{aligned}
\tag{8.100}
$$

From the definitions of ξ^+ and ξ^-, it follows that

$$
\xi^+=\frac{\xi+|\xi|\sigma}{2},\ |\xi^+|=|\xi|\cos\frac{\theta}{2},\ \frac{\xi}{|\xi|}\cdot\sigma=\cos\theta,0\leqslant\theta\leqslant\frac{\pi}{2},
$$

$$
\frac{|\xi|^2}{2}\leqslant|\xi^+|^2\leqslant|\xi|^2,|\xi|^2-|\xi^+|^2=|\xi^-|^2=|\xi|^2\sin^2\frac{\theta}{2}.
$$

From the integral in (8.100), we note that $2^{k-1}\leqslant|\xi|\leqslant 2^{k+1}$. From the definition of ξ^+, it follows that $2^{k-2}\leqslant|\xi^+|\leqslant 2^{k+1}$. Thus

$$
\hat f(\xi^+)=\widehat{\triangle_{k-2}f}(\xi^+)+\widehat{\triangle_{k-1}f}(\xi^+)+\widehat{\triangle_k f}(\xi^+)+\widehat{\triangle_{k+1}f}(\xi^+).
\tag{8.101}
$$

Plugging this expression of $f(\xi^+)$ in (8.100), and by the fact that $|\hat f(\xi^-)|\leqslant\|f\|_{L^1}\leqslant\|f_0\|_{L^1}$, (8.100) is bounded by

$$
\begin{aligned}
&\|f_0\|_{L^1}\int_{2^{k-1}\leqslant|\xi|\leqslant 2^{k+1}}\int_{S^{n-1}}b(\frac{\xi}{|\xi|}\cdot\sigma)|\widehat{\triangle_k f}(\xi^+)||A_\xi^k||\widehat{\triangle_k f}(\xi)|d\sigma d\xi\\
&+\|f_0\|_{L^1}\int_{2^{k-1}\leqslant|\xi|\leqslant 2^{k+1}}\int_{S^{n-1}}b(\frac{\xi}{|\xi|}\cdot\sigma)|\widehat{\triangle_{k-2}f}(\xi^+)||A_\xi^k||\widehat{\triangle_k f}(\xi)|d\sigma d\xi\\
&+\|f_0\|_{L^1}\int_{2^{k-1}\leqslant|\xi|\leqslant 2^{k+1}}\int_{S^{n-1}}b(\frac{\xi}{|\xi|}\cdot\sigma)|\widehat{\triangle_{k-1}f}(\xi^+)||A_\xi^k||\widehat{\triangle_k f}(\xi)|d\sigma d\xi\\
&+\|f_0\|_{L^1}\int_{2^{k-1}\leqslant|\xi|\leqslant 2^{k+1}}\int_{S^{n-1}}b(\frac{\xi}{|\xi|}\cdot\sigma)|\widehat{\triangle_{k+1}f}(\xi^+)||A_\xi^k||\widehat{\triangle_k f}(\xi)|d\sigma d\xi,
\end{aligned}
\tag{8.102}
$$

where $A_\xi^k=\varphi_k(\xi)-\varphi_k(\xi^+)$. By (8.91), we have

$$
\begin{aligned}
|A_\xi^k|&=|\varphi_k(\xi)-\varphi_k(\xi^+)|\leqslant|\tilde\varphi_k(|\xi|)-\tilde\varphi_k(|\xi^+|)|\\
&\lesssim\frac{1}{2^k}(|\xi|-|\xi^+|)\lesssim\frac{1}{2^k}\frac{|\xi|^2-|\xi^+|^2}{|\xi|+|\xi^+|}\lesssim\sin^2\frac{\theta}{2}.
\end{aligned}
\tag{8.103}
$$

We only estimate the first term in (8.102), the estimates of the other terms in (8.102) can be obtained similarly. By (8.103) and Hölder inequality, (8.102) is bounded by

$$\begin{aligned}
&\|f_0\|_{L^1}\int_{2^{k-1}\leqslant|\xi|\leqslant 2^{k+1}}\int_{S^{n-1}} b(\frac{\xi}{|\xi|}\cdot\sigma)|\widehat{\triangle_k f}(\xi^+)||A_\xi^k||\widehat{\triangle_k f}(\xi)|d\sigma d\xi\\
&\lesssim \|f_0\|_{L^1}\int_{\mathbb{R}^n}|\widehat{\triangle_k f}(\xi)|\int_{S^{n-1}} b(\frac{\xi}{|\xi|}\cdot\sigma)\sin^2\frac{\theta}{2}|\widehat{\triangle_k f}(\xi^+)|d\sigma d\xi\\
&\lesssim \|f_0\|_{L^1}\Big(\int|\widehat{\triangle_k f}(\xi)|^2 d\xi\Big)^{1/2}\Big\{\int\Big(\int b(\frac{\xi}{|\xi|}\cdot\sigma)\sin^2\frac{\theta}{2}|\widehat{\triangle_k f}(\xi^+)|d\sigma\Big)^2 d\xi\Big\}^{1/2}\\
&\lesssim \|f_0\|_{L^1}\|\triangle_k f\|_{L^2}^2.
\end{aligned}\tag{8.104}$$

which follows from the change of variables $\xi\to\xi^+$ as in Lemma 8.1. □

Theorem 8.1. *Under the hypothesis of Lemma 8.6, Let $f(t,v)$ be any weak non negative solution of the Cauchy problem* (8.79), *satisfying* (8.90). *Then for any $s\in\mathbb{R}^+$ and $t>0$, one has $f(t,v)\in H^s(\mathbb{R}^n)$.*

Proof. First, we have

$$\partial_t\|\triangle_k f\|_{L^2}=(\widehat{\triangle_k Q(f,f)},\widehat{\triangle_k f})_{L^2}+(\widehat{\triangle_k f},\widehat{\triangle_k Q(f,f)})_{L^2}.\tag{8.105}$$

From (8.51) it follows that

$$\begin{aligned}
&\Big(Q(f,\triangle_k f),\triangle_k f\Big)_{L^2}\\
&=\int_{\mathbb{R}^{2n}\times S^{n-1}} b(\frac{v-v_*}{|v-v_*|}\cdot\sigma)f(v_*)\triangle_k f(v)(\triangle_k f(v')-\triangle_k f(v))d\sigma dv_* dv\\
&=\frac{1}{2}\int_{\mathbb{R}^{2n}\times S^{n-1}} b(\frac{v-v_*}{|v-v_*|}\cdot\sigma)f(v_*)((\triangle_k f)^2(v')-(\triangle_k f)^2(v))d\sigma dv_* dv\\
&-\frac{1}{2}\int_{\mathbb{R}^{2n}\times S^{n-1}} b(\frac{v-v_*}{|v-v_*|}\cdot\sigma)f(v_*)(\triangle_k f(v')-\triangle_k f(v))^2 d\sigma dv_* dv\\
&=I_1-I_2.
\end{aligned}\tag{8.106}$$

For the first term in (8.106), using Lemma 8.1 in this chapter, we get

$$\begin{aligned}
I_1&=\int_{\mathbb{R}^{2n}\times S^{n-1}} b(k\cdot\sigma)((\triangle_k f)^2(v')-(\triangle_k f)^2(v))d\sigma dv dv_*\\
&=A\int_{\mathbb{R}^n}(\triangle_k f)^2(v_*)dv_*,
\end{aligned}\tag{8.107}$$

where

$$A=S^{n-2}\int_0^{\frac{\pi}{2}}\sin^{n-2}\theta\Big(\frac{1}{\cos^n\frac{\theta}{2}}-1\Big)b(\cos\theta)d\theta.\tag{8.108}$$

From $1-\cos^n\frac{\theta}{2}\leqslant n(1-\cos\frac{\theta}{2})=2n\sin^2\frac{\theta}{4}$ it follows that

$$I_1 \lesssim \|\triangle_k f\|_{L^2}^2. \tag{8.109}$$

For the second term in (8.106), using Corollary 8.1 and Lemma 8.3 in this chapter, we have

$$\begin{aligned} I_2 &= \frac{1}{2}\int_{\mathbb{R}^{2n}\times S^{n-1}} b(\frac{v-v_*}{|v-v_*|}\cdot\sigma)f(v_*)(\triangle_k f(v')-\triangle_k f(v))^2 d\sigma dv_* dv \\ &\geqslant C_{f_0}\int_{\mathbb{R}^n}(1+|\xi|^\nu)\varphi_k(\xi)\hat{f}(\xi)d\xi \geqslant C_{f_0}\|\triangle_k f\|_{H^{\frac{\nu}{2}}}^2. \end{aligned} \tag{8.110}$$

Thus collecting (8.105), (8.106), (8.109), (8.110) and Lemma 8.6, we get

$$\partial_t\|\triangle_k f\|_{L^2}+C_{f_0}\|\triangle_k f\|_{H^{\frac{\nu}{2}}}^2 \leqslant C\|f_0\|_{L^1}(\sum_{j=k-2}^{k+1}\|\triangle_j f\|_{L^2}^2). \tag{8.111}$$

Dividing(8.111) by 2^{kn}, we have

$$\begin{aligned} &\frac{\partial_t\|\triangle_k f\|_{L^2}}{2^{kn}}+(C_{f_0}2^{k\nu}-C\|f_0\|_{L^1})\frac{\|\triangle_k f\|_{L^2}^2}{2^{kn}} \\ &\leqslant C\|f_0\|_{L^1}\Big(\frac{\|\triangle_{k-2}f\|_{L^2}^2}{2^{(k-2)n}}+\frac{\|\triangle_{k-1}f\|_{L^2}^2}{2^{(k-1)n}}+\frac{\|\triangle_{k+1}f\|_{L^2}^2}{2^{(k+1)n}}\Big). \end{aligned} \tag{8.112}$$

Let $U_k(t)=\frac{\|\triangle_k f\|_{L^2}^2}{2^{kn}}$, $C_k=C_{f_0}2^{k\nu}-C\|f_0\|_{L^1}$ and $\beta=C\|f_0\|_{L^1}$. Then (8.112) can be rewritten by:

$$\partial_t U_k(t)+C_k U_k(t)\leqslant\beta(U_{k-2}(t)+U_{k-1}(t)+U_{k+1}(t)). \tag{8.113}$$

From Bernstein's inequality (Polyya Plancherel Nikoolski inequality), it follows that

$$\|\triangle_k f\|_{L^2}\leqslant C(\varphi)2^{nk}\|\triangle_k f\|_{L^1}\leqslant C(\varphi)2^{nk}\|f\|_{L^1}\leqslant C(\varphi)2^{nk}\|f_0\|_{L^1}.$$

For $k\geqslant 0, t\geqslant 0$,

$$U_k(t)\leqslant M,\quad\text{with}\quad M=C(\varphi)\|f_0\|_{L^1}. \tag{8.114}$$

By the definition of C_k, we show that C_k is nondecreasing, $C_k\geqslant 0$ with large enough k (without loss of generality, we can assume that $C_k\geqslant 0$ with $k\geqslant k_0\geqslant 3$). From (8.113) and (8.114), it follows that $U_k(t)$ satisfy the conditions of the following Lemma 8.7. Then for all integer $p\geqslant 1$, there exist constants A_p and D_p such that

$$U_k(t)\leqslant MA_p e^{-C_{k-2(p-1)}t}+MD_p\frac{1}{(C_{k-2(p-1)})^p}, t\geqslant 0 \tag{8.115}$$

for all $k \geqslant 2(p-1)+k_0+2$. Thus, for fixed $s>0$, $t>0$ and an integer $p \geqslant 1$ which we shall choose just below; using the definition of Sobolev spaces, we have

$$\begin{aligned}\|f\|_{H^s}^2 &\sim \sum_{k=0}^{\infty} 2^{2ks}\|\Delta_k f\|_{L^2}^2 \\ &= \sum_{k=0}^{2(p-1)+k_0+1} 2^{2ks}\|\Delta_k f\|_{L^2}^2 + \sum_{k=2(p-1)+k_0+2}^{\infty} 2^{2ks}\|\Delta_k f\|_{L^2}^2 \quad (8.116)\\ &= \sum_{k=0}^{2(p-1)+k_0+1} 2^{2ks}\|\Delta_k f\|_{L^2}^2 + \sum_{k=2(p-1)+k_0+2}^{\infty} 2^{k(2s+n)}U_k.\end{aligned}$$

In order to prove $f \in H^s$, it remains to show that the last series appearing above is a convergent one. From (8.115) it follows that

$$\begin{aligned}&\sum_{k=2(p-1)+k_0+2}^{\infty} 2^{k(2s+n)}U_k \\ &\leqslant \sum_{k=2(p-1)+k_0+2}^{\infty} 2^{k(2s+n)} M A_p e^{-C_{k-2(p-1)}t} \quad (8.117)\\ &+ \sum_{k=2(p-1)+k_0+2}^{\infty} 2^{k(2s+n)} M D_p \frac{1}{(C_{k-2(p-1)})^p}.\end{aligned}$$

Using the definition of C_k, we have

$$C_{k-2(p-1)} \sim C(f_0,p)2^{k\nu p}, \text{ if } k \text{ is large enough.} \quad (8.118)$$

Choosing p such that $\nu p \geqslant 2s+n+1$ for instance, yields that the two series on the right hand side of the last inequality are convergent. □

Lemma 8.7. (Iteration) *Let β, M be two non negative numbers. Given positive integer k_0, assume that $\{C_k\}_{k\geqslant k_0}$ is a sequence of positive numbers, and is non-decreasing in k, and satisfies: there exists a positive α such that $C_{k+1}-C_k \geqslant \alpha$ for all $k \geqslant k_0$. $\{U_k\}_{k\geqslant k_0} = \{U_k(t)\}_{k\geqslant k_0}$ is another sequence with $t \in \mathbb{R}^+$ satisfying:*

$$\partial_t U_k(t) + C_k U_k(t) \leqslant \beta(U_{k-2}(t) + U_{k-1}(t) + U_{k+1}(t)),$$
$$0 \leqslant U_k(t) \leqslant M, \ \forall k \geqslant k_0+2, \forall t \geqslant 0.$$

Then for any integer $p \geqslant 1$, there exist constants A_p, D_p such that, for all $k \geqslant 2(p-1)+k_0+2$, one has

$$U_k(t) \leqslant M A_p e^{-C_{k-2(p-1)}t} + M D_p \frac{1}{(C_{k-2(p-1)})^p}, t \geqslant 0.$$

Proof. It is done by iteration. First start from the fact that $U_k \leqslant M$, and using this in the right hand side of the differential inequality, leads to the above conclusion, for $p = 1$. Then, start from the above first iteration, and replace again in the rhs of the above inequality, and just repeat the process. □

Next, we will introduce another method of multiplier, in details, one can refer to [171]. For convenience, we assume $n = 3$. Let $f(t, v)$ be a weak solution of the Cauchy problem (8.79). For any fixed $T_0 > 0$, it follows that $f(t, v) \in L^1(\mathbb{R}^3) \subset H^{-2}(\mathbb{R}^3)$, $t \in [0, T_0]$. For $t \in [0, T_0], N > 0$ and $0 < \delta < 1$, define multiplier:

$$M_\delta(t,\xi) = (1 + |\xi|^2)^{\frac{Nt-4}{2}}(1 + \delta|\xi|^2)^{-N_0}, \ N_0 = \frac{NT_0 + 4}{2}$$

and the corresponding pseudo-differential operator $M_\delta(t, D_v)$ is defined by

$$M_\delta(t, D_v) = \mathcal{F}^{-1}M_\delta(t,\xi)\mathcal{F}.$$

Then for any $\delta \in (0, 1)$,

$$M_\delta(t, D_v)^2 f \in L^\infty([0, T_0]; W^{2,\infty}(\mathbb{R}^3)), \ M_\delta(t, D_v)f \in C([0, T_0]; L^2(\mathbb{R}^3)).$$

Lemma 8.8. *Under the hypothesis of Lemma 8.6, Let $f(t, v)$ be any weak non negative solution of the Cauchy problem* (8.79), *satisfying* (8.90). *Then*

$$(Q(f,f), M_\delta^2 f)_{L^2} - (Q(f, M_\delta f), M_\delta f)_{L^2} \leqslant C_f\|f_0\|_{L^1}\|M_\delta^2 f\|_{L^2}, \quad (8.119)$$

where the constant C_f is independent of $0 < \delta < 1$.

Proof. Similarly with the proof of Lemma 8.6, we have

$$\begin{aligned}
&(Q(f,f), M_\delta^2 f)_{L^2} - (Q(f, M_\delta f), M_\delta f)_{L^2} \\
&= (\widehat{Q(f,f)}, \widehat{M_\delta^2 f})_{L^2} - (\widehat{Q(f, M_\delta f)}, \widehat{M_\delta f})_{L^2} \\
&= \int_{\mathbb{R}^n}\int_{S^{n-1}} b(\frac{\xi}{|\xi|}\cdot\sigma)\hat{f}(\xi^-)\hat{f}(\xi^+)(M_\delta(t,\xi) - M_\delta(t,\xi^+))M_\delta(t,\xi)\bar{\hat{f}}(\xi)d\sigma d\xi.
\end{aligned} \quad (8.120)$$

It suffices to prove

$$M_\delta(t,\xi) - M_\delta(t,\xi^+) \leqslant N_0 2^{\frac{NT_0+4}{2}} M_\delta(t,\xi^+)\sin^2\frac{\theta}{2}. \quad (8.121)$$

Define

$$\tilde{M}_\delta(t, s) = (1 + s)^{\frac{Nt-4}{2}}(1 + \delta s)^{-N_0}, \quad s = |\xi|^2, s^+ = |\xi^+|^2,$$

so that

$$M_\delta(t,\xi) = \tilde{M}_\delta(t, |\xi|^2).$$

From the mean value theorem, it follows that there exists $s^+ < \tilde{s} < s$ such that

$$\tilde{M}_\delta(t,s) - \tilde{M}_\delta(t,s^+) = \frac{\partial \tilde{M}_\delta}{\partial s}(t,\tilde{s})(s-s^+).$$

By the definition of $\tilde{M}_\delta$,

$$\frac{\partial \tilde{M}_\delta}{\partial s}(t,s) = \Big(\frac{Nt-4}{2(1+s)} - \frac{\delta N_0}{1+\delta s}\Big)\tilde{M}_\delta(t,s).$$

By the following

$$\frac{s}{1+s}, \frac{\delta s}{1+\delta s} \leqslant 1; \Big|\frac{\tilde{M}_\delta(t,\tilde{s})}{\tilde{M}_\delta(t,s^+)}\Big| \leqslant 2^{\frac{NT_0+4}{2}}, s - s^+ = s\sin^2\frac{\theta}{2},$$

we have

$$\tilde{M}_\delta(t,s) - \tilde{M}_\delta(t,s^+) \leqslant N_0 2^{\frac{NT_0+4}{2}} \tilde{M}_\delta(t,s^+)\sin^2\frac{\theta}{2}.$$

Then

$$\begin{aligned}
&\int_{\mathbb{R}^n}\int_{S^{n-1}} b(\frac{\xi}{|\xi|}\cdot\sigma)\hat{f}(\xi^-)\hat{f}(\xi^+)(M_\delta(t,\xi) - M_\delta(t,\xi^+))M_\delta(t,\xi)\overline{\hat{f}}(\xi)d\sigma d\xi \\
&\leqslant \int_{\mathbb{R}^n}\int_{S^{n-1}} b(\frac{\xi}{|\xi|}\cdot\sigma)\sin^2\frac{\theta}{2}|\hat{f}(\xi^-)||\hat{f}(\xi^+)|M_\delta(t,\xi^+)M_\delta(t,\xi)|\overline{\hat{f}}(\xi)|d\sigma d\xi \\
&\leqslant C\|f_0\|_{L^1}\|M_\delta f\|_{L^2}^2.
\end{aligned}$$

□

Thus using Lemma 8.8, we give another proof of Theorem 8.1. We note that any weak solution f has the following properties: $M_\delta^2 f \in L^\infty([0,T_0];W^{2,\infty}(\mathbb{R}^3))$ and $M_\delta f \in C([0,T_0];L^2(\mathbb{R}^3))$. From the definition of weak solution, we choose $\psi(t,v) = M_\delta^2 f(t,v)$ as test function. For any $t \in (0,T_0)$,

$$\begin{aligned}
\int_{\mathbb{R}^3} f(t,v)M_\delta^2 f(t,v)dv &- \int_{\mathbb{R}^3} f_0(v)M_\delta^2 f(0,v)dv \\
&- \int_0^t d\tau \int_{\mathbb{R}^3} f(\tau,v)\partial_\tau(M_\delta^2 f(\tau,v))dv \qquad (8.122)\\
&= \int_0^t d\tau \int_{\mathbb{R}^3} Q(f,f)(\tau,v)M_\delta^2 f(\tau,v)dv.
\end{aligned}$$

Moreover, we have

$$\begin{aligned}
&\int_0^t d\tau \int_{\mathbb{R}^3} f(\tau,v)\partial_\tau(M_\delta^2 f(\tau,v))dv \\
&= \lim_{h\to 0}\int_0^t d\tau \int_{\mathbb{R}^3} (f(\tau,v) + f(\tau+h,v))\frac{M_\delta^2 f(\tau+h,v) - M_\delta^2 f(\tau,v)}{2h}dv
\end{aligned}$$

$$
\begin{aligned}
&= \lim_{h\to 0}\int_0^t d\tau \int_{\mathbb{R}^3} \\
&\qquad \Big\{\Big(M_\delta(\tau+h)f(\tau,v)+M_\delta(\tau+h)f(\tau+h,v)\Big)\frac{M_\delta f(\tau+h,v)}{2h} \\
&\qquad\qquad -\Big(M_\delta(\tau)f(\tau,v)+M_\delta(\tau)f(\tau+h,v)\Big)\frac{M_\delta f(\tau,v)}{2h}\Big\}dv \\
&= \lim_{h\to 0}\int_0^t d\tau \int_{\mathbb{R}^3} \frac{(M_\delta f)^2(\tau+h,v)-(M_\delta f)^2(\tau,v)}{2h}dv \\
&+\lim_{h\to 0}\int_0^t d\tau \int_{\mathbb{R}^3} \frac{1}{2h}\Big\{M_\delta(\tau+h)(f(\tau,v))M_\delta(\tau+h)(f(\tau+h,v)) \\
&\qquad\qquad - M_\delta(\tau)(f(\tau,v))M_\delta(\tau)(f(\tau+h,v))\Big\}dv \\
&= J_1+J_2. \qquad (8.123)
\end{aligned}
$$

For J_1, we have

$$
\begin{aligned}
J_1 &= \lim_{h\to 0}\int_0^t d\tau \int_{\mathbb{R}^3} \frac{(M_\delta f)^2(\tau+h,v)-(M_\delta f)^2(\tau,v)}{2h}dv \\
&= \lim_{h\to 0}\frac{1}{2h}\Big\{\int_0^{t+h} d\tau - \int_0^h d\tau\Big\}\int_{\mathbb{R}^3}(M_\delta f)^2(\tau,v)dv \\
&= \frac{1}{2}\int_{\mathbb{R}^3}(M_\delta f)^2(t,v)dv - \frac{1}{2}\int_{\mathbb{R}^3}(M_\delta f)^2(0,v)dv. \qquad (8.124)
\end{aligned}
$$

For J_2, we have

$$
\begin{aligned}
J_2 &= \lim_{h\to 0}\int_0^t d\tau \int_{\mathbb{R}^3} \frac{1}{2h}\Big\{M_\delta(\tau+h)(f(\tau,v))M_\delta(\tau+h,v)(f(\tau+h,v)) \\
&\qquad\qquad - M_\delta(\tau)(f(\tau,v))M_\delta(\tau)(f(\tau+h,v))\Big\}dv \\
&= \lim_{h\to 0}\int_0^t d\tau \int_{\mathbb{R}^3} \frac{1}{2h}\Big\{f(\tau,v)M_\delta^2(\tau+h,v)(f(\tau+h,v)) \\
&\qquad\qquad - f(\tau,v)M_\delta^2(\tau)(f(\tau+h,v))\Big\}dv \\
&= \lim_{h\to 0}\int_0^t d\tau \int_{\mathbb{R}^3} \frac{1}{2h}f(\tau,v)\Big\{\Big(M_\delta^2(\tau+h)-M_\delta^2(\tau)\Big)(f(\tau+h,v))\Big\}dv \\
&= \frac{1}{2}\int_0^t d\tau \int_{\mathbb{R}^3} f(\tau,v)(\partial_\tau M_\delta^2(\tau))(f(\tau,v))dv. \qquad (8.125)
\end{aligned}
$$

By (8.123), (8.124) and (8.125), (8.122) can be rewritten by

$$
\begin{aligned}
&\frac{1}{2}\int_{\mathbb{R}^3} f(t,v)M_\delta^2 f(t,v)dv - \frac{1}{2}\int_{\mathbb{R}^3} f_0(v)M_\delta^2 f(0,v)dv \\
&= \int_0^t d\tau\int_{\mathbb{R}^3} f(\tau,v)(\partial_\tau M_\delta^2)(f(\tau,v))dv + \int_0^t d\tau\int_{\mathbb{R}^3}(Q(f,f),M_\delta^2 f)_{L^2}(\tau,v)dv. \\
&\qquad (8.126)
\end{aligned}
$$

Similarly with the first proof of Theorem 8.1, or by the content of Section 3 in this chapter, we have

$$\|M_\delta f\|^2_{H^{\frac{\nu}{2}}} \leqslant C_f\Big\{(-Q(f, M_\delta f), M_\delta f)_{L^2} + \|M_\delta f\|^2_{L^2}\Big\}. \tag{8.127}$$

From Lemma 8.8, it follows that

$$\|M_\delta f\|^2_{H^{\frac{\nu}{2}}} \leqslant \widetilde{C}_f\Big\{(-Q(f, f), M^2_\delta f)_{L^2} + \|M_\delta f\|^2_{L^2}\Big\}. \tag{8.128}$$

Since

$$\partial_t M_\delta(t,\xi) = N M_\delta(t,\xi)\log\langle\xi\rangle,$$

we obtain

$$\int_0^t d\tau \int_{\mathbb{R}^3} f(\tau, v)(\partial_\tau M^2_\delta(\tau))(f(\tau, v))dv \leqslant 2N \int_0^t \|(\log\Lambda)^{\frac{1}{2}} M_\delta f(\tau)\|^2_{L^2} d\tau, \tag{8.129}$$

where $\log\Lambda = \mathcal{F}^{-1}\log\langle\xi\rangle\mathcal{F}$ and $\Lambda = \mathcal{F}^{-1}\langle\xi\rangle\mathcal{F}$. This together with (8.126) and (8.128), imply

$$\begin{aligned} &\|M_\delta f(t)\|^2_{L^2} + \frac{1}{2C_f}\int_0^t \|\Lambda^{\frac{\nu}{2}} M_\delta f(\tau)\|^2_{L^2} d\tau \\ &\leqslant \|M_\delta f(0)\|^2_{L^2} + 2N\int_0^t \|(\log\Lambda)^{\frac{1}{2}} M_\delta f(\tau)\|^2_{L^2} d\tau + \int_0^t \|M_\delta f(\tau)\|^2_{L^2} d\tau. \end{aligned} \tag{8.130}$$

Using $\log\langle\xi\rangle \leqslant \langle\xi\rangle$ and Gagliardo-Nirenberg interpolation inequality, for any $\varepsilon > 0$, we have

$$\begin{aligned} &\|M_\delta f(t)\|^2_{L^2} + \Big(\frac{1}{2C_f} - \varepsilon\Big)\int_0^t \|\Lambda^{\frac{\nu}{2}} M_\delta f(\tau)\|^2_{L^2} d\tau \\ &\leqslant \|M_\delta f(0)\|^2_{L^2} + C_{\varepsilon,N}\int_0^t \|M_\delta f(\tau)\|^2_{L^2} d\tau. \end{aligned} \tag{8.131}$$

By choosing $\varepsilon = \frac{1}{4C_f} > 0$, there exists a constant $C_{f,N}$ depending only on C_f, N, T_0 and being independent of $\delta \in (0,1)$, such that for any $t \in (0, T_0)$,

$$\|M_\delta f(t)\|^2_{L^2} \leqslant \|M_\delta f(0)\|^2_{L^2} + C_{f,N}\int_0^t \|M_\delta f(\tau)\|^2_{L^2} d\tau. \tag{8.132}$$

Then Gronwall inequality yields

$$\|M_\delta f(t)\|^2_{L^2} \leqslant e^{C_{f,N,t}} \|M_\delta(0) f_0)\|^2_{L^2}. \tag{8.133}$$

Since

$$\|M_\delta f(t)\|_{L^2} = \|(1-\delta\Delta)^{-N_0} f(t)\|_{H^{Nt-4}},$$

and

$$\|M_\delta(0)f_0\|_{L^2(\mathbb{R}^3)} = \|(1-\delta\Delta)^{-N_0}f_0\|_{H^{-4}(\mathbb{R}^3)} \leqslant \|f_0\|_{H^{-4}(\mathbb{R}^3)} \leqslant C\|f_0\|_{L^1(\mathbb{R}^3)},$$

we obtain

$$\|(1-\delta\Delta)^{-N_0}f(t)\|_{H^{Nt-4}} \leqslant \tilde{C}e^{C_{f,N,t}}\|f_0\|_{L^1(\mathbb{R}^3)},$$

where the constant $\tilde{C}$ is independent of $\delta \in (0,1)$. Finally, for any given $t > 0$ since N can be arbitrarily large, by letting $\delta \to 0$, we have $f(t) \in H^{+\infty}(\mathbb{R}^3)$.

Remark 8.3. Recently, the existence of the classical global solutions of the Boltzmann equation without angular cut-off was obtained by Gressman and Strain [83]. There are some recent works by Chen, Li and Xu [34; 36] which have been devoted to the study of the regularity for the Landau equation, which can be regarded as a limit of the Boltzmann equation when the collisions become grazing. Mouhot and Villani [172] for the first time establish the Landau damping in a nonlinear context.

Appendix A

Notations

We list some notations used in this book, most of them are familiar for the PDE readers.

(1) $\mathbb{R}$ = the set of real, $\mathbb{C}$ = the set of complex number, $\mathbb{N}$ = the set of natural number, $\mathbb{Z}$ = the set of integers, $\mathbb{Z}_+ = \mathbb{N} \cup \{0\}$, $\mathbb{R}_+ = [0, \infty)$.

(2) $a \vee b = \max(a, b)$; $a \wedge b = \min(a, b)$.

(3) p' denotes the duality number of p, i.e., $1/p + 1/p' = 1$, $\forall\, p \in [1, \infty]$.

(4) $C > 1$ and $0 < c < 1$ express universal constants that can be different at different places.

(5) $A \lesssim B$ means $A \leqslant CB$; $A \sim B$ means $A \lesssim B$ and $B \lesssim A$.

(6) $L^p := L^p(\mathbb{R}^n)$ is the Lebesgue space,

$$\|f\|_p := \|f\|_{L^p(\mathbb{R}^n)} = \left(\int_{\mathbb{R}^n} |f(x)|^p dx \right)^{1/p}$$

(7) Let $f = (f_1, ..., f_n)$ be a vector function,

$$\|f\|_p = (\|f_1\|_p^2 + ... + \|f_n\|_p^2)^{1/2}.$$

(8) For $x = (x_1, ..., x_n) \in \mathbb{R}^n$, we write $|x| = (|x_1|^2 + ... + |x_n|^2)^{1/2}$, and sometimes we also denote $|x| = |x_1| + ... + |x_n|$.

(9) Let $A \subset \mathbb{R}^n$, we denote by $|A|$ the measure of A.

(10) Let T and S be operators, we use the notation $T \sim S$ to express that T can be roughly regarded as S (when S is easily understood).

Appendix B

Definition of scattering operator

We give an exact definition of the scattering operator by the taking NLS (7.54) as an example.

Definition B.1. Let X be a Banach space, $S(t) = e^{it\Delta}$ be the evolution semigroup associated to (7.54), and $u(t)$ be the global solution of (7.54) with the initial datum $u_0 \in X$. If the limit

$$u_+ = \lim_{t\to\infty} S(-t)u(t)$$

exists in X, we say that u_+ is the *asymptotic state* of u_0 at $+\infty$. Also, if the limit

$$u_- = \lim_{t\to-\infty} S(-t)u(t)$$

exists in X, we say that u_- is the asymptotic state of u_0 at $-\infty$. In other words, $u(t)$ behaves as $t \to \pm\infty$ like the solutions $S(t)u_\pm$ of the linear Schrödinger equation (or the free Schrödinger equation, i.e. $iu_t + \Delta u = 0$). The inverse operators $\Omega_\pm = U_\pm^{-1}$ of the operators $U_\pm : u_0 \mapsto u_\pm$ are called the *forward/backward wave operators.* Note that the uniqueness aspect of the H^1-wellposedness theory ensures that the wave operators are injective. If they are also surjective, in other word, if every H^1-wellposed solution is global and scatters in H^1 as $t \to +\infty$, we say that we also have the *asymptotic completeness.* If the forward wave operator and the backward wave operator exist simultaneously, the mapping $\mathbf{S} = \Omega_+^{-1} \circ \Omega_- : u_- \mapsto u_+$ is called the *scattering operator.*

Remark B.1. The scattering theory involves two essential factors: the existence of the forward/backward wave operators and the asymptotic completeness. Generally speaking, the existence of wave operators is relatively easy to establish as long as the power p is suitable, and especially if a smallness condition is assumed, both in focusing and defocusing cases. However, the asymptotic completeness is a bit harder and requires some decay estimates even in the defocusing case.

Remark B.2. To reduce the number of cases slightly, we shall only consider scattering from $t = 0$ to $t = +\infty$ or vice versa. One can certainly consider scattering back and forth between $t = 0$ to $t = -\infty$, or between $t = -\infty$ and $t = +\infty$, but the theory is more or less the same in each of these cases.

In general, for the well-posedness and the scattering theory of nonlinear Schrödinger equations, we usually use the integral version of the equation, that is, the Duhamel formula

$$u(t) = S(t - t_0)u(t_0) - i \int_{t_0}^{t} S(t - \tau)(|u(\tau)|^{p-1}u(\tau))d\tau. \tag{B.1}$$

For the scattering theory of the equation (7.54) in H^1, we need to show the global solution of (7.54) with the initial datum $u(0) = u_0 \in H^1$ scatters to a solution $S(t)u_+$ of the associated linear equation as $t \to +\infty$ in H^1, that is,

$$\|u(t) - S(t)u_+\|_{H^1} \to 0, \quad \text{as } t \to +\infty,$$

or equivalently,

$$\|S(-t)u(t) - u_+\|_{H^1} \to 0, \quad \text{as } t \to +\infty.$$

In other words, we require that the function $S(-t)u(t)$ converges in H^1 as $t \to +\infty$. From the Duhamel formula (B.1), we know

$$S(-t)u(t) = u_0 - i \int_0^t S(-\tau)(|u(\tau)|^{p-1}u(\tau))d\tau.$$

Thus, u scatters in H^1 as $t \to +\infty$ if and only if the improper integral

$$\int_0^\infty S(-\tau)(|u(\tau)|^{p-1}u(\tau))d\tau \tag{B.2}$$

is conditionally convergent in H^1, in which case the asymptotic state u_+ is given by the formula

$$u_+ = u_0 - i \int_0^\infty S(-\tau)(|u(\tau)|^{p-1}u(\tau))d\tau. \tag{B.3}$$

We can regard the asymptotic state u_+ as a nonlinear perturbation of the initial state u_0. Comparing (B.1) with (B.3), we eliminate u_0 to obtain the identity

$$u(t) = S(t)u_+ + i \int_t^\infty S(t - \tau)(|u(\tau)|^{p-1}u(\tau))d\tau, \tag{B.4}$$

which can be viewed as the limiting case $t_0 = +\infty$ of (B.1).

Appendix C

Some fundamental results

C.1 Gagliardo-Nirenberg inequality in Sobolev spaces

The Gagliardo-Nirenberg inequality is a fundamental tool in the study of partial differential equations, some special cases of which were discovered by Gagliardo [75], Ladyzhenskaya [153] and Nirenberg [181]. The general version can be stated as follows.

Theorem C.1. *Let* $1 \leqslant p, p_0, p_1 \leqslant \infty$, $\ell, m \in \mathbb{N} \cup \{0\}$, $\ell < m$, $\ell/m \leqslant \theta \leqslant 1$,

$$\frac{1}{p} = \frac{\ell}{n} + \frac{1-\theta}{p_0} + \theta\left(\frac{1}{p_1} - \frac{m}{n}\right). \tag{C.1}$$

If $m - \ell - n/p_0$ *is an integer, we further assume that* $\ell/m \leqslant \theta < 1$. *Then for any* $u \in C_0^\infty(\mathbb{R}^n)$,

$$\sum_{|\alpha|=\ell} \|D^\alpha u\|_{L^p(\mathbb{R}^n)} \lesssim \|u\|_{L^{p_0}(\mathbb{R}^n)}^{1-\theta} \sum_{|\alpha|=m} \|D^\alpha u\|_{L^{p_1}(\mathbb{R}^n)}^{\theta}. \tag{C.2}$$

The proof of the Gagliardo-Nirenberg inequality is based on the global-derivative analysis in L^p spaces, which is rather complicated, see [74; 87], for instance.

C.2 Convexity Hölder inequality in sequence spaces ℓ_p^s

Analogous to the convexity Hölder inequality in Triebel-Lizorkin spaces, we have the following convexity Hölder inequality in sequence spaces ℓ_p^s.

Lemma C.1. *Let* $0 < q \leqslant \infty$, $-\infty < s_1$, $s_0 < \infty$ *with* $s_0 \neq s_1$, $0 < \theta < 1$, $s = (1-\theta)s_0 + \theta s_1$. *We have*

$$\|2^{sj} a_j\|_{\ell^q} \lesssim \|2^{s_0 j} a_j\|_{\ell^\infty}^{1-\theta} \|2^{s_1 j} a_j\|_{\ell^\infty}^{\theta}. \tag{C.3}$$

Proof. We can assume that $\{a_j\} = \{a_j\}_{j\geqslant 0}$, $a_j \geqslant 0$ and $s_1 < s_0$. Put $C_i = \sup_{j\geqslant 0} 2^{s_i j} a_j$, $i = 0, 1$. It suffices to consider the case $C_1 > 0$. We have $C_1 \leqslant C_0$. Take j_0 such that

$$\min(C_0/2^{s_0 j},\ C_1/2^{s_1 j}) = \begin{cases} C_0/2^{s_0 j}, \ j > j_0, \\ C_1/2^{s_1 j}, \ j \leqslant j_0. \end{cases}$$

It is easy to see that $C_0 \sim C_1 2^{(s_0-s_1)j_0}$. So,

$$\|2^{s_0 j} a_j\|_{\ell^\infty}^{1-\theta} \|2^{s_1 j} a_j\|_{\ell^\infty}^{\theta} \sim C_1 2^{(s_0-s_1)j_0(1-\theta)}. \tag{C.4}$$

On the other hand,

$$a_j \leqslant C_0/2^{s_0 j},\ j > j_0; \quad a_j \leqslant C_1/2^{s_1 j},\ 0 \leqslant j \leqslant j_0.$$

A simple calculation yields

$$\|2^{sj} a_j\|_{\ell^q} \lesssim C_1 2^{(s_0-s_1)j_0(1-\theta)}, \tag{C.5}$$

which implies the result, as desired. □

C.3 Inclusion between homogeneous Triebel-Lizorkin spaces

The inclusion among homogeneous Triebel-Lizorkin spaces is very useful but Triebel [224] only claimed that it is probably true. Here we give the details of the proof, which is analogous to that of Theorem 1.2.

Theorem C.2. *Let $1 \leqslant p_1 < p_2 < \infty$, $1 \leqslant r, q \leqslant \infty$ and $-\infty < s_2 < s_1 < \infty$ satisfy $s_1 - n/p_1 = s_2 - n/p_2$. Then we have*

$$\dot F^{s_1}_{p_1,q} \subset \dot F^{s_2}_{p_2,r}. \tag{C.6}$$

Proof. By $\ell^r \subset \ell^q$ for $q \geqslant r$, it suffices to show that

$$\dot F^{s_1}_{p_1,\infty} \subset \dot F^{s_2}_{p_2,1}. \tag{C.7}$$

We can assume that $\|f\|_{\dot F^{s_1}_{p_1,\infty}} = 1$. Recalling the equivalent norm on L^p, we have

$$\|f\|^{p_2}_{\dot F^{s_2}_{p_2,1}} \sim \int_0^\infty t^{p_2-1} \left| \left\{ x : \sum_{k\in\mathbb{Z}} 2^{ks_2} |(\triangle_k f)(x)| > t \right\} \right| dt, \tag{C.8}$$

where $|\{\cdots\}|$ denotes the measure of the set $\{\cdots\}$. It is easy to see that

$$\sum_{k=K+1}^{\infty} 2^{ks_2} |\triangle_k f| \lesssim 2^{K(s_2-s_1)} \sup_k 2^{ks_1} |\triangle_k f|. \tag{C.9}$$

By Corollary 1.1,

$$\|\triangle_k f\|_\infty \lesssim 2^{kn/p_1}\|\triangle_k f\|_{p_1} \lesssim 2^{k(n/p_1-s_1)}\|f\|_{F^{s_1}_{p_1,\infty}}. \tag{C.10}$$

Hence, for any $K \in \mathbb{Z}$,

$$\sum_{k=-\infty}^{K} 2^{ks_2}|\triangle_k f| \lesssim \sum_{k=-\infty}^{K} 2^{k(s_2-s_1+n/p_1)} \lesssim 2^{Kn/p_2}. \tag{C.11}$$

Choosing $K \in \mathbb{Z}$ such that $C2^{Kn/p_2} \sim t/2$, we have $2^K \sim t^{p_2/n}$. If $\sum_{k\in\mathbb{Z}} 2^{ks_2}|(\triangle_k f)(x)| > t$, then it follows from (C.9) and (C.11) that

$$C2^{K(s_2-s_1)} \sup_{k\in\mathbb{Z}} 2^{ks_1}|\triangle_k f| \geqslant \sum_{k=K+1}^{\infty} 2^{ks_2}|\triangle_k f| > t/2. \tag{C.12}$$

Collecting (C.8) and (C.12), we have

$$\begin{aligned}\|f\|^{p_2}_{\dot F^{s_2}_{p_2,1}} &\lesssim \int_0^\infty t^{p_2-1}\left|\left\{x : \sup_k 2^{ks_1}|(\triangle_k f)(x)| > ct^{p_2/p_1}\right\}\right| dt\\ &\lesssim \int_0^\infty \tau^{p_1-1}\left|\left\{x : \sup_k 2^{ks_1}|(\triangle_k f)(x)| > \tau\right\}\right| d\tau\\ &\lesssim 1,\end{aligned} \tag{C.13}$$

which implies the result, as desired. □

C.4 Riesz-Thorin interpolation theorem

Theorem C.3. *Let $1 \leqslant p_i, q_i \leqslant \infty$, $p_0 \neq p_1$, $q_0 \neq q_1$ satisfy*

$$T: L^{p_i} \to L^{q_i}, \quad i = 0, 1.$$

Assume that $\theta \in (0,1)$ satisifies

$$\frac{1}{p} = \frac{1-\theta}{p_0} + \frac{\theta}{p_1}, \quad \frac{1}{q} = \frac{1-\theta}{q_0} + \frac{\theta}{q_1}.$$

Then we have $T : L^p \to L^q$ and

$$\|T\|_{L^p\to L^q} \leqslant \|T\|^{1-\theta}_{L^{p_0}\to L^{q_0}}\|T\|^{\theta}_{L^{p_1}\to L^{q_1}}.$$

C.5 Hardy-Littlewood-Sobolev inequality

Up to now, the singular integration is still a difficult problem in the theory of harmonic analysis. Hardy-Littlewood-Sobolev inequality is a fundamental tool in this subject, cf. [202]. Let $0 < \alpha < n$,

$$I_\alpha f(x) = \int_{\mathbb{R}^n} \frac{f(y)}{|x-y|^{n-\alpha}} dy.$$

Proposition C.1. *Let* $1 < p, q < \infty$ *and* $0 < \alpha < n$ *satisfy* $1/p = 1/q + \alpha/n$. *Then we have*

$$\|I_\alpha f\|_q \lesssim \|f\|_p. \tag{C.14}$$

C.6 Van der Corput lemma

Lemma C.2. *Let* $\varphi \in C_0^\infty(\mathbb{R})$. *Assume that* $P \in C^2(\mathbb{R})$ *satisfies that for any* $\xi \in \operatorname{supp} \varphi$, $|P^{(k)}(\xi)| \geqslant 1$. *Moreover, we assume that the following alternative condition holds*

(i) $k \geqslant 2$;
(ii) $k = 1$ *and* $P'(x)$ *is a monotone function.*

Then we have

$$\left| \int e^{i\lambda P(\xi)} \varphi(\xi) d\xi \right| \lesssim \lambda^{-1/k} (\|\varphi\|_\infty + \|\varphi'\|_1).$$

C.7 Littlewood-Paley square function theorem

The Littlewood-Paley square function theorem is one of the most important results in the early stage of harmonic analysis. It seems that Triebel-Lizorkin spaces are also inspired by this theorem.

Proposition C.2. *Let* $1 < p < \infty$, $s \in \mathbb{R}$. *Then*

$$\|u\|_{\dot{H}^s_p} \sim \|u\|_{\dot{F}^s_{p,2}}, \quad \|u\|_{H^s_p} \sim \|u\|_{F^s_{p,2}}. \tag{C.15}$$

In particular,

$$\|u\|_{L^p} \sim \|u\|_{\dot{F}^0_{p,2}}, \quad \|u\|_{L^p} \sim \|u\|_{F^0_{p,2}}. \tag{C.16}$$

The proof of the Littlewood-Paley square function theorem can be found in [202].

C.8 Complex interpolation in modulation spaces

We have the following

Theorem C.4. *Let* $0 < p, q, p_i, q_i \leqslant \infty$, $s, s_i \in \mathbb{R}$ *with* $i = 0, 1$ *and*

$$s = (1-\theta)s_0 + \theta s_1, \quad \frac{1}{p} = \frac{1-\theta}{p_0} + \frac{\theta}{p_1}, \quad \frac{1}{q} = \frac{1-\theta}{q_0} + \frac{\theta}{q_1}. \tag{C.17}$$

Then we have

$$(M^{s_0}_{p_0,q_0}, M^{s_1}_{p_1,q_1})_\theta = M^s_{p,q}.$$

Recall that the result of Theorem C.4 is quite similar to Besov spaces, indeed, if (C.17) holds, then we have

$$(B^{s_0}_{p_0,q_0}, B^{s_1}_{p_1,q_1})_\theta = B^s_{p,q}. \tag{C.18}$$

See [245].

C.9 Christ-Kiselev lemma

The Christ-Kiselev lemma (cf. [46]) is very useful for the study of nonlinear dispersive equations. There are a series of generalizations to the Christ-Kiselev lemma in recent years, see Molinet-Ribaud [166] and Smith-Sogge [201]. The following result is due to Wang-Han-Huang [243]. Denote

$$Tf(t) = \int_{-\infty}^{\infty} K(t,t')f(t')dt', \quad T_{re}f(t) = \int_0^t K(t,t')f(t')dt'. \tag{C.19}$$

If $T: Y_1 \to X_1$ implies that $T_{re}: Y_1 \to X_1$, then $T: Y_1 \to X_1$ is said to be a well restricted operator.

Proposition C.3. *Let T be as in* (C.19). *We have the following conclusions.*

(1) *If* $\wedge_{i=1}^3 p_i > (\vee_{i=1}^3 q_i) \vee (q_1q_3/q_2)$, *then* $T: L^{q_1}_{x_1} L^{q_2}_{x_2} L^{q_3}_t(\mathbb{R}^3) \to L^{p_1}_{x_1} L^{p_2}_{x_2} L^{p_3}_t(\mathbb{R}^3)$ *is a well restricted operator.*
(2) *If* $q_1 < \wedge_{i=1}^3 p_i$, *then* $T: L^{q_1}_t L^{q_2}_{x_1} L^{q_3}_{x_2}(\mathbb{R}^3) \to L^{p_1}_{x_1} L^{p_2}_{x_2} L^{p_3}_t(\mathbb{R}^3)$ *is a well restricted operator.*
(3) *If* $p_1 > (\vee_{i=1}^3 q_i) \vee (q_1q_3/q_2)$, *then* $T: L^{q_1}_{x_1} L^{q_2}_{x_2} L^{q_3}_t(\mathbb{R}^3) \to L^{p_1}_t L^{p_2}_{x_1} L^{p_3}_{x_2}(\mathbb{R}^3)$ *is a well restricted operator.*
(4) *If* $\wedge_{i=1}^3 p_i > (\vee_{i=1}^3 q_i) \vee (q_1q_3/q_2)$, *then* $T: L^{q_1}_{x_1} L^{q_2}_{x_2} L^{q_3}_t(\mathbb{R}^3) \to L^{p_2}_{x_2} L^{p_1}_{x_1} L^{p_3}_t(\mathbb{R}^3)$ *is a well restricted operator.*

In higher spatial dimensions, we have similar results.

Proposition C.4. *Let T be as in* (C.19). *We have the following results.*

(1) *If* $\min(p_1, p_2, p_3) > \max(q_1, q_2, q_3, \ q_1q_3/q_2)$, *then*

$$T : L^{q_1}_{x_1} L^{q_2}_{x_2,\ldots,x_n} L^{q_3}_t(\mathbb{R}^{n+1}) \to L^{p_1}_{x_1} L^{p_2}_{x_2,\ldots,x_n} L^{p_3}_t(\mathbb{R}^{n+1})$$

is a well restriction operator.

(2) *If* $p_0 > (\vee^3_{i=1} q_i) \vee (q_1q_3/q_2)$, *then*

$$T : L^{q_1}_{x_1} L^{q_2}_{x_2,\ldots,x_n} L^{q_3}_t(\mathbb{R}^{n+1}) \to L^{p_0}_t L^{p_1}_{x_1} \ldots L^{p_n}_{x_n}(\mathbb{R}^{n+1})$$

is a well restriction operator.

(3) *If* $q_0 < \min(p_1, p_2, p_3)$, *then*

$$T : L^{q_0}_t L^{q_1}_{x_1} \ldots L^{q_n}_{x_n}(\mathbb{R}^{n+1}) \to L^{p_1}_{x_1} L^{p_2}_{x_2,\ldots,x_n} L^{p_3}_t(\mathbb{R}^{n+1})$$

is a well restriction operator.

(4) *If* $\min(p_1, p_2, p_3) > \max(q_1, q_2, q_3, \ q_1q_3/q_2)$, *then*

$$T : L^{q_1}_{x_2} L^{q_2}_{x_1,x_3,\ldots,x_n} L^{q_3}_t(\mathbb{R}^{n+1}) \to L^{p_1}_{x_1} L^{p_2}_{x_2,\ldots,x_n} L^{p_3}_t(\mathbb{R}^{n+1})$$

is a well restriction operator.

Bibliography

[1] M. J. Ablowitz, R. Haberman, Nonlinear evolution equations in two and three dimensions, Phys. Rev. Lett., **35** (1975), 1185–1188.

[2] R. Alexandre, L. Desvillettes, C. Villani and B. Wennberg, Entropy dissipation and long range interactions, Arch. Rat. Mech. Anal., **152** (2000), 327–355.

[3] R. Alexandre and M. El Safadi, Littlewood–Paley theory and regularity issues in Boltzmann homogeneous equations. I. Non-cutoff case and Maxwellian molecules, Math. Models Methods Appl. Sci., **15** (2005), 907–920.

[4] R. Alexandre and M. El Safadi, Littlewood–Paley theory and regularity issues in Boltzmann homogeneous equations. II. Non cutoff case and non Maxwellian molecules, Discrete Contin. Dyn. Syst., **24** (2009), 1–11.

[5] R. Aulaskari, J. Xiao, R. H. Zhao, On subspaces and subsets of $BMOA$ and UBC, Analysis, **15** (1995), 101–121.

[6] H. Bahouri and J. Shatah, Decay estimates for the critical semilinear wave equations, Ann. Inst. H. Poincaré, Anal. Non Linéaire, **15** (1997), 783–789.

[7] M. Beals, Self-spreading and strength of singularities for solutions to semilinear wave equation, Ann. of Math., **118** (1983), 187–214.

[8] I. Bejenaru, S. Herr, J. Holmer, D. Tataru, On the 2D Zakharov system with L^2-Schrödinger data, Nonlinearity, **22** (2009), 1063–1089.

[9] I. Bejenaru and T. Tao, Sharp well-posedness and ill-posedness results for a quadratic non-linear Schrödinger equation, J. Funct. Anal., **233** (2006), 228–259.

[10] T. B. Benjamin, Internal waves of permanent form in fluids of great depth, J. Fluid Mech., **29** (1967), 559–592.

[11] A. Bényi, K.A. Okoudjou, Local well-posedness of nonlinear dispersive equations on modulation spaces, Bull. London Math. Soc., **41** (2009), 549–558.

[12] A. Bényi, K. Gröchenig, K.A. Okoudjou and L.G. Rogers, Unimodular Fourier multiplier for modulation spaces, J. Funct. Anal., **246** (2007), 366–384.

[13] J. Bergh and J. Löfström, *Interpolation Spaces,* Springer–Verlag, 1976.

[14] J. Bourgain, Fourier transform restriction phenomena for certain lattice subsets and applications to nonlinear evolution equations I, Schrödinger equations, Geom. Funct. Anal., **3** (1993), 107–156.

[15] J. Bourgain, Fourier transform restriction phenomena for certain lattice subsets and applications to nonlinear evolution equations II, The KdV-equation, Geom. Funct. Anal., **3** (1993), 209–262.

[16] J. Bourgain, On the Cauchy problem for the Kadomtsev–Petviashvili equation, Geom. Funct. Anal., **3** (1993), 315–341.

[17] J. Bourgain, Refinements of Strichartz' inequality and applications to 2D-NLS with critical nonlinearity, Int. Math. Res. Not., (1998), no. 5, 253–283.

[18] J. Bourgain, *Global solutions of nonlinear Schrödinger equation,* Amer. Math. Soc., 1999.

[19] J. Bourgain, Global well posedness of defocusing critical nonlinear Schrödinger equation in the radial case, J. Amer. Math. Soc., **12** (1999), 145–171.

[20] J. Bourgain, J. Colliander, On well-posedness of the Zakharov system, Int. Math. Res. Not., **11** (1996), 515–546.

[21] P. Brenner, On space-time means and everywhere defined scattering operators for nonlinear Klein–Gordon equations, Math. Z., **186** (1984), 383–391.

[22] P. Brenner, On scattering and everywhere-defined scattering operators of nonlonear Klein-Gordon equations, J. Differential Equations, **56** (1985), 310–344.

[23] H. Brezis and P. Mironescu, Gagliardo–Nirenberg, composition and products in fractional Sobolev spaces, J. Evol. Equ., **1** (2001), 387–404.

[24] L. Carleson, *Some analytical problems related to statistical mechanics*, Euclidean Harmonic Analysis, Lecture Notes in Math., Vol. **779**, Springer, Berlin, 1980, 5–45.

[25] T. Cazenave, F. B. Weissler, Some remarks on the nonlinear Schrödinger equation in the critical case, Lecture Notes in Math., **1394** (1989), 18–29.

[26] T. Cazenave and F.B. Weissler, The Cauchy problem for the critical nonlinear Schrödinger equation in H^s, Nonlinear Anal. TMA, **14** (1990), 807–836.

[27] T. Cazenave, *Semilinear Schrödinger equations*, Courant Lecture Notes in Mathematics, Vol. **10**, 2003.

[28] C. Cercignani, *The Boltzmann Equation and Its Applications,* Springer–Verlag, 1988.

[29] J. Y. Chemin, Théormes d'unicité pour le système de Navier–Stokes tridimensionnel, Journal d'Anal. Math. **77** (1999), 27–50.

[30] J. Y. Chemin, I. Gallagher, On the global wellposedness of the 3D Navier–Stokes equations with large initial data. Ann. Sci. École Norm. Sup. **39** (2006), 679–698.

[31] J. Y. Chemin, I. Gallagher, Wellposedness and stability results for the Navier-Stokes equations in $\mathbb{R}^3$, Ann. Inst. H. Poincaré, Anal. Non Linéaire, **26** (2009), 599–624.

[32] J. Y. Chemin, I. Gallagher, Large, global solutions to the Navier–Stokes equations, slowly varying in one direction, Tran. Amer. Math. Soc., **362** (2010), 2859–2873.

[33] J. Y. Chemin, P. Zhang, On the global wellposedness to the 3-D incompressible anisotropic Navier–Stokes equations. Comm. Math. Phys., **272** (2007), 529–566.

[34] H. Chen, W. X. Li, C. J. Xu, Propagation of Gevrey regularity for solutions of Landau equations, Kinet. Relat. Models, **1** (2008), 355–368.

[35] H. Chen, W. X. Li, C. J. Xu, Gevrey hypoellipticity for linear and non-linear Fokker–Planck equations, J. Differential Equations, **246** (2009), 320–339.

[36] H. Chen, W. X. Li, C. J. Xu, Analytic smoothness effect of solutions for spatially homogeneous Landau equation, J. Differential Equations, **248** (2010), 77–94.

[37] H. Chen, W. X. Li, C. J. Xu, Gevrey hypoellipticity for a class of kinetic equations, Commun. Part. Diff. Equations, **36** (2011), 693–728.

[38] J. C. Chen, D. S. Fan and L. Sun, Asymptotic estimates For unimodular Fourier multipliers on modulation spaces, Preprint.

[39] Cheskidov A., Shvydkoy R, The regularity of weak solutions of the 3D Navier-Stokes equations in $B^{-1}_{\infty,\infty}$, Arch. Ration. Mech. Anal., **195** (2010), 159–169.

[40] J. Colliander, *Nonlinear Schrödinger equations,* Lecture notes, 2004.

[41] J. E. Colliander, J. M. Delort, C. E. Kenig and G. Staffilani, Bilinear estimates and applications to 2D NLS, Trans. Amer. Math. Soc., **353** (2001), 3307–3325.

[42] J. Colliander, M. Keel, G. Staffilani, H. Takaoka and T. Tao, Sharp global well-posedness for KdV and modified KdV on $\mathbb{R}$ and $\mathbb{T}$, J. Amer. Math. Soc., **16** (2003), 705–749.

[43] J. Colliander, M. Keel, G. Staffilani, H. Takaoka and T. Tao, Viriel, Morawetz, and interaction Morawetz inequalities, Lecture notes, 2004.

[44] J. Colliander, M. Keel, G. Staffilani, H. Takaoka, and T. Tao, Global well-posedness and scattering for the energy-critical nonlinear Schrödinger equation in $\mathbb{R}^3$, Ann. of Math., **167** (2008), 767–865.

[45] J. Colliander, M. Keel, G. Staffilani, H. Takaoka and T. Tao, Resonant decompositions and the I-method for cubic nonlinear Schrödinger equation on $\mathbb{R}^2$, arXiv:0704.2730.

[46] M. Christ, A. Kiselev, Maximal functions associated to filtrations, J. Funct. Anal., **179** (2001), 406–425.

[47] M. Christ, J. Colliander and T. Tao, Asymptotics, frequency modulation and low regularity ill-posedness for canonical defocusing equations, Amer. J. Math., **125** (2003), 1235–1293.

[48] P. A. Clarkson and J. A. Tuszyriski, Exact solutions of the multidimensional derivative nonlinear Schrödinger equation for many-body systems near criticality, J. Phys. A: Math. Gen. **23** (1990), 4269–4288.

[49] E. Cordero, F. Nicola, Some new Strichartz estimates for the Schrödinger equation. J. Differential Equations, **245** (2008), 1945–1974.

[50] E. Cordero, F. Nicola, Remarks on Fourier multipliers and applications to the wave equation, J. Math. Anal. Appl., **353** (2009), 583–591.

[51] E. Cordero, F. Nicola, Sharpness of some properties of Wiener amalgam and modulation spaces. Bull. Aust. Math. Soc., **80** (2009), 105–116.

[52] S. B. Cui, Pointwise estimates for a class of oscillatory integrals and related L^p–L^q estimates, J. Fourier Anal. Appl., **11** (2005), 441–457.

[53] S. B. Cui, Pointwise estimates for oscillatory integrals and related L^p-L^q estimates, II Multidimensional case, J. Fourier Anal. Appl., **12** (2006), 605–627.

[54] J. M. Delort, Existence globale et comportement asymptotique pour léquation de Klein–Gordon quasi-linéaire à données petites en dimension 1, Ann. Sci. École Norm. Sup., **34** (2001), 1–61.

[55] J. M. Delort, D. Y. Fang, R. Y. Xue, Global existence of small solutions for quadratic quasilinear Klein-Gordon systems in two space dimensions. J. Funct. Anal., **211** (2004), 288–323.

[56] C. Deng, J. H. Zhao, S. B. Cui, Well-posedness of a dissipative nonlinear electrohydrodynamic system in modulation spaces, Nonlinear Anal., TMA, **73** (2010), 2088–2100.

[57] L. Desvillettes, About the regularization properties of the non cut-off Kac equation, Comm. Math. Phys., **168** (1995), 417–440.

[58] L. Desvillettes, Regularization properties of the 2-dimensional non radially symmetric non cutoff spatially homogeneous Boltzmann equation for Maxwellian molecules, Trans. Theory Stat. Phys., **26** (1997), 341–357.

[59] L. Desvillettes, *About the use of the Fourier transform for the Boltzmann equation,* Summer School on Methods and Models of Kinetic Theory (MMKT 2002). Riv. Mat. Univ. Parma, **7** (2003), 1–99.

[60] L. Desvillettes and B. Wennberg, Smoothness of the solution of the spatially homogeneous Boltzmann equation without cutoff, Comm. Partial Differential Equations, **29** (2004), 133–155.

[61] W. Y. Ding, Y. D. Wang, Schrödinger flow of maps into symplectic manifolds, Science in China Ser. A, **41** (1998), 746–755.

[62] W. Y. Ding and Y. D. Wang, Local Schrödinger flow into Kähler manifolds, Sci. China Ser. A, **44** (2001), 1446–1464.

[63] J. M. Dixon and J. A. Tuszynski, Coherent structures in strongly interacting many-body systems: II, Classical solutions and quantum fluctuations, J. Phys. A: Math. Gen., **22** (1989), 4895–4920.

[64] B. Dodson, Global well-posedness and scattering for the defocusing, L^2-critical, nonlinear Schrödinger equation when $d \geqslant 3$, arXiv:0912.2467.

[65] B. Dodson, Global well-posedness and scattering for the defocusing, L^2-critical, nonlinear Schrödinger equation when $d = 2$, arXiv:1006.1375.

[66] B. Dodson, Global well-posedness and scattering for the defocusing, L^2-critical, nonlinear Schrödinger equation when $d = 1$, arXiv:1010.0040.

[67] H. Dong, D. Du, On the local smoothness of solutions of the Navier–Stokes equations, J. Math. Fluid Mech., **9** (2007), 139–152.

[68] H. Dong, D. Du, The Navier–Stokes equations in the critical Lebesgue space, Comm. Math. Phys., **292** (2009), 811–827.

[69] X. T. Duong, L. X. Yan, New function spaces of BMO type, the John-Nirenberg inequality, interpolation, and applications, Comm. Pure Appl. Math. **58** (2005), 1375–1420.

[70] L. Escauriaza, G. A. Seregin, and V. Šverák, $L^{3,\infty}$-solutions of Navier–Stokes equations and backward uniqueness, Uspekhi Mat. Nauk, **58** (2003), 3–44.

[71] D. Y. Fang, H. Pecher, S. J. Zhong, Low regularity global well-posedness for the two-dimensional Zakharov system, Analysis (Munich), **29** (2009), 265–281.

[72] D. Y. Fang, J. Xie and T. Cazenave, Scattering for the focusing energy-subcritical NLS, Science in China, Ser A, Math., in press.

[73] H. G. Feichtinger, Modulation spaces on locally compact Abelian group, Technical Report, University of Vienna, 1983. Published in: "Proc. Internat. Conf. on Wavelet and Applications", 99–140. New Delhi Allied Publishers, India, 2003.

[74] A. Friedmann, *Partial Differential Equations*, Holt, Rinehart and Winston, New York, 1969.

[75] E. Gagliardo, Proprieta di alcune classi di funzioni in pia variabili, Richerche Mat., **7** (1958), 102–137; **9** (1959), 24–51.

[76] V. Georgiev, P. Popivanov, Global solution to the two-dimensional Klein–Gordon equation, Comm. Partial Differential Equations **16** (1991), 941–995.

[77] Y. Giga, Solutions for semilinear parabolic equations in L^p and regularity of weak solutions of the Navier–Stokes system, J. Differential Equations,**61** (1986), 186–212.

[78] J. Ginibre and G. Velo, On a class of nonlinear Schrödinger equations II, Scattering theory J. Funct. Anal., **32** (1979), 33–71.

[79] J. Ginibre and G. Velo, Time decay of finite energy solutions of the nonlinear Klein-Gordon and Schrödinger equations, Ann. Inst. H. Poincaré, Phys. Theor., **43** (1985), 399–442.

[80] J. Ginibre and G. Velo, Generalized Strichartz inequalities for the wave equation, J. Funct. Anal., **133** (1995), 50–68.

[81] J. Ginibre, Y. Tsutsumi and G. Velo, On the Cauchy problem for the Zakharov System, J. Funct. Anal., **151** (1997), 384–436.

[82] L. Grafakos, *Classical and modern Fourier analysis,* Pearson/Prentice Hall, 2004.

[83] P. T. Gressman, R. M. Strain, Global classical solutions of the Boltzmann equation without angular cut-off, arXiv:1011.5441.

[84] M. Grillakis, On nonlinear Schrödinger equations, Comm. Partial Differential Equations, **25** (2000), 1827–1844.

[85] P. Gröbner, *Banachräume Glatter Funktionen und Zerlegungsmethoden*, Doctoral thesis, University of Vienna, 1992.

[86] K. Gröchenig, *Foundations of Time–Frequency Analysis*, Birkhäuser, Boston, MA,

2001.

[87] B. L. Guo, *Viscosity Elimination Method and the Viscosity of Difference Scheme*, Chinese Sci. Publisher, 2004.

[88] B. L. Guo, S. J. Ding, *Landau–Lifshitz equations*, Frontiers of Research with the Chinese Academy of Sciences, 1. World Scientific, 2008.

[89] B. L. Guo and B. X. Wang, The Cauchy problem for the Davey–Stewartson systems, Comm. Pure Appl. Math., **52** (1999), 1477–1490.

[90] Z. H. Guo, Global Well-posedness of Korteweg-de Vries equation in $H^{-3/4}(\mathbb{R})$, J. Math. Pures Appl., **91** (2009), 583–597.

[91] Z. H. Guo, Local well-posedness and a priori bounds for the modified Benjamin–Ono equation without using a gauge transformation, arXiv:0807.3764.

[92] Z. H. Guo, Local Well-posedness for dispersion generalized Benjamin–Ono equations in Sobolev spaces, arXiv:0812.1825.

[93] Z. H. Guo, The Cauchy problem for a class of derivative nonlinear dispersive equations, Doctoral thesis, Peking Univ., (2009).

[94] Z. H. Guo, L. Z. Peng, B. X. Wang, Decay estimates for a class of wave equations, J. Funct. Anal., **254** (2008), 1642–1660.

[95] Z. H. Guo, L. Z. Peng and B. X. Wang, On the local regularity of the KP-I equation in anisotropic Sobolev space, J. Math. Pure Appl., **94** (2010), 414–432.

[96] Z. H. Guo and B. X. Wang, Global well posedness and inviscid limit for the Korteweg–de Vries–Burgers equation, J. Differential Equations, **246** (2009), 3864–3901.

[97] M. Hadac, Well-posedness for the Kadomtsev–Petviashvili-II equation and generalizations, Trans. Amer. Math. Soc., **360** (2008), 6555–6572.

[98] M. Hadac, S. Herr and H. Koch, Well-posedness and scattering for the KP-II equation in a critical space, Ann. Inst. H. Poincaré, Anal. Non Linéaire, **26** (2009), 917–941.

[99] H. Hajaiej, L. Molinet, T. Ozawa, B. X. Wang, Necessary and sufficient conditions for the fractional Gagliardo–Nirenberg inequalities and applications to the Navier–Stokes and generalized boson equations, arXiv:1004.4287.

[100] C. C. Hao, L. Hsiao and B. X. Wang, Wellposedness for the fourth order nonlinear Schrödinger equations, J. Math. Anal. Appl., **320** (2006), 246–265.

[101] C. C. Hao, L. Hsiao and B. X. Wang, Wellposedness of Cauchy problem for the fourth order nonlinear Schrödinger equations in multi-dimensional spaces, J. Math. Anal. Appl., **328** (2007), 58–83.

[102] N. Hayashi, The initial value problem for the derivative nonlinear Schrödinger equation in the energy space. Nonlinear Anal., TMA, **20** (1993), 823–833.

[103] N. Hayashi, E.I. Kaikina, P.I. Naumkin, Damped wave equation in the subcritical case, J. Differential Equations, **207** (2004), 161–194.

[104] N. Hayashi, E.I. Kaikina, P.I. Naumkin, Damped wave equation with a critical nonlinearity, Trans. Amer. Math. Soc., **358** (2006), 1117–1163.

[105] N. Hayashi and T. Ozawa, Finite energy solutions of nonlinear Schrödinger equations of derivative type, SIAM J. Math. Anal., **25** (1994), 1488–1503.

[106] N. Hayashi, T. Ozawa, Modified wave operators for the derivative nonlinear Schrödinger equation. Math. Ann., **298** (1994), 557–576.

[107] N. Hayashi, T. Ozawa, Remarks on nonlinear Schrödinger equations in one space dimension. Differential Integral Equations **7** (1994), 453–461.

[108] S. Herr, A. Ionescu, C. Kenig and H. Koch, A para-differential renormalization technique for nonlinear dispersive equations, arXiv:0907.4649.

[109] Z. H. Huo and Y. L. Jia, The Cauchy problem for the fourth-order nonlinear

Schrödinger equation related to the vortex filament, J. Differential Equations, **214** (2005), 1–35.

[110] Z. H. Huo and Y. L. Jia, A refined well-posedness for the fourth-order nonlinear Schrödinger equation related to the vortex filament, Comm. Partial Differential Equations, **32** (2007), 1493–1510.

[111] Z. H. Huo, Y. Morimoto, S. Ukai and T. Yang, Regularity of solutions for spatially homogeneous Boltzmann equation without angular cutoff, Kinet. Relat. Models, **1** (2008), 453–489.

[112] R. Ikehata, K. Nishihara, H. J. Zhao, Global asymptotics of solutions to the Cauchy problem for the damped wave equation with absorption, J. Differential Equations, **226** (2006), 1–29.

[113] A. D. Ionescu and C. E. Kenig, Global well-posedness of the Benjamin–Ono equation in low-regularity spaces, J. Amer. Math. Soc., **20** (2007), 753–798.

[114] A. Ionescu and C. E. Kenig, Low-regularity Schrödinger maps, II: Global well posedness in dimensions $d \geqslant 3$, Commun. Math. Phys., **271** (2007), 523–559.

[115] A. D. Ionescu, C. E. Kenig and D. Tataru, Global well-posedness of the KP-I initial-value problem in the energy space, Invent. Math., **173** (2008), 265–304.

[116] T. Iwabuchi, Navier-Stokes equations and nonlinear heat equations in modulation spaces with negative derivative indices. J. Differential Equations, **248** (2010), 1972–2002.

[117] F. John, *Plane waves and spherical means, Applied to partial differential equations,* Springer, 1981.

[118] L.V. Kapitanskii, Weak and yet weak solutions of semilinear wave equations, Comm. Partial Differential Equations, **19** (1994), 1629–1676.

[119] T. Kato, Strong L^p solutions of the Navier–Stokes equations in $\mathbb{R}^m$ with applications to weak solutions, Math. Z., **187** (1984), 471–480.

[120] T. Kato, On nonlinear Schrödinger equations, Ann.Inst.H.Poincaré, Phys. Theor. **46** (1987), 113–129.

[121] M. Keel and T. Tao, End point Strichartz estimates, Amer. J. Math., **120** (1998), 955–980.

[122] C. E. Kenig, On the local and global well-posedness theory for the KP-I equation, Ann. Inst. H. Poincaré, Anal. Non Linéaire, **21** (2004), 827–838.

[123] C. E. Kenig. G. S. Koch, An alternative approach to regularity for the Navier–Stokes equations in critical spaces, Preprint.

[124] C.E. Kenig, and F. Merle, Global well-posedness, scattering and blow-up for the energy-critical, focusing, non-linear Schrödinger equation in the radial case. Invent. Math., **166** (2006), 645–675.

[125] C. E. Kenig, G. Ponce, C. Rolvent, L. Vega, The genreal quasilinear untrahyperbolic Schrodinger equation, Adv. Math., **206** (2006), 402–433.

[126] C. E. Kenig, G. Ponce and L. Vega, Oscillatory integrals and regularity of dispersive equations, Indiana Univ. Math. J., **40** (1991), 33–69.

[127] C. E. Kenig, G. Ponce and L. Vega, Well-posedness of the initial value problem for the Korteweg-de Vries equation, J. Amer. Math. Soc., **4** (1991), 323–347.

[128] C. E. Kenig, G. Ponce and L. Vega, The Cauchy problem for the Korteweg-de Vries equation in Sobolev spaces of negative indices, Duke Math. J., **71** (1993), 1–21.

[129] C.E. Kenig, G. Ponce and L. Vega, Well-posedness and scattering results for generalized KdV equation via contraction principles, Comm. Pure Appl. Math., **46** (1993), 527–620.

[130] C. E. Kenig, G. Ponce, L. Vega, Small solutions to nonlinear Schrödinger equation, Ann. Inst. Henri Poincaré, Sect C, **10** (1993), 255–288.

[131] C. E. Kenig, G. Ponce and L. Vega, Quadratic forms for the 1-D semilinear Schrödinger equation, Trans. Amer. Math. Soc., **346** (1996), 3323–3353.

[132] C. E. Kenig, G. Ponce and L. Vega, A bilinear estimate with applications to the KdV equation, J. Amer. Math. Soc., **9** (1996), 573–603.

[133] C. E. Kenig, G. Ponce and L. Vega, Smoothing effects and local existence theory for the generalized nonlinear Schrödinger equations, Invent. Math., **134** (1998), 489–545.

[134] C. E. Kenig, G. Ponce and L. Vega, On the ill-posedness of some canonical dispersive equations, Duke Math. J., **106** (2001), 617–633.

[135] C. E. Kenig, G. Ponce, L. Vega, The Cauchy problem for quasi-linear Schrödinger equations, Invent. Math., **158** (2004), 343–388.

[136] C. E. Kenig and H. Takaoka, Global wellposedness of the modified Benjamin–Ono equation with initial data in $H^{1/2}$, Int. Math. Res. Not., **2006**, Art. ID 95702, 1–44.

[137] R. Killip, D. Li, M. Visan and X. Y. Zhang, Characterization of minimal-mass blowup solutions to the focusing mass-critical NLS, SIAM J. Math. Anal., **41** (2009), 219–236.

[138] R. Killip, T. Tao, and M. Visan, The cubic nonlinear Schrödinger equation in two dimensions with radial data, J. Euro. Math. Soc., **11** (2009), 1203–1258.

[139] R. Killip, M. Visan, The focusing energy-critical nonlinear Schrödinger equation in dimensions five and higher, Amer. J. Math. **132** (2010), no. 2, 361–424..

[140] R. Killip, M. Visan, and X. Y. Zhang, The mass-critical nonlinear Schrödinger equation with radial data in dimensions three and higher, Anal. of PDE, **1** (2008), 229–266.

[141] N. Kishimoto, Well-posedness of the Cauchy problem for the Korteweg-de Vries equation at the critical regularity, Preprint.

[142] N. Kishimoto, *Low-regularity Local Well-posedness for Quadratic Nonlinear Schrödinger Equations*, Master Thesis, Kyoto University, 2008.

[143] S. Klainerman, Long-time behavior of solutions to nonlinear evolution equations, Arch. Rational Mech. Anal., **78** (1982), 73–98.

[144] S. Klainerman, Global existence of small amplitude solutions to nonlinear Klein–Gordon equations in four space-time dimensions, Comm. Pure Appl. Math., **38** (1985), 631–641.

[145] S. Klainerman and M. Machedon, Space-time estimates for null forms and the local existence theorem, Comm. Pure Appl. Math., **46** (1993), no. 9, 1221–1268.

[146] S. Klainerman and M. Machedon, Smoothing estimates for null forms and applications, Duke Math. J., **81** (1995), 99–133.

[147] S. Klainerman, G. Ponce, Global small amplitude solutions to nonlinear evolution equations, Commun. Pure Appl. Math., **36** (1983), 133–141.

[148] M. Kobayashi, Modulation spaces $M^{p,q}$ for $0 < p, q \leqslant \infty$, J. of Funct. Spaces and Appl., **4** (2006), no.3, 329–341.

[149] M. Kobayashi, Dual of modulation spaces, J. of Funct. Spaces and Appl., **5** (2007), 1–8.

[150] M. Kobayashi, Y. Sawano, Molecular decomposition of the modulation spaces, Osaka J. Math., **47** (2010), 1029–1053.

[151] H. Koch and D. Tataru, Well-posedness for the Navier–Stokes equations, Adv. Math., **157** (2001), 22–35.

[152] B. G. Konopelchenko, B. T. Matkarimov, On the inverse scattering transform of the Ishimori equations, Phys. Lett., **135** (1989), 183–189.

[153] O. Ladyzhenskaya, The solvability "in the large" of the boundary value problem for the Navier-Stokes equation in the case of two space variables, Comm. Pure Appl. Math., **12** (1959), 427–433.

[154] O. Ladyzhenskaya, *The Mathematical Theory of Viscous Incompressible Flow,* Second English edition, Mathematics and its Applications, Vol. **2**, Gordon and Breach, Science Publishers, New York–London–Paris, 1969.

[155] P. G. Lemarié-Rieusset, *Recent developments in the Navier–Stokes problem,* A CRC Press Company, 2002.

[156] S.P. Levandosky, Decay estimates for fourth order wave equations, J. Differential Equations, **143** (1998), 360–413.

[157] F. Linares and G. Ponce, On the Davey–Stewartson systems, Ann. Inst. H. Poincaré, Anal. NonLinéaire, **10** (1993), 523–548.

[158] H. Lindblad and C.D. Sogge, On existence and scattering with minimal regularity for semilinear wave equations, J. Funct. Anal., **130** (1995), 357–426.

[159] J. L. Lions, *Some Methods for Solving Nonlinear Boundary Value Problems,* Translated by B. L. Guo and L. R. Wang, Sun Yat-sen University Press, 1992.

[160] P. L. Lions, Regularity and compactness for Boltzmann collision operator without cut-off. C. R. Acad. Sci. Paris Series I, **326** (1998), 37–41.

[161] W. Littman, Fourier transforms of surface-carried measures and differentiability of surface averages, Bull. of Amer. Math. Soc., **69** (1963), 766–770.

[162] S. Machihara and T. Ozawa, Interpolation inequalities in Besov spaces, Proc. Amer. Math. Soc., **131** (2002), 1553–1556.

[163] R. May, Rôle de l'space de Besov $B^{-1}_{\infty,\infty}$ dans le contrôle de l'explosion évetuelle en temps fini des solutions réguli'eres des équations de Navier–Stokes, C. R. Acad. Sci. Paris, **323** (2003), 731–734.

[164] A. Miyachi, F. Nicola, S. Riveti, A. Taracco and N. Tomita, Estimates for unimodular Fourier multipliers on modulation spaces, Proc. Amer. Math. Soc., **137** (2009), 3869–3883.

[165] L. Molinet and F. Ribaud, On the low regularity of the Korteweg–de Vries–Burgers equation, Int. Math. Res. Not., **37** (2002), 1979–2005.

[166] L. Molinet, F. Ribaud, Well-posedness results for the generalized Benjamin–Ono equation with small initial data, J. Math. Pures Appl., **83** (2004), 277–311.

[167] L. Molinet, J.C. Saut and N. Tzvetkov, Ill-posedness issues for the Benjamin–Ono and related equations, SIAM J. Math. Anal., **33** (2001), 982–988.

[168] L. Molinet, J. C. Saut and N. Tzvetkov, Global well-posedness for the KP-I equation, Math. Ann., **324** (2002) 255–275, **328** (2004), 707–710.

[169] L. Molinet, J. C. Saut and N. Tzvetkov, Well-posedness and ill-posedness results for the Kadomtsev–Petviashvili-I equation, Duke Math. J., **115** (2002), 353–384.

[170] C. S. Morawetz and W. A. Strauss, Decay and scattering of solutions of a nonlinear relativistic wave equation, Comm. Pure Appl. Math.,**25** (1972), 1–31.

[171] Y. Morimoto, S. Ukai, C. J. Xu and T. Yang, Regularity of the solution to the spatially homogeneous Boltzmann equation without angular cutoff, Discrete Contin. Dyn. Syst., **24** (2009), 187–212.

[172] C. Mouhot and C. Villani, On the Landau damping, arXiv:0904.2760.

[173] M. Nakamura and T. Ozawa, Nonlinear Schrödinger equations in the Sobolev space of critical order, J. Funct. Anal., **155** (1998), 364–380

[174] M. Nakamura and T. Ozawa, The Cauchy problem for nonlinear Klein–Gordon equations in the Sobolev spaces, Publ. Res. Inst. Math. Sci., **37** (2001), 255–293.

[175] K. Nakanishi, Energy scattering for Hartree equations, Math. Res. Lett., **6** (1999), 107–118.

[176] K. Nakanishi, Energy scattering for nonlinear Klein–Gordon and Schrödinger equations in spatial dimensions 1 and 2, J. Funct. Anal., **169** (1999), 201–225.

[177] K. Nakanishi, Scattering theory for nonlinear Klein–Gordon equation with Sobolev critical power, Int. Math. Res. Notices, **1999**, 31–60.

[178] K. Nakanishi, Unique global existence and asymptotic behavior of solutions for wave equations with non-coercive nonlinearity, Comm. Partial Differential Equations, **24** (1999), 185–221.

[179] K. Nakanishi, Remarks on the energy scattering for nonlinear Klein–Gordon and Schrödinger equations, Tohoku Math. J., **53** (2001), 285–303.

[180] K. Nakanishi, H. Takaoka and Y. Tsutsumi, Counterexamples to bilinear estimates related to the KdV equation and the nonlinear Schrödinger equation, Methods of Appl. Anal., **8** (2001) 569–578.

[181] L. Nirenberg, On elliptic partial differential equations, Ann. Sc. Norm. Sup. Pisa, Ser. III, **13**, (1959), 115–162.

[182] H. Ono, Algebraic solitary waves in stratified fluids, J. Phys. Soc. Japan, **39** (1975) 1082–1091.

[183] F. Oru, *Rôle des oscillations dans quelques probl'emes d'nalyse non-linéaire*, Doctorat de Ecole Normale Supérieure de Cachan, 1998.

[184] T. Ozawa, On the nonlinear Schrödinger equations of derivative type, Indiana Univ. Math. J., **45** (1996), 137–163.

[185] T. Ozawa, K. Tsutaya, Y. Tsutsumi, Global existence and asymptotic behavior of solutions for the Klein–Gordon equations with quadratic nonlinearity in two space dimensions, Math. Z., **222** (1996), 341–362.

[186] T. Ozawa and J. Zhai, Global existence of small classical solutions to nonlinear Schrödinger equations, Ann. I. H. Poincaré, Anal. Non Linéaire, **25** (2008), 303–311.

[187] M. Paicu, Équation anisotrope de Navier–Stokes dans des espaces critiques, Rev. Mat. Iberoamericana, **21** (2005), 179–235.

[188] Y. P. Pao, Boltzmann collision operator with inverse power intermolecular potential,, I, II. Commun. Pure Appl. Math., **27** (1974), 407–428, 559–581.

[189] H. Pecher, Low energy scattering for nonlinear Klein–Gordon equations, J. Funct. Anal., **63** (1985), 101–122.

[190] H. Pecher, L^p-Abschätzungen und klassische Lösungen für nichtlineare Wellengleichungen. I, Math. Z., **150** (1976), 159–183.

[191] H. Pecher, Nonlinear small data scattering for the wave and Klein–Gordon equations, Math.Z., **185** (1984), 261–270.

[192] J. Peetre, Applications de la théorie des espaces d'interpolation dans l'analyse harmonique, Ricerche Mat., **15** (1966), 1–34.

[193] J. Rauch, *Partial Differential Equations*, Graduate Texts in Math., Springer–Verlag, 1991.

[194] J. Rauch and M. Reed, Nonlinear microlocal analysis of semilinear hyperbolic systems in one spatial dimension, Duke Math. J., **49** (1982), 397–475.

[195] M. Ruzhansky, J. Smith, *Dispersive and Strichartz estimates for hyperbolic equations with constant coefficients*, MSJ Memoirs, **22**, Mathematical Society of Japan, Tokyo, 2010.

[196] E. Ryckman, M. Visan, Global well-posedness and scattering for the defocusing energy-critical nonlinear Schrödinger equation in $\mathbb{R}^{1+4}$, Amer. J. Math., **129** (2007), 1–60.

[197] I.E. Segal, Dispersion for nonlinear relativistic equation II, Ann. Sc. Ec. Norm Sci., **4** (1968), 459–497.

[198] I.E. Segal, Space-time decay for solutions of wave equations, Adv. Math., **22** (1976), 304–311.

[199] J. Shatah, Global existence of small classical solutions to nonlinear evolution equations, J. Differential Equations, **46** (1982), 409–423.

[200] J. Shatah, Normal forms and quadratic nonlinear Klein–Gordon equations, Comm. Pure Appl. Math., **38** (1985), 685–696.

[201] H. F. Smith, C. D. Sogge, Global Strichartz estimates for nontrapping perturbations of Laplacian, Comm. Partial Differential Equations, **25** (2000), 2171–2183.

[202] E. M. Stein, *Singular Integrals and Differentiability Properties of Functions*, Princeton Univ. Press, Princeton, New Jersey, 1970.

[203] E. M. Stein, *Harmonic Analysis,* Princeton Univ. Press, Princeton, New Jersey, 1992.

[204] W. Strauss, Nonlinear scattering theory at low energy, J. Funct. Anal., **42** (1981), 110–133 and **43** (1981), 281–293.

[205] W. Strauss, *Nonlinear Wave Equations,* Lecture Notes, Vol. **73**, Amer. Math. Soc., Providence, RI. 1989.

[206] R. S. Strichartz, Restriction of Fourier transform to quadratic surfaces and decay of solutions of wave equations, Duke Math. J., **44** (1977), 705–714.

[207] M. Sugimoto, L^p–$L^{p'}$-estimates for hyperbolic equations, Math. Res. Lett., **2** (1995), 171–178.

[208] M. Sugimoto, Estimates for hyperbolic equations with non-convex characteristics, Math. Z., **222** (1996), 521–531.

[209] M. Sugimoto, M. Ruzhansky, A smoothing property of Schrödinger equations and a global existence result for derivative nonlinear equations, Advances in analysis, 315–320, World Sci. Publ., Hackensack, NJ, 2005.

[210] M. Sugimoto amd N. Tomita, The dilation property of modulation spaces and their inclusion relation with Besov spaces, J. Funct. Anal., **248** (2007), 79–106.

[211] C. Sulem and P.L. Sulem, *The Nonlinear Schrödinger Equation: Self-Focusing and Wave Collaps*, Appl. Math. Sci., **139**, Springer–Verlag, 1999.

[212] H. Takaoka, Well-posedness for the Kadomtsev–Petviashvili-II equation, Adv. Differential Equations, **5** (2000), 1421–1443.

[213] H. Takaoka, Well-posedness for the one-dimensional nonlinear Schrödinger equation with the derivative nonlinearity, Adv. Differential Equations, **4** (1999), 561–580.

[214] H. Takaoka and N. Tzvetkov, On the local regularity of the Kadomtsev–Petviashvili-II equation, Int. Math. Res. Not., **2001** (2001) 77–114.

[215] T. Tao, Spherically averaged endpoint Strichartz estimates for the two-dimensional Schrödinger equation, Comm. Partial Differential Equations, **25** (2000), 1471–1485.

[216] T. Tao, Multilinear weighted convolution of L^2-functions, and applications to nonlinear dispersive equations, Amer. J. Math., **123** (2001), 839–908.

[217] T. Tao, Global well-posedness and scattering for the higher-dimensional energy-critical non-linear Schrödinger equation for radial data, New York J. Math., **11** (2005), 57–80.

[218] T. Tao, *Nonlinear dispersive equations*, volume 106 of *CBMS Regional Conference Series in Mathematics*, Published for the Conference Board of the Mathematical Sciences, Washington, DC, 2006.

[219] T. Tao, Scattering for the quantic generalized Korteweg-de Vries equation, J. Differential Equations, **232** (2007), 623–651.

[220] T. Tao, M. Visan and X. Y. Zhang, The nonlinear Schrödinger equation with combined power-type nonlinearities, Comm. Partial Differential Equations, **32** (2007), 1281–1343.

[221] D. Tataru, Local and global results for wave maps I, Comm. Partial Differential Equations, **23** (1998), 1781–1793.

[222] J. Toft, Continuity properties for modulation spaces, with applications to pseudo-differential calculus, I, J. Funct. Anal., **207** (2004), 399–429.

[223] P. Tomas, A restriction theorem for the Fourier transform, Bull. Amer. Math. Soc., **81** (1975), 477–478.

[224] H. Triebel, *Theory of Function Spaces,* Birkhäuser–Verlag, 1983.

[225] H. Triebel, Modulation spaces on the Euclidean n-spaces, Z. Anal. Anwendungen, **2** (1983), 443–457.

[226] Y. Tsutsumi, L^2-solutions for nonlinear Schrödinger equations and nonlinear groups. Funkcial. Ekvac., **30** (1987), 115–125.

[227] J. A. Tuszynski and J. M. Dixon, Coherent structures in strongly interacting many-body systems: I, Derivation of dynamics, J. Phys. A: Math. Gen., **22** (1989), 4877–4894.

[228] S. Ukai, Local solutions in Gevrey classes to the nonlinear Boltzmann equation without cutoff, Japan J. Appl. Math., **1** (1984), 141–156.

[229] C. Villani, *A Review of Mathematical Topics in Collisional Kinetic Theory,* Handbook of Fluid Mechanics, eds. S. Friedlander and D. Serre (North–Holland, 2002).

[230] C. Villani, On a new class of weak solutions to the spatially homogeneous Boltzmann and Landau equations, Arch. Rational Mech. Anal., **143** (1998), no. 3, 273–307.

[231] C. Villani, Regularity estimates via entropy dissipation for the spatially homogeneous Boltmann equation, Rev. Mat. Iberoamericana, **15** (1999), 335–352.

[232] M. Visan, The defocusing energy-critical nonlinear Schrödinger equation in higher dimensions, Duke Math. J., **138** (2007), 281–374.

[233] H. Wadade, Remarks on the Gagliardo–Nirenberg type inequality in the Besov and the Triebel-Lizorkin spaces in the limiting case, J. Fourier Anal. Appl., **15** (2009), 857–870.

[234] B. X. Wang, *The Cauchy Problem for the Nonlinear Schrödinger, Klein–Gordon Equations and Their Coupled Equations*, Doctoral Thesis, Inst. Appl. Phys. and Comput. Math., Dec. 1993, 1–224.

[235] B. X. Wang, Bessel (Riesz) potentials on Banach function spaces and their applications I, Theory, Acta Math. Sinica (N.S.), **14** (1998), 327–340.

[236] B. X. Wang, On existence and scattering for critical and subcritical nonlinear Klein–Gordon equations in H^s, Nonlinear Analysis, TMA, **31** (1998), 173–187.

[237] B. X. Wang , On scattering of solutions for the critical and subcritical nonlinear Klein-Gordon equations, Discrete Contin. Dyn. Syst., **5** (1999), 753–763.

[238] B. X. Wang, The limit behavior of solutions for the complex Ginzburg–Landau equation, Commun. Pure Appl. Math., **55** (2002), no.4, 481–508.

[239] B. X. Wang, The smoothness of scattering operators for the sinh–Gordon and nonlinear Schrödinger equations, Acta Math. Sinica (Engl. Ser.), **18** (2002), 549–564.

[240] B. X. Wang, Exponential Besov spaces and their applications to certain evolution equations with dissipation, Commun. Pure and Appl. Anal., **3** (2004), 883–919.

[241] B. X. Wang, Nonlinear scattering theory for a class of wave equations in H^s, J. Math. Anal. Appl., **296** (2004), 74–96.

[242] B. X. Wang, Concentration Phenomenon for the L^2 Critical and Super Critical Nonlinear Schrödinger Equation in Energy Spaces, Commun. Contemp. Math., **8** (2006), 309–330.

[243] B. X. Wang, L. J. Han, C. Y. Huang, Global well-Posedness and scattering for the derivative nonlinear Schrödinger equation with small rough data, Ann. I. H. Poincaré, Anal. Non Linéaire, **26** (2009), 2253–2281.

[244] B. X. Wang, C. C. Hao and H. Hudzik, Energy scattering for the nonlinear Schrödinger equations with exponential growth in lower spatial dimensions, J. Differential Equations, **228** (2006), 311–338.

[245] B. X. Wang and C. Y. Huang, Frequency-uniform decomposition method for the generalized BO, KdV and NLS equations, J. Differential Equations, **239** (2007), 213–250.

[246] B. X. Wang and H. Hudzik, The global Cauchy problem for the NLS and NLKG with small rough data, J. Differential Equations, **231** (2007), 36–73.

[247] B. X. Wang, L. F. Zhao, B. L. Guo, Isometric decomposition operators, function spaces $E^{\lambda}_{p,q}$ and their applications to nonlinear evolution equations, J. Funct. Anal., **233** (2006), 1–39.

[248] W. K. Wang, T. Yang, Point-wise estimates and L_p convergence rates to diffusion waves for p-system with damping, J. Differential Equations **187** (2003), 310–336.

[249] W. K. Wang, W. J. Wang, The pointwise estimates of solutions for semilinear dissipative wave equation in multi-dimensions. J. Math. Anal. Appl., **366** (2010), 226–241.

[250] F. B. Weissler, Existence and nonexistence of global solutions for a semilinear heat equation, Israel Math. J. **39** (1981), 29–40.

[251] F. B. Weissler, The Navier–Stokes initial value problem in L^p, Arch. Rational Mech. Anal. **74** (1980), 219–230.

[252] N. Wiener, Tauberian theorems, Ann. of Math., **33** (1932), 1–100.

[253] J. Xiao, Homothetic variant of fractional Sobolev space with application to Navier–Stokes system, Dyn. Partial Differ. Equ. **4** (2007), 227–245.

[254] D. C. Yang, W. Yuan, A new class of function spaces connecting Triebel–Lizorkin spaces and Q spaces, J. Funct. Anal., **255** (2008), 2760–2809.

[255] D. C. Yang, W. Yuan, New Besov-type spaces and Triebel–Lizorkin-type spaces including Q spaces, Math. Z., **265** (2010), 451–480.

[256] K. Yosida, *Functional Analysis,* Springer–Verlag, 1981.

[257] W. Yuan, W. Sickel, D. C. Yang *Morrey and Campanato Meet Besov, Lizorkin and Triebel,* Lect. Note in Math., **2005** (2010), 1-281.

[258] V. E. Zakharov, E. A. Kuznetson, Multi-scale expansions in the theory of systems integrable by inverse scattering method, Physica D, **18** (1986), 455–463.

[259] V. E. Zakharov, E. I. Schulman, Degenerated dispersion laws, motion invariant and kinetic equations, Physica D, **1** (1980), 185–250.

[260] Y. L. Zhou, B. L. Guo, Weak solution of system of ferromagnetic chain with several variables. Sci. Sinica Ser. A, **30** (1987), 1251–1266.

[261] Y. L. Zhou, B. L. Guo and S. Tan, Existence and uniqueness of smooth solution for system of ferro-magnetic chain. Sci. China Ser. A **34** (1991), 257–266.

Index